The interacting boson–fermion model has become in recent years the standard model for the description of atomic nuclei with an odd number of protons and/or neutrons. This books describes the mathematical framework on which the interacting boson–fermion model is built and presents applications to a variety of situations encountered in nuclei.

The book addresses both the analytical and the numerical aspects of the problem. The analytical aspect requires the introduction of rather complex group theoretic methods, including the use of graded (or super) Lie algebras. The first (and so far only) example of supersymmetry occurring in nature is also discussed.

The book is the first comprehensive treatment of the subject and will appeal to both theoretical and experimental physicists. The large number of explicit formulas for level energies, electromagnetic transition rates and intensities of transfer reactions presented in the book provide a simple but detailed way to analyse experimental data. This book can also be used as a text book for advanced graduate students.

T0268996

CAMBRIDGE MONOGRAPHS ON
MATHEMATICAL PHYSICS

General editors: P. V. Landshoff, D. R. Nelson, D. W. Sciama, S. Weinberg

THE INTERACTING BOSON–FERMION MODEL

Cambridge Monographs on Mathematical Physics

THE INTERACTING
BOSON–FERMION MODEL

F. IACHELLO

Center for Theoretical Physics
Yale University

P. VAN ISACKER

SERC Daresbury Laboratory

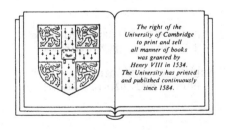

The right of the
University of Cambridge
to print and sell
all manner of books
was granted by
Henry VIII in 1534.
The University has printed
and published continuously
since 1584.

CAMBRIDGE UNIVERSITY PRESS

Cambridge

New York   *Port Chester*   *Melbourne*   *Sydney*

CAMBRIDGE UNIVERSITY PRESS
Cambridge, New York, Melbourne, Madrid, Cape Town, Singapore, São Paulo

Cambridge University Press
The Edinburgh Building, Cambridge CB2 2RU, UK

Published in the United States of America by Cambridge University Press, New York

www.cambridge.org
Information on this title: www.cambridge.org/9780521380928

First published 1991
This digitally printed first paperback version 2005

A catalogue record for this publication is available from the British Library

Library of Congress Cataloguing in Publication data
Iachello, F.
The interacting Boson–Fermion model / F. Iachello, P. Van Isacker.
p.   cm. – (Cambridge monographs on mathematical physics)
Includes bibliographical references (p.   ) and index
ISBN 0-521-38092-8
1. Interacting boson-fermion models. I. Van Isacker, P.
II. Title. III. Series
QC793.5.B622I23   1990
539.7′21 – dc20   90–2395
CIP

ISBN-13 978-0-521-38092-8 hardback
ISBN-10 0-521-38092-8 hardback

ISBN-13 978-0-521-02164-7 paperback
ISBN-10 0-521-02164-2 paperback

Contents

Part III: The interacting boson–fermion model-k

Part IV: High-lying collective modes

Preface

The interacting boson model has emerged in the last fifteen years as a unified framework for the description of the collective properties of nuclei. The key ingredients of this model are its algebraic structure based on the powerful methods of group theory, the possibility it gives to perform calculations in all nuclei and its direct connection with the shell model that allows one to derive its properties from microscopic interactions.

The interacting boson model deals with nuclei with an even number of protons and neutrons. However, more than half of the nuclear species have an odd number of protons and/or neutrons. In these nuclei there is an interplay between collective (bosonic) and single-particle (fermionic) degrees of freedom. The interacting boson model was extended to cover these situations by introducing the interacting boson–fermion model. This book, which is the second in a series of three, describes the interacting boson–fermion model and its applications. It has two aspects, an algebraic (group-theoretic) aspect and a numerical one. The algebraic aspect describes the coupling of bosons and fermions. The situation here is by far more complex than in the case of even–even nuclei and, for this reason, it is described in greater detail. The study of coupled Bose–Fermi systems is a novel application of algebraic methods and as such has a wider scope than that presented here. It has been used recently in other fields of physics, as for example in the coupling of electronic and rotation–vibration degrees of freedom in molecules. The discussion of coupled Bose–Fermi systems (Chapter 3) may, at times, seem tedious but it is a necessary ingredient for a detailed study and classification of the large variety of observed odd–even and odd–odd nuclei.

The theoretical difficulty that one encounters in the study of odd–even and odd–odd nuclei is reflected in a similar experimental difficulty. Quite often, spin and parity assignments of levels are difficult to make and thus a comparison between theory and experiment is not as straightforward as in the case of even–even nuclei. This book is intended as a guide to sort out from this complex situation the key features of collective properties of odd–even and odd–odd nuclei.

The development of the interacting boson–fermion model has been a cooperative effort of many researchers and we want to take this opportunity to thank *all* the persons who have contributed to it. In particular we wish to thank Baha Balantekin, Roelof Bijker, Jan Jolie, Serdar Kuyucak, Amiram Leviatan and Olaf Scholten, who have developed a good fraction of the formalism reported here. Another special thanks goes to Jose Arias, Clara Alonso, Pieter Brussaard, Alex Dieperink, Jan Dobeš, Jorge Dukelski, Da-Hsuan Feng, Alejandro Frank, Fritz Hahne, Kris Heyde, Michael Kirson, Brahman Kota, Dimitri Kusnezov, Georghi Kyrchev, Pertti Lipas, Manolo Lozano, Giuseppe Maino, Iain Morrison, Vladimir Paar, Stuart Pittel, Frederick Scholtz, Hong-Zhou Sun, Stanislav Szpikowski, Alberto Ventura, Andrea Vitturi, Nobuaki Yoshida and Lina Zuffi, who have contributed to several aspects of the model. Among the experimenters who have analyzed the predictions of the model and compared it with the data, we wish to thank Nives Blasi, Noemi Benczer-Koller, Richard Casten, Jolie Cizewski, Adrian Gelberg, Giovanni Lo Bianco, Michel Vergnes, Jean Vervier, Peter von Brentano, David Warner and John Wood. Finally, we wish to thank Klaas Allaart, Akito Arima, Bruce Barrett, J. Phillip Elliott, Joseph Ginocchio, Takaharu Otsuka, Igal Talmi and Li-Ming Yang for their contribution to the microscopic aspects of the interacting boson–fermion model, and Herman Feshbach and D. Allan Bromley for their constant interest and encouragement.

New Haven, Connecticut
September 1989

Part I

THE INTERACTING
BOSON–FERMION MODEL-1

[PART I]

THE INTERACTING
BOSON-FERMION MODEL

1

Operators

1.1 Introduction

In many cases in physics, one has to deal simultaneously with collective and single-particle excitations of the system. The collective excitations are usually bosonic in nature while the single-particle excitations are often fermionic. One is therefore led to consider a system which includes bosons and fermions. In this book we discuss applications of a general algebraic theory of mixed Bose–Fermi systems to atomic nuclei. The collective degrees of freedom here can be described in terms of a system of interacting bosons as discussed in a previous book (Iachello and Arima, 1987), henceforth referred to as Volume 1. The single-particle degrees of freedom represent the motion of individual nucleons in the average nuclear field. They are described in terms of a system of interacting fermions. The coupling of fermions and bosons leads to the interacting boson–fermion model which has been used extensively in recent years to discuss the properties of nuclei with an odd number of nucleons.

The interacting boson–fermion model was introduced by Arima and one of us in 1975 (Arima and Iachello, 1975). It was subsequently expanded by Iachello and Scholten (1979) and cast into a form more readily amenable to calculations. As in the corresponding case of even-mass systems, the algebra of creation and annihilation operators can be realized in several ways. One of these is the Holstein–Primakoff realization which leads to a slightly different version of the interacting boson–fermion model called the truncated quadrupole phonon–fermion model (Paar, 1980; Paar and Brant, 1981), based on the boson realization introduced by Janssen, Jolos and Dönau in 1974 and discussed in Sect. 1.4.6 of

Volume 1. In this book we discuss only the algebraic and geometric properties of the interacting boson–fermion model. The
microscopic origin and justification will be dealt with in a subsequent book. As in the case of even-mass systems, there are
several versions of the model which differ in their treatment of
the proton and neutron degrees of freedom. In the first version,
called the interacting boson–fermion model-1 (IBFM-1) and discussed in Part I of this book, no distinction is made between
protons and neutrons. In the other versions of the model they are
treated explicitly. The interacting boson–fermion model-2 (IBFM-
2) applies to nuclei where protons and neutrons occupy different
valence shells, while the interacting boson–fermion model-3 and
4 (IBFM-3 and IBFM-4) deal with lighter nuclei where protons
and neutrons occupy the same valence shell in which case isospin
becomes important. These will be discussed in Parts II and III.

1.2 Boson and fermion operators

In the interacting boson–fermion model the collective degrees of
freedom are described by boson operators. The properties of these
operators were discussed in great detail in Volume 1 and will be
only briefly reviewed here. To lowest order of approximation only
bosons with angular momentum and parity $J^P = 0^+$ and 2^+
are retained (s and d bosons). The corresponding creation and
annihilation operators are written as

$$b_{l,m}^\dagger; \qquad b_{l,m}; \qquad (l = 0, 2; -l \leq m \leq l), \qquad (1.1)$$

or

$$b_\alpha^\dagger; \qquad b_\alpha; \qquad (\alpha = 1, \dots, 6), \qquad (1.2)$$

and satisfy the commutation relations

$$[b_{l,m}, b_{l',m'}^\dagger] = \delta_{ll'}\delta_{mm'},$$
$$[b_{l,m}, b_{l',m'}] = [b_{l,m}^\dagger, b_{l',m'}^\dagger] = 0. \qquad (1.3)$$

or

$$[b_\alpha, b_{\alpha'}^\dagger] = \delta_{\alpha\alpha'}; \qquad [b_\alpha, b_{\alpha'}] = [b_\alpha^\dagger, b_{\alpha'}^\dagger] = 0. \qquad (1.4)$$

In addition to collective degrees of freedom, one wants to describe single-particle degrees of freedom. In nuclei, the single particles are protons and neutrons. These are fermions. The angular momentum and parity of these particles depends on the allowed orbits as will be discussed in more detail in Part II. Here we shall denote the angular momentum by j and its z-component by m. An interacting boson–fermion model is specified by the number and the values of angular momenta retained. In treating the single-particle degrees of freedom, it is also convenient to use the formalism of second quantization and introduce the fermion creation and annihilation operators

$$
\begin{aligned}
a^\dagger_{j,m}, \quad (m = \pm\tfrac{1}{2}, \pm\tfrac{3}{2}, \ldots, \pm j), \\
a_{j,m}, \quad (m = \pm\tfrac{1}{2}, \pm\tfrac{3}{2}, \ldots, \pm j).
\end{aligned}
\tag{1.5}
$$

These operators satisfy anticommutation relations

$$
\begin{aligned}
\{a_{j,m}, a^\dagger_{j',m'}\} = \delta_{jj'}\delta_{mm'}, \\
\{a_{j,m}, a_{j',m'}\} = \{a^\dagger_{j,m}, a^\dagger_{j',m'}\} = 0,
\end{aligned}
\tag{1.6}
$$

where the curly brackets denote an anticommutator, $\{A, B\} = AB + BA$, for any two operators A and B. These have to be contrasted with the commutation relations satisfied by the boson operators, (1.3). There the square brackets denote a commutator, $[A, B] = AB - BA$.

Instead of the double label j, m we shall use, at times, a single index i and denote the operators by

$$
a^\dagger_i; \quad a_i; \quad (i = 1, \ldots, n),
\tag{1.7}
$$

with anticommutation relations

$$
\{a_i, a^\dagger_{i'}\} = \delta_{ii'}; \quad \{a_i, a_{i'}\} = \{a^\dagger_i, a^\dagger_{i'}\} = 0.
\tag{1.8}
$$

Finally, it is assumed that boson and fermion operators commute:

$$
[b_{l,m}, a_{j',m'}] = [b_{l,m}, a^\dagger_{j',m'}] = [b^\dagger_{l,m}, a_{j',m'}] = [b^\dagger_{l,m}, a^\dagger_{j',m'}] = 0.
\tag{1.9}
$$

This is a natural assumption if bosons and fermions are elementary particles. In nuclei, where bosons are composite particles (fermion pairs), it is a model assumption. The effects of the compositeness of the bosons are introduced through an additional interaction (exchange interaction).

Spherical tensors can be constructed from the creation and annihilation operators in the usual way. The creation operators already transform in the appropriate way. The annihilation operators do not but one can introduce the operators

$$\tilde{a}_{j,m} = (-)^{j-m} a_{j,-m}, \qquad (1.10)$$

that transform appropriately under rotations. With these operators one can form tensor products as discussed in Volume 1. The phase convention (1.10), $(-)^{j-m}$, is chosen to conform with the majority of articles written on the interacting boson–fermion model. This phase is still consistent with that used for the boson operators, $(-)^{l+m}$, Eq. (1.9) of Volume 1, since for bosons (integer l) either choice, $(-)^{l+m}$ or $(-)^{l-m}$, gives the same result.

1.3   Basis states

In the formalism of second quantization, basis states can be constructed by repeated application of creation and annihilation operators on a vacuum state. For bosons the basis is:

$$\mathcal{B}: \qquad b_\alpha^\dagger b_{\alpha'}^\dagger \ldots |\text{o}\rangle, \qquad (1.11)$$

while for fermions it is:

$$\mathcal{F}: \qquad a_i^\dagger a_{i'}^\dagger \ldots |\text{o}\rangle. \qquad (1.12)$$

Due to their commutation relations, a major difference between boson and fermion operators is that, while one can put any number of bosons in a certain state, one can place only one fermion in the same state. This implies that

$$\left(a_i^\dagger\right)^2 |\text{o}\rangle = 0, \qquad (1.13)$$

that is, all indices in (1.12) must be different. The basis of the interacting boson–fermion model is the product of (1.11) and (1.12), usually written as

$$\mathcal{BF}: \qquad a_i^\dagger a_{i'}^\dagger \ldots b_\alpha^\dagger b_{\alpha'}^\dagger \ldots |0\rangle. \qquad (1.14)$$

Whether fermion operators are written to the left or to the right of boson operators is not relevant since they commute with each other.

It is also here convenient to construct states with good angular momentum by coupling the boson and fermion operators appropriately,

$$\mathcal{BF}: \qquad [[a_j^\dagger \times a_{j'}^\dagger \times \cdots]^{(L_F)} \times [b_l^\dagger \times b_{l'}^\dagger \times \cdots]^{(L_B)}]_M^{(L)}|0\rangle. \qquad (1.15)$$

Since the angular momentum alone is, in general, not sufficient to characterize the states uniquely, one needs extra labels. These will be discussed in Chapter 2.

1.4   Physical operators

1.4.1 The Hamiltonian operator

The model Hamiltonian contains a part that describes the bosons, H_B, a part that describes the fermions, H_F, and a part that describes the interaction between bosons and fermions, V_{BF},

$$H = H_B + H_F + V_{BF}. \qquad (1.16)$$

In the interacting boson–fermion model it is assumed that the Hamiltonian conserves separately the number of bosons, N_B, and the number of fermions, N_F. The structure of the various parts of the Hamiltonian operator is then as in Eq. (1.19) of Volume 1,

$$H_B = E_0 + \sum_{\alpha\beta} \epsilon_{\alpha\beta} b_\alpha^\dagger b_\beta + \sum_{\alpha\alpha'\beta\beta'} \tfrac{1}{2} u_{\alpha\alpha'\beta\beta'} b_\alpha^\dagger b_{\alpha'}^\dagger b_\beta b_{\beta'} + \cdots,$$

$$H_F = \mathcal{E}_0 + \sum_{ik} \eta_{ik} a_i^\dagger a_k + \sum_{ii'kk'} \tfrac{1}{2} v_{ii'kk'} a_i^\dagger a_{i'}^\dagger a_k a_{k'} + \cdots,$$

$$V_{BF} = \sum_{\alpha i\beta k} w_{\alpha i\beta k} b_\alpha^\dagger a_i^\dagger b_\beta a_k + \cdots. \qquad (1.17)$$

This Hamiltonian can be rewritten in such a way that its invariance under rotations becomes evident,

$$H_{\rm B} = E_0 + \sum_l \epsilon_l \sqrt{2l+1} [b_l^\dagger \times \tilde{b}_l]_0^{(0)}$$

$$+ \sum_{L_{\rm B},ll'l''l'''} \tfrac{1}{2} u_{ll'l''l'''}^{(L_{\rm B})} [[b_l^\dagger \times b_{l'}^\dagger]^{(L_{\rm B})} \times [\tilde{b}_{l''} \times \tilde{b}_{l'''}]^{(L_{\rm B})}]_0^{(0)} + \cdots,$$

$$H_{\rm F} = \mathcal{E}_0 - \sum_j \eta_j \sqrt{2j+1} [a_j^\dagger \times \tilde{a}_j]_0^{(0)}$$

$$+ \sum_{L_{\rm F},jj'j''j'''} \tfrac{1}{2} v_{jj'j''j'''}^{(L_{\rm F})} [[a_j^\dagger \times a_{j'}^\dagger]^{(L_{\rm F})}$$

$$\times [\tilde{a}_{j''} \times \tilde{a}_{j'''}]^{(L_{\rm F})}]_0^{(0)} + \cdots,$$

$$V_{\rm BF} = - \sum_{J,ljl'j'} w_{ljl'j'}^{(J)} \sqrt{2J+1} [[b_l^\dagger \times a_j^\dagger]^{(J)}$$

$$\times [\tilde{b}_{l'} \times \tilde{a}_{j'}]^{(J)}]_0^{(0)} + \cdots. \tag{1.18}$$

The coefficients $w_{ljl'j'}^{(J)}$ in (1.18) are the boson–fermion interaction matrix elements,

$$w_{ljl'j'}^{(J)} = \langle b_l a_j; J | V_{\rm BF} | b_{l'} a_{j'}; J \rangle. \tag{1.19}$$

Hermiticity of the Hamiltonian imposes further restrictions on the parameters in (1.18). For instance, assuming the matrix elements (1.19) to be real, one finds for the boson–fermion interaction that $w_{ljl'j'}^{(J)} = w_{l'j'lj}^{(J)}$. Other parametrizations of the boson–fermion interaction are possible. Two of them have been frequently used in calculations with the interacting boson–fermion model. They are referred to as the multipole expansion,

$$V_{\rm BF} = \sum_{L,ll'jj'} w_{ll'jj'}^{\prime(L)} (-)^L \sqrt{2L+1} [[b_l^\dagger \times \tilde{b}_{l'}]^{(L)} \times [a_j^\dagger \times \tilde{a}_{j'}]^{(L)}]_0^{(0)} + \cdots,$$

$$\tag{1.20}$$

and the exchange expansion,

$$V_{\rm BF} = \sum_{J,ljl'j'} w_{ljl'j'}^{\prime\prime(J)} \sqrt{2J+1} : [[b_l^\dagger \times \tilde{a}_j]^{(J)} \times [\tilde{b}_{l'} \times a_{j'}^\dagger]^{(J)}]_0^{(0)} : + \cdots,$$

$$\tag{1.21}$$

where the colons (: $\cdots$:) denote normal ordering. Normal ordering in this case implies that $a_{j'}^\dagger$ should stand on the left of $\tilde{a}_j$ with a minus sign. For these parametrizations, Hermiticity implies the relations $w_{ll''jj'}^{\prime(L)} = w_{l''ljj'}^{\prime(L)}$ and $w_{ll''jj'}^{\prime(L)} = (-)^{j-j'} w_{ll''j'j}^{\prime(L)}$ in (1.20) and $w_{ljl'j'}^{\prime\prime(J)} = w_{l'j'lj}^{\prime\prime(J)}$ in (1.21). The coefficients $w_{ll''jj'}^{\prime(L)}$ and $w_{ljl'j'}^{\prime\prime(J)}$ are related to the matrix elements (1.19) in the following way:

$$w_{ll''jj'}^{\prime(L)} = -\sum_J (-)^{j+l''+J}(2J+1) \left\{ \begin{array}{ccc} l & j & J \\ j' & l'' & L \end{array} \right\} w_{ljl''j'}^{(J)},$$

$$w_{ljl'j'}^{\prime\prime(J)} = \sum_{J'} (2J'+1) \left\{ \begin{array}{ccc} l & j' & J' \\ l' & j & J \end{array} \right\} w_{lj'l'j}^{(J')}. \tag{1.22}$$

In this expansion, the quantity in curly brackets denotes a Wigner 6j-symbol (de-Shalit and Talmi, 1963).

In most calculations, only terms up to two creation and two annihilation operators have been retained. In that case, the Hamiltonian H_B has been written down explicitly in Volume 1. In order to write down the parts H_F and V_{BF} one needs to know the values of j. As an example, we consider the case in which j can take only one value, $j = 3/2$. Omitting the index $j = 3/2$ from the fermion operators, one has

$$H_F = \mathcal{E}_0 - \eta\sqrt{4}[a^\dagger \times \tilde{a}]_0^{(0)} + \sum_{L_F=0,2} \tfrac{1}{2} v^{(L_F)}[[a^\dagger \times a^\dagger]^{(L_F)} \times [\tilde{a} \times \tilde{a}]^{(L_F)}]_0^{(0)},$$

$$\begin{aligned} V_{BF} &= w_{ss}^{\prime(0)}[[s^\dagger \times \tilde{s}]^{(0)} \times [a^\dagger \times \tilde{a}]^{(0)}]_0^{(0)} + w_{dd}^{\prime(0)}[[d^\dagger \times \tilde{d}]^{(0)} \times [a^\dagger \times \tilde{a}]^{(0)}]_0^{(0)} \\ &+ w_{dd}^{\prime(1)}[[d^\dagger \times \tilde{d}]^{(1)} \times [a^\dagger \times \tilde{a}]^{(1)}]_0^{(0)} \\ &+ w_{dd}^{\prime(2)}[[d^\dagger \times \tilde{d}]^{(2)} \times [a^\dagger \times \tilde{a}]^{(2)}]_0^{(0)} \\ &+ w_{sd}^{\prime(2)}[[s^\dagger \times \tilde{d} + d^\dagger \times \tilde{s}]^{(2)} \times [a^\dagger \times \tilde{a}]^{(2)}]_0^{(0)} \\ &+ w_{dd}^{\prime(3)}[[d^\dagger \times \tilde{d}]^{(3)} \times [a^\dagger \times \tilde{a}]^{(3)}]_0^{(0)}. \end{aligned} \tag{1.23}$$

Here we have also used the fact that the Hamiltonian H is an Hermitian operator, $H^\dagger = H$. There are thus three parameters, η, $v^{(0)}$ and $v^{(2)}$, specifying the fermion Hamiltonian H_F and six parameters, $w_{ss}^{\prime(0)}$, $w_{dd}^{\prime(0)}$, $w_{dd}^{\prime(1)}$, $w_{sd}^{\prime(2)}$, $w_{dd}^{\prime(2)}$ and $w_{dd}^{\prime(3)}$, specifying the boson–fermion interaction V_{BF} in its multipole form.

1.4.2 Transition operators

Operators inducing electromagnetic transitions of multipolarity
L also contain a part describing the bosons, $T_B^{(L)}$, and a part
describing the fermions, $T_F^{(L)}$,

$$T^{(L)} = T_B^{(L)} + T_F^{(L)}. \tag{1.24}$$

The structure of each term is:

$$T_B^{(L)} = t_0^{(0)} \delta_{L0} + \sum_{\alpha\beta} t_{\alpha\beta}^{(L)} b_\alpha^\dagger b_\beta + \cdots,$$

$$T_F^{(L)} = f_0^{(0)} \delta_{L0} + \sum_{ik} f_{ik}^{(L)} a_i^\dagger a_k + \cdots. \tag{1.25}$$

In principle, the transition operator $T^{(L)}$ contains also a boson–
fermion part, $T_{BF}^{(L)}$, of the form

$$T_{BF}^{(L)} = \sum_{ik\alpha\beta} r_{ik\alpha\beta}^{(L)} a_i^\dagger a_k b_\alpha^\dagger b_\beta + \cdots. \tag{1.26}$$

This part, however, contains at least two creation and two annihilation operators and is usually neglected. Again, since the
operators $T^{(L)}$ must transform as tensors of rank L under rotations, it is more convenient to rewrite (1.25) in coupled-tensor
form,

$$T_{B,\mu}^{(L)} = t_0^{(0)} \delta_{L0} + \sum_{ll'} t_{ll'}^{(L)} [b_l^\dagger \times \tilde{b}_{l'}]_\mu^{(L)} + \cdots,$$

$$T_{F,\mu}^{(L)} = f_0^{(0)} \delta_{L0} + \sum_{jj'} f_{jj'}^{(L)} [a_j^\dagger \times \tilde{a}_{j'}]_\mu^{(L)} + \cdots. \tag{1.27}$$

In addition to being tensors under rotations, the electromagnetic
transition operators have a definite character under parity. If only
bosons with $J^P = 0^+$ and 2^+ are considered, the parity of the
boson part $T_B^{(L)}$ is always positive. The fermion part $T_F^{(L)}$ instead
can have either positive or negative parity. As will be discussed in
Part II, for each fermion the angular momentum j is built from an
orbital angular momentum l_j and a spin $s = 1/2$. Its parity is thus
$(-)^{l_j}$. When combining fermion creation, $a_j^\dagger$, and annihilation, $\tilde{a}_{j'}$,

operators one must make sure that the combined parity is equal to that of the transition. This implies that the coefficients $f_{jj'}^{(L)}$ vanish unless $(-)^{l_j+l'_j+L}$ is $+1$ for electric transitions and -1 for magnetic transitions.

Usually, only terms containing one creation and one annihilation operator are retained. The explicit form of $T_B^{(L)}$ is then given in Eq. (1.24) of Volume 1. In order to write down $T_F^{(L)}$ one needs to know the values of j. As an example, we consider again the case $j = 3/2$ for which one obtains operators with multipolarity $L = 0, 1, 2, 3$,

$$
\begin{aligned}
T_{\mathrm{F},0}^{(\mathrm{E0})} &= f^{(0)} + f'^{(0)}[a^\dagger \times \tilde{a}]_0^{(0)}, \\
T_{\mathrm{F},\mu}^{(\mathrm{M1})} &= f^{(1)}[a^\dagger \times \tilde{a}]_\mu^{(1)}, \\
T_{\mathrm{F},\mu}^{(\mathrm{E2})} &= f^{(2)}[a^\dagger \times \tilde{a}]_\mu^{(2)}, \\
T_{\mathrm{F},\mu}^{(\mathrm{M3})} &= f^{(3)}[a^\dagger \times \tilde{a}]_\mu^{(3)}.
\end{aligned}
\tag{1.28}
$$

1.4.3 Independent parameters

Some of the parameters in the Hamiltonian and the transition operators can be eliminated, by using the condition that the number of bosons, N_B, and fermions, N_F, is conserved. For example, one of the terms in V_{BF}, (1.23), can be eliminated to yield

$$
H_\mathrm{F} = \mathcal{E}_0 - \eta'\sqrt{4}[a^\dagger \times \tilde{a}]_0^{(0)}
$$
$$
+ \sum_{L_\mathrm{F}=0,2} \tfrac{1}{2}v^{(L_\mathrm{F})}[[a^\dagger \times a^\dagger]^{(L_\mathrm{F})} \times [\tilde{a} \times \tilde{a}]^{(L_\mathrm{F})}]_0^{(0)}
$$

$$
\begin{aligned}
V_{\mathrm{BF}} = \; &w_{dd}''^{(0)}[[d^\dagger \times \tilde{d}]^{(0)} \times [a^\dagger \times \tilde{a}]^{(0)}]_0^{(0)} + w_{dd}'^{(1)}[[d^\dagger \times \tilde{d}]^{(1)} \times [a^\dagger \times \tilde{a}]^{(1)}]_0^{(0)} \\
&+ w_{dd}'^{(2)}[[d^\dagger \times \tilde{d}]^{(2)} \times [a^\dagger \times \tilde{a}]^{(2)}]_0^{(0)} \\
&+ w_{sd}'^{(2)}[[s^\dagger \times \tilde{d}+d^\dagger \times \tilde{s}]^{(2)} \times [a^\dagger \times \tilde{a}]^{(2)}]_0^{(0)} \\
&+ w_{dd}'^{(3)}[[d^\dagger \times \tilde{d}]^{(3)} \times [a^\dagger \times \tilde{a}]^{(3)}]_0^{(0)},
\end{aligned}
\tag{1.29}
$$

with

$$
\eta' = \eta - \frac{N_\mathrm{B}}{\sqrt{4}}w_{ss}'^{(0)}, \qquad w_{dd}''^{(0)} = w_{dd}'^{(0)} - \sqrt{5}w_{ss}'^{(0)}.
\tag{1.30}
$$

This elimination reduces the number of parameters in V_{BF} by one, and is similar to that discussed in Volume 1.

1.4.4 Special forms of the boson–fermion interaction

The form (1.18) of V_{BF} is too general for a purely phenomenological analysis. On the basis of microscopic considerations, to be discussed in Volume 3, it has been suggested (Iachello and Scholten, 1979) that a simpler form may account for most of the observed properties. In this simpler form only three terms are retained. First, a monopole term, written in the form

$$V_{\text{BF}}^{\text{MON}} = \sum_j A_j [[d^\dagger \times \tilde{d}]^{(0)} \times [a_j^\dagger \times \tilde{a}_j]^{(0)}]_0^{(0)}; \qquad (1.31)$$

second, a quadrupole term, written in the form

$$V_{\text{BF}}^{\text{QUAD}} = \sum_{jj'} \Gamma_{jj'} [[(s^\dagger \times \tilde{d} + d^\dagger \times \tilde{s}) + \chi(d^\dagger \times \tilde{d})]^{(2)} \times [a_j^\dagger \times \tilde{a}_{j'}]^{(2)}]_0^{(0)};$$

$$(1.32)$$

and finally, an interaction, called exchange interaction, that takes into account the fact that the bosons are fermion pairs (Talmi, 1981; Scholten and Dieperink, 1981; Otsuka *et al.*, 1987). This interaction can be written as

$$V_{\text{BF}}^{\text{EXC}} = \sum_{jj'j''} \Lambda_{jj'}^{j''} : [[d^\dagger \times \tilde{a}_j]^{(j'')} \times [\tilde{d} \times a_{j'}^\dagger]^{(j'')}]_0^{(0)} :, \qquad (1.33)$$

and has been shown (Yoshida *et al.*, 1988) to be of crucial importance in reproducing the signature dependence of electromagnetic transitions in odd–even nuclei. Thus, two terms are taken from the multipole expansion (1.20) and one term from the exchange expansion (1.21). They can be converted to two-body matrix elements (1.19) using the relations (1.22). In the exchange interaction one could also have the terms

$$V_{\text{BF}}^{\text{EXC}'} = \sum_{jj'} \Lambda_{jj'}^{'j'} : \{[[d^\dagger \times \tilde{a}_j]^{(j')} \times [\tilde{s} \times a_{j'}^\dagger]^{(j')}]_0^{(0)}$$

$$+ [[s^\dagger \times \tilde{a}_{j'}]^{(j')} \times [\tilde{d} \times a_j^\dagger]^{(j')}]_0^{(0)}\} :, \quad (1.34)$$

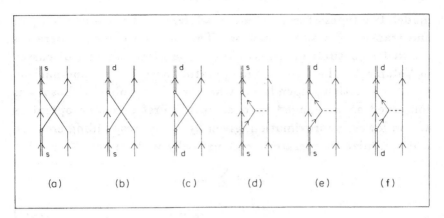

Fig. 1-1   Schematic representation of the boson–fermion interaction. Graphs (a), (b) and (c) represent the exchange interaction, while graphs (d), (e) and (f) represent the direct interaction.

and

$$V_{\text{BF}}^{\text{EXC}''} = \sum_j \Lambda_{jj}''^{j} : [[s^\dagger \times \tilde{a}_j]^{(j)} \times [\tilde{s} \times a_j^\dagger]^{(j)}]_0^{(0)} : . \qquad (1.35)$$

If only one single-particle orbit j is taken into account, these two terms can be eliminated and included in (1.32) and (1.33), respectively. When several j orbits are taken into account, these terms must be retained since their dependence on the indices j and j' is different from that of (1.32) and (1.33). Microscopic calculations also indicate the presence of a direct interaction of the type (1.19) between a fermion and a d boson (Talmi, 1981; Gelberg, 1983).

The boson–fermion interaction can be displayed graphically. This is usually done by denoting the bosons by a double line, since they are fermion pairs, the fermion by a single line and the interaction by a wavy line. The direct and exchange interactions are then displayed as in Fig. 1.1.

1.4.5 Transfer operators

Another set of operators of particular interest in nuclear physics is formed by transfer operators. In the interacting boson–fermion model, two types of transfer are possible, transfer of a nucleon pair and transfer of a single nucleon. Two-nucleon transfer operators are written entirely in terms of boson operators and were discussed in Volume 1. The form of the operators describing a one-nucleon transfer reaction depends on whether the number of bosons is conserved or is changed by one. In the first case, the operators are, in lowest approximation, given by the corresponding creation and annihilation operators, schematically written as

$$P_+^{(j)} = \sum_i p_i^{(j)} a_i^\dagger,$$
$$P_-^{(j)} = \sum_i p_i^{(j)} a_i, \qquad (1.36)$$

or, more explicitly, as

$$P_{+,m}^{(j)} = p_j a_{j,m}^\dagger,$$
$$P_{-,m}^{(j)} = p_j \tilde{a}_{j,m}, \qquad (1.37)$$

where j denotes the transferred angular momentum. However, for the transfer operators (1.37), it has been found that the lowest order is not sufficient to describe the experimental situation, since it does not take into account the composite nature of the bosons. Consequently, one needs to introduce higher-order terms. To next order, these are written as

$$P_{3,+}^{(j)} = \sum_{\alpha\beta i} q_{\alpha\beta i}^{(j)} b_\alpha^\dagger b_\beta a_i^\dagger,$$
$$P_{3,-}^{(j)} = \sum_{\alpha\beta i} q_{\alpha\beta i}^{(j)} b_\beta^\dagger b_\alpha a_i, \qquad (1.38)$$

or, in coupled-tensor form,

$$P_{3,+,m}^{(j)} = \sum_{ll',k,j'} q_{ll',k,j'}^{(j)} [[b_l^\dagger \times \tilde{b}_{l'}]^{(k)} \times a_{j'}^\dagger]_m^{(j)},$$
$$P_{3,-,m}^{(j)} = \sum_{ll',k,j'} (-)^{l+l'-k} q_{ll',k,j'}^{(j)} [[b_{l'}^\dagger \times \tilde{b}_l]^{(k)} \times \tilde{a}_{j'}]_m^{(j)}, \quad (1.39)$$

where we have explicitly introduced the phase $(-)^{l+l'-k}$ for the operators to transform as spherical tensors under rotations. In the second case, when the number of bosons is changed by one, the transfer operators are in lowest order

$$P_+^{\prime(j)} = \sum_{\alpha i} p_{\alpha i}^{\prime(j)} b_\alpha a_i^\dagger,$$

$$P_-^{\prime(j)} = \sum_{\alpha i} p_{\alpha i}^{\prime(j)} b_\alpha^\dagger a_i, \qquad (1.40)$$

or, more explicitly,

$$P_{+,m}^{\prime(j)} = \sum_{l j'} p_{l j'}^{\prime(j)} [\tilde{b}_l \times a_{j'}^\dagger]_m^{(j)},$$

$$P_{-,m}^{\prime(j)} = \sum_{l j'} p_{l j'}^{\prime(j)} [b_l^\dagger \times \tilde{a}_{j'}]_m^{(j)}. \qquad (1.41)$$

The operators with index + describe a transfer reaction from an even–even to an even–odd nucleus, while the operators with index − describe the inverse reaction.

1.4.6 Special forms of the transfer operators

It has been suggested (Scholten, 1980) that a special form of the higher-order terms in the transfer operators is often sufficient to describe the experimental situation. In this form only two terms are retained in (1.39). The first is a monopole term,

$$P_{3,+,m}^{\mathrm{MON}(j)} = q_0^{(j)} [[d^\dagger \times \tilde{d}]^{(0)} \times a_j^\dagger]_m^{(j)}. \qquad (1.42)$$

The second is a quadrupole term,

$$P_{3,+,m}^{\mathrm{QUAD}(j)} = \sum_{j'} q_{2,j'}^{(j)} [[(s^\dagger \times \tilde{d} + d^\dagger \times \tilde{s}) + \chi(d^\dagger \times \tilde{d})]^{(2)} \times a_{j'}^\dagger]_m^{(j)}. \qquad (1.43)$$

Similar expressions hold for the substraction operators $P_{3,-,m}^{(j)}$. More realistic but more complicated transfer operators have recently been proposed by Sofia and Vitturi (1989) on the basis of microscopic theory.

2

Algebras

2.1 Introduction

In the previous chapter we described the structure of the Hamiltonian and other operators of the interacting boson–fermion model. Properties of the model could then be found by numerical methods. However, due to the complexity of the model, it is of particular interest here to use the powerful techniques of algebraic methods in order to find solutions in closed form. For problems with both fermionic and bosonic degrees of freedom, the study of the algebraic properties of the system requires algebras built with both fermion and boson operators. In Volume 1 algebras built with boson operators were discussed. In this volume, we discuss algebras built with fermion operators and the combination of these with algebras built with boson operators. The study here is somewhat more complex than that of Volume 1. The reader is required to have a good knowledge of the theory of Lie algebras and to be familiar with the discussion of Volume 1.

2.2 Fermion algebras

Consider the set of bilinear products of fermion creation $(a_i^\dagger)$ and annihilation (a_i) operators

$$g: \qquad A_{ik} = a_i^\dagger a_k, \qquad (i, k = 1, \ldots, n). \qquad (2.1)$$

These operators satisfy the commutation relations

$$[A_{ik}, A_{st}] = A_{it}\delta_{ks} - A_{sk}\delta_{it}, \qquad (2.2)$$

together with the appropriate Jacobi identities. They are generators of the unitary algebra in n dimensions, u(n). For reasons which will become apparent in future sections, they have been denoted by A_{ik} rather than G_{ik}. As in Volume 1, we will not distinguish between an algebra, g, and its associated group G, and denote both by a capital letter, U(n). However, since we have here algebras formed both with boson and fermion operators, when needed, we shall attach a superscript B or F to distinguish the two possibilities.

It is convenient to rewrite the generators of U(n) in a coupled-tensor notation (Racah, 1949),

$$g: \qquad A^{(\lambda)}_{\mu}(j,j') = [a^{\dagger}_{j} \times \tilde{a}_{j'}]^{(\lambda)}_{\mu}. \qquad (2.3)$$

The operators (2.3) satisfy the commutation relations

$$[A^{(\lambda)}_{\mu}(j,j'), A^{(\lambda')}_{\mu'}(j'',j''')]$$

$$= - \sum_{\lambda''\mu''} \sqrt{(2\lambda+1)(2\lambda'+1)}(\lambda\mu\ \lambda'\mu'|\lambda''\mu'')$$

$$\times \left[(-)^{\lambda''+j+j'''} \left\{ \begin{array}{ccc} \lambda & \lambda' & \lambda'' \\ j''' & j & j' \end{array} \right\} \delta_{j'j''} A^{(\lambda'')}_{\mu''}(j,j''') \right.$$

$$\left. -(-)^{\lambda+\lambda'+j'+j''} \left\{ \begin{array}{ccc} \lambda & \lambda' & \lambda'' \\ j'' & j' & j \end{array} \right\} \delta_{jj'''} A^{(\lambda'')}_{\mu''}(j'',j') \right], \qquad (2.4)$$

where j, j', j'' and j''' are half-integers.

The algebraic structure of bilinear products of fermion operators is much richer than that of bilinear products of boson operators discussed in Volume 1. We shall therefore analyze it in somewhat greater detail. We divide the discussion in two parts, the first being devoted to the case in which j can assume only one value (single j) and the second being devoted to the case in which j can assume more than one value (multiple j).

2.3 Single j

In the single-j case , the value of n is

$$n = 2j + 1, \qquad (2.5)$$

as can be seen from (1.5). The n^2 operators

$$A_\mu^{(\lambda)}(j,j) = [a_j^\dagger \times \tilde{a}_j]_\mu^{(\lambda)}, \qquad (2.6)$$

satisfy the commutation relations (2.4) with $j = j' = j'' = j'''$ and are the generators of $U(2j + 1)$. If one excludes the operator with $\lambda = \mu = 0$, one has the algebra $SU(2j + 1)$. It is, in general, possible to find a further subset which is closed with respect to commutation. As can be seen from (2.4), the operators with odd λ form a closed algebra. They generate the algebra $Sp(n)$ or, to be more precise, the compact algebra $Sp(n, C)$. However, following standard notation, we shall delete the letter C and denote this algebra simply by $Sp(n)$. As one can see from Table 2.2 of Volume 1, this algebra has $\frac{1}{2}n(n + 1)$ generators. A further subalgebra is that of the three-dimensional rotations, $SU(2) \approx O(3)$, which is generated by the operators (2.6) with $\lambda = 1$. (The symbol $\approx$ denotes isomorphic algebras, which will be discussed in Sect. 2.4. For orthogonal algebras we also delete the letter S (special) since it is not essential for the discussion in this book.) Finally, one can consider the algebra of rotations around the z-axis, $O(2)$, generated by the operator (2.6) with $\lambda = 1$ and $\mu = 0$. Thus, in general, a chain of algebras is:

$$U(2j + 1) \supset SU(2j + 1) \supset Sp(2j + 1) \supset SU(2) \supset O(2). \qquad (2.7)$$

We now consider specific examples of (2.7).

2.3.1 The case $j = 1/2$

In this case there are $2^2 = 4$ operators of the type (2.6). These can be written as

$$\begin{aligned}
A_\mu^{(1)}(\tfrac{1}{2}, \tfrac{1}{2}) &= [a_{1/2}^\dagger \times \tilde{a}_{1/2}]_\mu^{(1)} & & 3 \\
A_0^{(0)}(\tfrac{1}{2}, \tfrac{1}{2}) &= [a_{1/2}^\dagger \times \tilde{a}_{1/2}]_0^{(0)} & & 1 \\
& & & \overline{} \\
& & & 4,
\end{aligned} \qquad (2.8)$$

where we have indicated to the right the number of operators. The four operators (2.8) generate the algebra $U(2)$. The three operators $A_\mu^{(1)}(\frac{1}{2},\frac{1}{2})$ generate SU(2) $\approx$ O(3) $\approx$ Sp(2),

$$A_\mu^{(1)}(\tfrac{1}{2},\tfrac{1}{2}) \;=\; [a_{1/2}^\dagger \times \tilde{a}_{1/2}]_\mu^{(1)}, \tag{2.9}$$

while the single operator

$$A_0^{(1)}(\tfrac{1}{2},\tfrac{1}{2}) \;=\; [a_{1/2}^\dagger \times \tilde{a}_{1/2}]_0^{(1)} \tag{2.10}$$

generates the algebra O(2) of rotations around the z-axis. This yields the chain of algebras

$$\text{U(2)} \supset \text{Sp(2)} \supset \text{O(2)}. \tag{2.11}$$

2.3.2   The case $j = 3/2$

In this case, there are $4^2 = 16$ operators generating the Lie algebra U(4),

$$
\begin{aligned}
A_\mu^{(3)}(\tfrac{3}{2},\tfrac{3}{2}) &= [a_{3/2}^\dagger \times \tilde{a}_{3/2}]_\mu^{(3)} & &7 \\
A_\mu^{(2)}(\tfrac{3}{2},\tfrac{3}{2}) &= [a_{3/2}^\dagger \times \tilde{a}_{3/2}]_\mu^{(2)} & &5 \\
A_\mu^{(1)}(\tfrac{3}{2},\tfrac{3}{2}) &= [a_{3/2}^\dagger \times \tilde{a}_{3/2}]_\mu^{(1)} & &3 \\
A_0^{(0)}(\tfrac{3}{2},\tfrac{3}{2}) &= [a_{3/2}^\dagger \times \tilde{a}_{3/2}]_0^{(0)} & &1 \\
& & &\overline{} \\
& & &16.
\end{aligned}
\tag{2.12}
$$

Deleting the last operator we have the algebra of SU(4),

$$
\begin{aligned}
A_\mu^{(3)}(\tfrac{3}{2},\tfrac{3}{2}) &= [a_{3/2}^\dagger \times \tilde{a}_{3/2}]_\mu^{(3)} & &7 \\
A_\mu^{(2)}(\tfrac{3}{2},\tfrac{3}{2}) &= [a_{3/2}^\dagger \times \tilde{a}_{3/2}]_\mu^{(2)} & &5 \\
A_\mu^{(1)}(\tfrac{3}{2},\tfrac{3}{2}) &= [a_{3/2}^\dagger \times \tilde{a}_{3/2}]_\mu^{(1)} & &3 \\
& & &\overline{} \\
& & &15.
\end{aligned}
\tag{2.13}
$$

Subalgebras of SU(4) are that of Sp(4), with 10 generators,

$$
\begin{aligned}
A_\mu^{(3)}(\tfrac{3}{2},\tfrac{3}{2}) &= [a_{3/2}^\dagger \times \tilde{a}_{3/2}]_\mu^{(3)} & &7 \\
A_\mu^{(1)}(\tfrac{3}{2},\tfrac{3}{2}) &= [a_{3/2}^\dagger \times \tilde{a}_{3/2}]_\mu^{(1)} & &3 \\
& & &\overline{} \\
& & &10,
\end{aligned}
\tag{2.14}
$$

of SU(2) $\approx$ O(3),

$$A_\mu^{(1)}(\tfrac{3}{2}, \tfrac{3}{2}) \;=\; [a_{3/2}^\dagger \times \tilde{a}_{3/2}]_\mu^{(1)} \qquad\qquad 3, \qquad\qquad (2.15)$$

and of O(2),

$$A_0^{(1)}(\tfrac{3}{2}, \tfrac{3}{2}) \;=\; [a_{3/2}^\dagger \times \tilde{a}_{3/2}]_0^{(1)} \qquad\qquad 1. \qquad\qquad (2.16)$$

This yields a chain of algebras,

$$\text{U}(4) \supset \text{SU}(4) \supset \text{Sp}(4) \supset \text{SU}(2) \supset \text{O}(2). \qquad (2.17)$$

2.3.3 The case $j = 5/2$

Here, there are $6^2 = 36$ operators generating U(6),

$$A_\mu^{(\lambda)}(\tfrac{5}{2}, \tfrac{5}{2}) \;=\; [a_{5/2}^\dagger \times \tilde{a}_{5/2}]_\mu^{(\lambda)}, \qquad \lambda = 0, 1, 2, 3, 4, 5. \quad (2.18)$$

Deleting the operator with $\lambda = 0$ we obtain the algebra SU(6). Retaining only terms with $\lambda = 1, 3, 5$, we obtain the 21 generators of Sp(6). Finally, retaining only the three operators with $\lambda = 1$, we have SU(2) and the single operator $A_0^{(1)}(\tfrac{5}{2}, \tfrac{5}{2})$ generates O(2). We thus have the group chain

$$\text{U}(6) \supset \text{SU}(6) \supset \text{Sp}(6) \supset \text{SU}(2) \supset \text{O}(2). \qquad (2.19)$$

Algebras for half-integer $j > 5/2$ can be constructed in a similar way.

2.4   Isomorphic Lie algebras

A concept which plays an important role in applications is that of isomorphic Lie algebras. These are algebras that have the same Lie commutation relations. All the isomorphisms of real simple Lie algebras are known (Wybourne, 1974). The isomorphisms of interest for applications in this book are shown in Table 2.1. Isomorphisms will be denoted here by the symbol $\approx$.

Table 2-1 Isomorphisms of real simple Lie algebras

Isomorphic Lie algebras
$su(2) \approx so(3) \approx sp(2)$
$so(5) \approx sp(4)$
$so(4) \approx su(2) \oplus su(2) \approx so(3) \oplus so(3) \approx sp(2) \oplus sp(2)$
$su(4) \approx so(6)$

2.5 Multiple *j*

The classification of chains of algebras associated with multiple j is rather complex and we present here only some selected cases. If the values of j are $j_1, j_2, \ldots$, the unitary group generated by the operators (2.1) is U(n) with

$$n = \sum_{j_i} (2j_i + 1), \qquad i = 1, 2, \ldots . \tag{2.20}$$

In cases of practical interest n may be rather large, for example 20 or 30. There are two cases of interest.

2.5.1 Class F-1. *Spinor algebras*

This class occurs whenever the set of values $j_1, j_2, \ldots$, forms a spinor representation of an orthogonal algebra. Spinor representations will be discussed below in Sect. 2.6. Particularly important is the case in which the values of j form a representation of the algebra of Spin(6) $\approx$ SU(4). The spinor representation of lowest dimensionality is here formed by the single value $j = 3/2$. This case was treated in Sect. 2.3.2. The spinor representation of next highest dimensionality has $j = 1/2, 3/2, 5/2, 7/2$ with $n = 20$. There are in total $20^2 = 400$ generators. Rather than writing them explicitly, we give here only the chain of subgroups

$$\text{U}(20) \supset \text{Spin}(6) \approx \text{SU}(4) \supset \text{Spin}(5) \approx \text{Sp}(4) \supset$$
$$\text{Spin}(3) \approx \text{SU}(2) \supset \text{Spin}(2) \approx \text{O}(2). \tag{2.21}$$

Because of the properties of spinor groups described in Sect. 2.6, this chain is identical to (2.17) except for the fact that it starts from U(20) rather than U(4).

2.5.2 Class F-2. Pseudo-spin algebras

This class is obtained in the following way. Divide the angular momentum j into two pieces, one integer, called pseudo-orbital, k, and one half-integer, called pseudo-spin, s (Hecht and Adler, 1969; Arima *et al.*, 1969). A simple example of this breaking is the case $j = 1/2, 3/2$. This can be viewed as a pseudo-orbital angular momentum $k = 1$, coupled to a pseudo-spin, $s = 1/2$, with $n = 6$. Breaking U(6) in this way corresponds to introducing the algebras $U_k(3)$ and $U_s(2)$. In general an algebra $U(n_j)$ can be broken into

$$U(n_j) \supset U_k(n_k) \otimes U_s(n_s), \qquad (2.22)$$

where

$$n_j = \sum_{j_i}(2j_i + 1),$$

$$n_k = \sum_{k_i}(2k_i + 1),$$

$$n_s = \sum_{s_i}(2s_i + 1). \qquad (2.23)$$

In (2.22) we have used the product sign $\otimes$. Actually, the algebra $u(n_j)$ is broken into the direct sum $\oplus$ of $u_k(n_k)$ and $u_s(n_s)$,

$$u(n_j) \supset u_k(n_k) \oplus u_s(n_s). \qquad (2.24)$$

However, since direct sums of algebras correspond to direct products of groups, it has become customary to use the product sign $\otimes$ when using the capital letters U and the sum sign $\oplus$ when using lowercase letters u. The generators of the unitary groups $U_k(n_k)$ and $U_s(n_s)$ can be obtained from those of $U(n_j)$ by transforming from $j-j$ to $k-s$ coupling (Bijker, 1984):

$$K_\mu^{(\lambda)}(k, k') = - \sum_s \sum_{jj'} \sqrt{(2j + 1)(2j' + 1)}(-)^{j'+\lambda+k+s}$$

$$\times \left\{ \begin{matrix} j & j' & \lambda \\ k' & k & s \end{matrix} \right\} A_\mu^{(\lambda)}(j, j'),$$

$$S_\mu^{(\lambda)}(s, s') = \sum_k \sum_{jj'} \sqrt{(2j + 1)(2j' + 1)}(-)^{s'+\lambda+j+k}$$

$$\times \left\{ \begin{matrix} s & s' & \lambda \\ j' & j & k \end{matrix} \right\} A_\mu^{(\lambda)}(j, j'), \qquad (2.25)$$

Table 2-2 Some algebras corresponding to class F-2

j	k	s	Algebra
1/2,3/2	1	1/2	$u(6) \supset u_k(3) \oplus u_s(2)$
1/2,3/2,5/2	0,2	1/2	$u(12) \supset u_k(6) \oplus u_s(2)$
	1	3/2	$u(12) \supset u_k(3) \oplus u_s(4)$
1/2,3/2,5/2,7/2	1,3	1/2	$u(20) \supset u_k(10) \oplus u_s(2)$
	2	3/2	$u(20) \supset u_k(5) \oplus u_s(4)$

and they satisfy the commutation relations

$$[K_\mu^{(\lambda)}(k,k'), K_{\mu'}^{(\lambda')}(k'',k''')]$$

$$= \sum_{\lambda''\mu''} \sqrt{(2\lambda+1)(2\lambda'+1)}(\lambda\mu\,\lambda'\mu'|\lambda''\mu'')$$

$$\times \left[(-)^{\lambda''+k+k'''} \left\{ \begin{matrix} \lambda & \lambda' & \lambda'' \\ k''' & k & k' \end{matrix} \right\} \delta_{k'k''} K_{\mu''}^{(\lambda'')}(k,k''')\right.$$

$$\left. -(-)^{\lambda+\lambda'+k'+k''} \left\{ \begin{matrix} \lambda & \lambda' & \lambda'' \\ k'' & k' & k \end{matrix} \right\} \delta_{kk'''} K_{\mu''}^{(\lambda'')}(k'',k')\right],$$

$$[S_\mu^{(\lambda)}(s,s'), S_{\mu'}^{(\lambda')}(s'',s''')]$$

$$= \sum_{\lambda''\mu''} \sqrt{(2\lambda+1)(2\lambda'+1)}(\lambda\mu\,\lambda'\mu'|\lambda''\mu'')(-)^{s+s'+s''+s'''}$$

$$\times \left[(-)^{\lambda''} \left\{ \begin{matrix} \lambda & \lambda' & \lambda'' \\ s''' & s & s' \end{matrix} \right\} \delta_{s's''} S_{\mu''}^{(\lambda'')}(s,s''')\right.$$

$$\left. -(-)^{\lambda+\lambda'} \left\{ \begin{matrix} \lambda & \lambda' & \lambda'' \\ s'' & s' & s \end{matrix} \right\} \delta_{ss'''} S_{\mu''}^{(\lambda'')}(s'',s')\right],$$

$$[K_\mu^{(\lambda)}(k,k'), S_{\mu'}^{(\lambda')}(s,s')] = 0. \qquad (2.26)$$

With the help of these commutation relations, one can form subalgebras in the usual way. In Table 2.2 some cases of interest are listed.

Table 2-3 Number of half-
integers that characterize
the spinor representations of
orthogonal groups

Group	Number
$O(n), n =$ even	$n/2$
$O(n), n =$ odd	$(n-1)/2$

Table 2-4 Isomorphisms
of spinor algebras

Isomorphic algebras
$\mathrm{spin}(2) \approx \mathrm{so}(2)$
$\mathrm{spin}(3) \approx \mathrm{su}(2)$
$\mathrm{spin}(4) \approx \mathrm{su}(2) \oplus \mathrm{su}(2)$
$\mathrm{spin}(5) \approx \mathrm{sp}(4)$
$\mathrm{spin}(6) \approx \mathrm{su}(4)$

2.6   Spinor algebras and groups

In Sect. 2.4 of Volume 1 the tensor representations of Lie groups
were discussed. These are characterized by a set of integers.
However, for the orthogonal groups one can also have spinor representations (Wybourne, 1974) characterized by a set of half-integer
values. The number of half-integers that characterize the spinor
representations of orthogonal groups is the same as for tensor representations and is given in Table 2.3. The spinor algebras, up to
Spin(6), are isomorphic to classical Lie algebras, as shown in Table 2.4 (Gilmore, 1974). Another peculiarity of orthogonal groups
$O(n)$ is that, for even n, there are two equivalent representations,
one with Young numbers

$$[\lambda_1, \lambda_2, \ldots, \lambda_{n/2}], \tag{2.27}$$

and one with

$$[\lambda_1, \lambda_2, \ldots, -\lambda_{n/2}]. \tag{2.28}$$

2.7 Basis states for fermions

2.7.1 Single j

Tensor representations of Lie algebras are characterized by a set of integers arranged in a Young tableau:

$$[\lambda_1, \lambda_2, \ldots, \lambda_n] = \overbrace{\square\square\square\cdots\square}^{\lambda_1}$$
$$\overbrace{\square\square\cdots\square}^{\lambda_2}$$
$$\vdots$$
$$\overbrace{\square\cdots\square}^{\lambda_n}. \qquad (2.29)$$

In the case of identical bosons discussed in Volume 1, the total wave function must be symmetric. This implies that the allowed Young tableaux are those with only one row. The situation is different for fermions. Here the total wave function must be antisymmetric. This implies that the allowed Young tableaux are those with only one column, characterized by the number of fermions, $N_{\rm F}$,

$$\{N_{\rm F}\} \equiv [\overbrace{1, 1, \ldots, 1}^{N_{\rm F}}, 0, 0, \ldots] \equiv \left.\begin{array}{c} \square \\ \square \\ \vdots \\ \square \end{array}\right\} N_{\rm F}. \qquad (2.30)$$

The construction of a basis for fermions amounts to the decomposition of the representations (2.30) of $U(2j+1)$ into those of its subgroups.

We begin by noting that we can construct only states up to $N_{\rm F} = 2j + 1$, since this is the maximum allowed number of rows in the Young tableau. The determination of the number of labels that characterize uniquely totally antisymmetric states of $U(n)$ is rather complex. The algebras in the chain (2.7) provide some labels, which may or may not be sufficient to characterize the states. If they are not sufficient one needs extra (or missing)

Table 2-5 Classification of antisymmetric states j^{N_F}
(Flowers, 1952)

$j = 1/2$	U(2)		J	
	[0]		0	
	[1]		1/2	
	$[1,1] \equiv [0]$		0	

$j = 3/2$	U(4)		Sp(4)	J
	[0]		(0,0)	0
	[1]		(1,0)	3/2
	[1,1]		(0,0)	0
			(1,1)	2
	$[1,1,1] \equiv [1]$		(1,0)	3/2
	$[1,1,1,1] \equiv [0]$		(0,0	0

$j = 5/2$	U(6)		Sp(6)	J
	[0]		(0,0,0)	0
	[1]		(1,0,0)	5/2
	[1,1]		(0,0,0)	0
			(1,1,0)	2,4
	[1,1,1]		(1,0,0)	5/2
			(1,1,1)	3/2,9/2
	$[1,1,1,1] \equiv [1,1]$		(0,0,0)	0
			(1,1,0)	2,4
	$[1,1,1,1,1] \equiv [1]$		(1,0,0)	5/2
	$[1,1,1,1,1,1] \equiv [0]$		(0,0,0)	0

labels, ν, giving rise to the classification scheme

$$\left| \begin{array}{cccc} U(n) & \supset \quad Sp(n) & \supset \; SU(2) & \supset \; O(2) \\ \downarrow & \downarrow & \downarrow & \downarrow \\ \{N_F\} & (n_1, n_2, \ldots, n_{n/2}) & \nu, J & M_J \end{array} \right\rangle . \quad (2.31)$$

Up to $j = 7/2$ no extra labels are needed. We now consider the classification of the three cases discussed in Sect. 2.3.

(i) The case $j = 1/2$

In this case the group chain is (2.11). The number of fermions can be $N_F = 0, 1, 2$. For fermion systems there is an equivalence between the representation with N_F boxes and the one with $2j + 1 - N_F$ boxes (particle–hole conjugation). Thus, for $j = 1/2$, the representations [1,1] and [0] are equivalent, $[1,1] \equiv [0]$. The number of fermions uniquely defines the states,

$$\left| \begin{array}{cccc} U(2) & \supset & Sp(2) \approx SU(2) & \supset & O(2) \\ \downarrow & & & & \downarrow \\ \{N_F\} & & & & M_J \end{array} \right\rangle. \qquad (2.32)$$

(ii) The case $j = 3/2$

In this case the group chain is (2.17). The decomposition of representations of U(4) into representations of Sp(4) is given by Flowers (1952). The quantum numbers can be written as

$$\left| \begin{array}{cccccccc} U(4) & \supset & Sp(4) & \supset & SU(2) & \supset & O(2) \\ \downarrow & & \downarrow & & \downarrow & & \downarrow \\ \{N_F\} & & (n_1, n_2) & & J & & M_J \end{array} \right\rangle. \qquad (2.33)$$

It would seem that five quantum numbers are needed to label the states. However, the representations (n_1, n_2) of Sp(4) are restricted and can actually be labelled by one quantum number, namely the number of 1s in the Young tableau, ν_F. This number is given by $N_F, N_F - 2, \ldots, 1$ or 0.

(iii) The case $j = 5/2$

In this case the group chain is (2.19). The quantum numbers can be written as

$$\left| \begin{array}{cccccccc} U(6) & \supset & Sp(6) & \supset & SU(2) & \supset & O(2) \\ \downarrow & & \downarrow & & \downarrow & & \downarrow \\ \{N_F\} & & (n_1, n_2, n_3) & & J & & M_J \end{array} \right\rangle. \qquad (2.34)$$

Also here the representations (n_1, n_2, n_3) of Sp(6) are restricted and there is only one quantum number, ν_F. The four quantum numbers N_F, ν_F, J, M_J are sufficient to determine the states uniquely. This situation persists up to and including $j = 7/2$. For larger js further quantum numbers are needed.

The complete classification of all states is shown in Table 2.5 for $j \leq 5/2$.

2.7.2 Multiple j. Spinor algebras

We consider first the case of spinor algebras. A case of interest here is $j = 1/2, 3/2, 5/2, 7/2$, to be classified using the group

Table 2-6 Partial classification of antisymmetric
states $(j = 1/2, 3/2, 5/2, 7/2)^{N_{\mathrm{F}}}$

U(20)	Spin(6)	Spin(5)	Spin(3)
N_{F}	$(\sigma_1, \sigma_2, \sigma_3)$	(τ_1, τ_2)	J
1	$(\frac{3}{2}, \frac{1}{2}, \frac{1}{2})$	$(\frac{3}{2}, \frac{1}{2})$	1/2,5/2,7/2
		$(\frac{1}{2}, \frac{1}{2})$	3/2
2	(2,2,1)	(2,2)	0,2,3,4,6
		(2,1)	1,2,3,4,5
	(3,0,0)	(3,0)	0,3,4,6
		(2,0)	2,4
		(1,0)	2
		(0,0)	0
	(2,1,0)	(2,1)	1,2,3,4,5
		(2,0)	2,4
		(1,1)	1,3
		(1,0)	2
	(1,0,0)	(1,0)	2
		(0,0)	0

chain (2.21). We begin by noting that these values of j transform
as the [1] representation of U(20) and as the $(\frac{3}{2}, \frac{1}{2}, \frac{1}{2})$ representation of Spin(6). The latter is identical to the representation
[2,1] of SU(4). Antisymmetric states for more than one parti-
cle can be constructed by taking appropriate products of these
representations and result in basis states that can be written as

$$\left| \begin{array}{ccccc} \mathrm{U}(20) \supset & \mathrm{Spin}(6) & \supset \mathrm{Spin}(5) \supset & \mathrm{Spin}(3) \supset & \mathrm{Spin}(2) \\ \downarrow & \downarrow & \downarrow & \downarrow & \downarrow \\ \{N_{\mathrm{F}}\} & \nu, (\sigma_1, \sigma_2, \sigma_3) & (\tau_1, \tau_2) & \nu', J & M_J \end{array} \right\rangle,$$
(2.35)

where ν and ν' represent missing labels, if needed. The
classification for $N_{\mathrm{F}} \leq 2$ is shown in Table 2.6.

2.7.3 Multiple j. Pseudo-spin algebras

For pseudo-spin algebras the construction of basis states is slightly
simpler. Antisymmetric states of $\mathrm{U}(n_j)$ are constructed by taking products of states of $\mathrm{U}_k(n_k)$ and $\mathrm{U}_s(n_s)$. It was shown by

Table 2-7 Partial classification of antisymmetric
states $(j = 1/2, 3/2)^{N_F}$

U(6)	$SU_k(3)$	$SU_s(2)$	$O_k(3)$	$SU_j(2)$
N_F	(λ,μ)	S	L	J
1	(1,0)	1/2	1	1/2,3/2
2	(2,0)	0	0,2	0,2
	(0,1)	1	1	0,1,2

Flowers (1952) that, in order to preserve the overall antisymmetry under $U(n_j)$, the Young tableaux of $U_k(n_k)$ and $U_s(n_s)$ must have conjugate symmetry (i.e. rows and columns interchanged). We consider here as an example the case $j = 1/2, 3/2$, which is classified according to

$$
\left|
\begin{array}{l}
\text{U(6)} \supset \text{SU}_k(3) \otimes \text{SU}_s(2) \supset O_k(3) \otimes \text{SU}_s(2) \\
\quad\downarrow \qquad\quad \downarrow \qquad\qquad \downarrow \qquad\qquad \downarrow \\
\{N_F\} \qquad (\lambda,\mu) \qquad S \qquad\quad L \\[2ex]
\supset \text{SU}_j(2) \supset O_j(2) \\
\qquad \downarrow \qquad\qquad \downarrow \\
\qquad J \qquad\quad M_J
\end{array}
\right\rangle ,
$$

$$(2.36)$$

where the Elliott quantum numbers $\lambda = \lambda_1 - \lambda_2$ and $\mu = \lambda_2$ are used instead of the Young tableau numbers λ_1 and λ_2 themselves (Elliott, 1958). The overall antisymmetry under U(6) imposes a relation between the $SU_k(3)$ and $SU_s(2)$ representations. This leads to the classification of states shown in Table 2.7. The other cases listed in Table 2.2 can be dealt with in a similar way.

2.8 Coupled Bose–Fermi algebras

The algebraic structure of coupled Bose–Fermi systems is very complex. One starts from the algebraic structure of the bosons described as in Volume 1 in terms of bilinear products of boson operators

$$g^{\mathrm{B}}: \qquad B_{\alpha\beta} = b_\alpha^\dagger b_\beta, \qquad (\alpha, \beta = 1, \ldots, n_{\mathrm{B}}), \qquad (2.37)$$

where we have used the notation $B_{\alpha\beta}$ rather than $G_{\alpha\beta}$ for reasons which will become apparent later. The operators (2.37) generate the algebra of $\mathrm{u}^{\mathrm{B}}(n_{\mathrm{B}})$. In the case of the interacting boson model-1, $n_{\mathrm{B}} = 6$. Similarly, one introduces the algebraic structure of the fermions described by the bilinear products of fermion operators

$$g^{\mathrm{F}}: \qquad A_{ik} = a_i^\dagger a_k, \qquad (i, k = 1, \ldots, n_{\mathrm{F}}). \qquad (2.38)$$

The operators A_{ik} generate the Lie algebra of $\mathrm{u}^{\mathrm{F}}(n_{\mathrm{F}})$. The algebraic structure of the coupled system is the direct sum of the two, $\mathrm{u}^{\mathrm{B}}(n_{\mathrm{B}}) \oplus \mathrm{u}^{\mathrm{F}}(n_{\mathrm{F}})$, denoted by

$$\mathrm{U}^{\mathrm{B}}(n_{\mathrm{B}}) \otimes \mathrm{U}^{\mathrm{F}}(n_{\mathrm{F}}), \qquad (2.39)$$

when capital letters are used, as discussed in Sect. 2.5.2. In (2.39) a superscript B or F is added in order to distinguish between boson and fermion algebras and groups.

In general, bosons are classified by a chain of algebras,

$$G^{\mathrm{B}} \supset G'^{\mathrm{B}} \supset G''^{\mathrm{B}} \supset \cdots, \qquad (2.40)$$

while fermions are classified by another chain,

$$G^{\mathrm{F}} \supset G'^{\mathrm{F}} \supset G''^{\mathrm{F}} \supset \cdots. \qquad (2.41)$$

If some of the algebras in the two chains coincide or if they are isomorphic, one can couple them. This coupling will be pictorially described by a lattice of algebras. Often, there are several ways in which the algebras can be coupled leading to several routes within the same lattice,

$$
\begin{array}{ccccc}
G^{\mathrm{B}} & & \otimes & & G^{\mathrm{F}} \\
\downarrow & \searrow{\scriptstyle a} & & \nearrow{\scriptstyle a} & \downarrow \\
G'^{\mathrm{B}} & & G^{\mathrm{BF}} & & G'^{\mathrm{F}} \\
\downarrow & \searrow{\scriptstyle b} & \downarrow & \nearrow{\scriptstyle b} & \downarrow \\
\vdots & & G'^{\mathrm{BF}} & & \vdots \\
& & \downarrow & & \\
& & \vdots & &
\end{array}
\qquad (2.42)
$$

denoted by $a, b, \ldots$.

In the previous sections we have considered two classes of fermion algebras, spinor algebras and pseudo-spin algebras. In the case of spinor algebras the chain is as in (2.41) and leads to a lattice of the type (2.42). This lattice will be called *twofold*. For pseudo-spin algebras, G^{F} is further split into $G_k^{\mathrm{F}} \otimes G_s^{\mathrm{F}}$. This leads to *threefold* lattices,

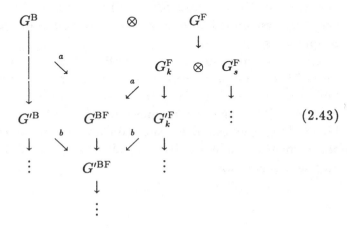

$$(2.43)$$

The freedom in choosing from different coupling routes has a positive and a negative aspect. The positive aspect is that it leads to many situations that can be solved in closed form. The negative aspect is that it is not clear from the beginning which particular route is appropriate to classify the states of a particular nucleus.

The generators of coupled algebras are a linear combination of the generators (2.37) of the boson algebra and those of the fermion algebra, (2.38). In the notation adopted here, these will be mostly written in coupled-tensor form as

$$B_\mu^{(\lambda)}(l, l') = [b_l^\dagger \times \tilde{b}_{l'}]_\mu^{(\lambda)},$$
$$A_\mu^{(\lambda)}(j, j') = [a_j^\dagger \times \tilde{a}_{j'}]_\mu^{(\lambda)}. \qquad (2.44)$$

For spinor algebras (twofold lattices), the generators of the coupled algebras will be a combination of the boson and fermion generators (2.44). When this combination occurs at the earliest possible stage in the chain, it will be called *maximal*. It is clear that a combination at the level of the rotation group is always possible. However, this usually gives trivial results. The earlier the combination of boson and fermion algebras occurs, the more constraints are imposed on the form of the Hamiltonian (discussed in Sect. 2.11).

The situation for pseudo-spin algebras is slightly more complicated. One must first go from the operators $A_\mu^{(\lambda)}(j,j')$ to the operators $K_\mu^{(\lambda)}(k,k')$ and $S_\mu^{(\lambda)}(s,s')$ via (2.25). The B-operators are then first added to the K-operators and subsequently to the S-operators.

The entire procedure for the coupling of algebras is a generalization of the familiar concept of the coupling of angular momenta. For example, starting from the boson algebra $O^B(3)$, generated by the orbital angular momentum, $\vec{L}$, and the fermion algebra $SU^F(2)$, generated by the spin angular momentum, $\vec{S}$, one can form the combined $SU^{BF}(2)$ algebra, generated by the total angular momentum, $\vec{J}$,

$$\vec{J} = \vec{L} + \vec{S}. \tag{2.45}$$

2.9 Particle–hole conjugation. Automorphism

An interesting aspect of algebras built with fermion operators is that there exists an automorphism among the generators of the algebra that preserves the commutation relations. Under this automorphism the generators transform as

$$A_\mu^{(\lambda)}(j,j') \to \bar{A}_\mu^{(\lambda)}(j,j')$$
$$= (-)^{j-j'-\lambda+1} A_\mu^{(\lambda)}(j',j) - \sqrt{2j+1}\,\delta_{\lambda 0}\delta_{\mu 0}\delta_{jj'},$$
$$K_\mu^{(\lambda)}(k,k') \to \bar{K}_\mu^{(\lambda)}(k,k') = (-)^{k-k'-\lambda+1} K_\mu^{(\lambda)}(k',k) + C\delta_{\lambda 0}\delta_{\mu 0}\delta_{kk'},$$
$$S_\mu^{(\lambda)}(s,s') \to \bar{S}_\mu^{(\lambda)}(s,s')$$
$$= (-)^{s-s'-\lambda+1} S_\mu^{(\lambda)}(s',s) + C'\delta_{\lambda 0}\delta_{\mu 0}\delta_{ss'}, \tag{2.46}$$

where C and C' are some c-numbers depending on the pseudo-orbital and pseudo-spin angular momenta involved. This automorphism corresponds to a transformation of the fermion operators

$$a_{j,m}^\dagger \to \tilde{a}_{j,m}, \qquad \tilde{a}_{j,m} \to -a_{j,m}^\dagger, \tag{2.47}$$

which exchanges creation and annihilation operators (particle-hole conjugation). Although the transformed operators (2.46) generate the same algebra as the original operators, it is customary to place a bar over the corresponding algebra, $\overline{G}$, to stress the fact that the relevant representations are different. For orthogonal groups the particle–hole conjugation has no effect and there is no need to introduce transformed algebras.

Using the barred algebras one can construct lattices similar to those of Sect. 2.8. For example, the threefold lattice (2.43) now becomes

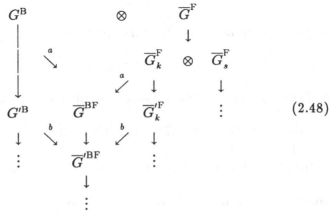

$$(2.48)$$

Note that for SU(2) the particle–hole conjugation has no effect.

2.10   Basis states

Once a particular algebraic structure has been chosen, the next step is to construct the corresponding basis. This basis is provided by the representations of the algebras in the chain. For the algebras G^{B} and G^{F}, the representations are characterized by the number of bosons, N_{B}, and the number of fermions, N_{F}. The appropriate representations of G^{B} are totally symmetric, $[N_{\mathrm{B}}]$, while those of G^{F} are totally antisymmetric, $\{N_{\mathrm{F}}\}$, and given by (2.30). The problem of finding a basis amounts to decomposing representations of an algebra into those of a subalgebra and eventually, at the stage where the algebras are coupled, it amounts to combining the representations. For example, for the chain

$$G^B \otimes G^F \supset G'^B \otimes G'^F \supset G'^{BF} \supset G''^{BF}, \tag{2.49}$$

one has to decompose first

$$G^B \supset G'^B, \qquad G^F \supset G'^F, \tag{2.50}$$

then combine G'^B with G'^F,

$$G'^B \otimes G'^F \supset G'^{BF}, \tag{2.51}$$

and finally decompose G'^{BF} into G''^{BF},

$$G'^{BF} \supset G''^{BF}. \tag{2.52}$$

This procedure provides a certain number of labels

$$\left| \begin{array}{cccccc} G^B & \otimes & G^F & \supset & G'^B & \otimes & G'^F & \supset & G'^{BF} & \supset & G''^{BF} \\ \downarrow & & \downarrow & & \downarrow & & \downarrow & & \downarrow & & \downarrow \\ [N_B] & & \{N_F\} & & \text{labels} & & \text{labels} & & \text{labels} & & \text{labels} \end{array} \right\rangle, \tag{2.53}$$

which may or may not be sufficient to characterize the states uniquely. If they are sufficient, (2.53) is the end of the procedure. If not, one must find *missing labels*.

The combination of representations is best done by using Young tableaux with appropriate multiplication rules (Hamermesh, 1962). One can also take advantage of some of the isomorphisms of algebras discussed in Sect. 2.4. In one of the cases that will be discussed below, one combines $O^B(6)$ with $SU^F(4)$ into $Spin^{BF}(6)$. All three algebras are isomorphic and the calculations are best done using $SU^F(4)$. Simple conversion formulas exist which translate a set of labels of an algebra into labels of an isomorphic algebra. These will be discussed case by case.

2.11 Dynamic symmetries

The basis states constructed in the way described in the previous section can be used for two purposes. First, they form a basis

Table 2-8 Eigenvalues of some Casimir operators of spinor groups

Group	Labels	Order	$\langle \mathcal{C} \rangle$
$\mathrm{Spin}(2n+1)$	$(f_1, f_2, \ldots, f_n)$	2	$\sum\limits_{i=1}^{n} 2f_i(f_i + 2n + 1 - 2i)$
$\mathrm{Spin}(2n)$	$(f_1, f_2, \ldots, f_n)$	2	$\sum\limits_{i=1}^{n} 2f_i(f_i + 2n - 2i)$

in which numerical calculations can be performed. The computer code ODDA (Scholten, 1979) performs calculations in one of the bases of the type (2.53), namely the one in which the coupling occurs at the level of the rotation group. The second purpose is that they provide solutions that can be written in closed form. These solutions, called dynamic symmetries, are particularly important in the present situation which is rather complex.

The technique used to find analytic solutions is the same as that discussed in Volume 1. One writes the Hamiltonian in terms of Casimir invariants of the algebras contained in the chain in which one is interested. For example, for (2.49),

$$H = E_0 + \alpha \mathcal{C}(G^{\mathrm{B}}) + \beta \mathcal{C}(G^{\mathrm{F}}) + \alpha' \mathcal{C}(G'^{\mathrm{B}}) + \beta' \mathcal{C}(G'^{\mathrm{F}})$$
$$+ \gamma \mathcal{C}(G'^{\mathrm{BF}}) + \delta \mathcal{C}(G''^{\mathrm{BF}}), \tag{2.54}$$

where $\mathcal{C}(G)$ denotes a Casimir invariant of G. As discussed in Volume 1, each algebra has, in general, several Casimir invariants of different orders. If only two-body interactions are included in the Hamiltonian, then H will contain at most quadratic invariants. The expectation value of H in the basis (2.53) gives the energies of the states,

$$E(\text{labels}) = E_0 + \alpha \langle \mathcal{C}(G^{\mathrm{B}}) \rangle + \beta \langle \mathcal{C}(G^{\mathrm{F}}) \rangle + \alpha' \langle \mathcal{C}(G'^{\mathrm{B}}) \rangle$$
$$+ \beta' \langle \mathcal{C}(G'^{\mathrm{F}}) \rangle + \gamma \langle \mathcal{C}(G'^{\mathrm{BF}}) \rangle + \delta \langle \mathcal{C}(G''^{\mathrm{BF}}) \rangle. \tag{2.55}$$

The eigenvalues of the linear and quadratic invariants of $U(n)$, $SU(n)$, $O(2n)$, $O(2n+1)$ and $Sp(2n)$ are given in Volume 1, Table 2.8. For the spinor algebras, they are identical to those of the corresponding orthogonal algebras and are given in Table 2.8. The only difference is that now the f_is are half-integer numbers.

Table 2-9 Eigenvalues of some Casimir operators of unitary groups

Group	Representation	Order	$\langle \mathcal{C} \rangle$
$U^B(n_B)$	$\overbrace{\square\square\cdots\square}^{N_B}$	1	N_B
		2	$N_B(N_B + n_B - 1)$
$U^F(n_F)$	$N_F \left\{ \begin{array}{c} \square \\ \square \\ \vdots \\ \square \end{array} \right.$	1	N_F
		2	$N_F(n_F + 1 - N_F)$

Returning to (2.55) one notes that the Casimir invariants of G^B and G^F contribute equally to all states since their expectation values depend only on N_B and N_F which are constant for a given nucleus. They are important only when one calculates binding energies. In Table 2.9 we quote the expectation values of the linear and quadratic invariants of $U^B(n_B)$ and $U^F(n_F)$ for totally symmetric and totally antisymmetric states.

2.12   Wave functions. Isoscalar factors

The final step in the use of dynamic symmetries for problems in structure physics is the evaluation of matrix elements of operators, such as the electromagnetic transition operators. In order to do this evaluation in closed form, one needs the full power of algebraic methods. One must first construct wave functions explicitly for the relevant chain, for example (2.49). Since one first combines the representations of G'^B and G'^F into G'^{BF} and subsequently decomposes those into representations of G''^{BF}, one needs coupling coefficients. The concept of coupling coefficients is described in the book of Wybourne (1974). It is a generalization of the familiar concept of Clebsch–Gordan coefficients for the rotation group, $(lm_l\, sm_s | jm_j)$. For nested chains such as those in (2.49), it must be enlarged to that of *isoscalar factors*. The calculation of isoscalar factors for a nested chain

$$G'^{\mathrm{B}} \otimes G'^{\mathrm{F}} \supset G'^{\mathrm{BF}} \supset G''^{\mathrm{BF}} \supset G'''^{\mathrm{BF}} \supset \cdots, \qquad (2.56)$$

is simplified by Racah's factorization lemma (Racah, 1949), which states that the isoscalar factors for $G'^{\mathrm{BF}} \supset G''^{\mathrm{BF}} \supset G'''^{\mathrm{BF}} \supset \cdots$ can be factorized at each step into those for $G'^{\mathrm{BF}} \supset G''^{\mathrm{BF}}$, those for $G''^{\mathrm{BF}} \supset G'''^{\mathrm{BF}}$, etc. Isoscalar factors for each step will be denoted by

$$\left\langle \begin{array}{cc|c} \Lambda_1 & \Lambda_2 & \Lambda \\ \lambda_1 & \lambda_2 & \lambda \end{array} \right\rangle, \qquad (2.57)$$

where $\Lambda_1, \lambda_1, \Lambda_2, \lambda_2$ and Λ, λ denote the labels of the groups in question. For example, in the isoscalar factor for $O(6) \supset O(5)$, Λ_1, Λ_2 and Λ are $O(6)$ labels, while λ_1, λ_2 and λ are $O(5)$ labels. This notation will become clearer later when applied to the explicit computation of isoscalar factors. The wave functions for the combined groups can then be written with the help of isoscalar factors as

$$|\Lambda\lambda\rangle = \sum_{\lambda_1\lambda_2} \left\langle \begin{array}{cc|c} \Lambda_1 & \Lambda_2 & \Lambda \\ \lambda_1 & \lambda_2 & \lambda \end{array} \right\rangle |\Lambda_1\lambda_1\rangle \, |\Lambda_2\lambda_2\rangle, \qquad (2.58)$$

which is again a generalization of the familiar Clebsch–Gordan relation for the rotation group.

The knowledge of isoscalar factors allows one to compute the matrix elements of transition operators T and transfer operators P of Sect. 1.4 between states of the basis (2.53). Examples of such calculations will be given later.

3

Bose–Fermi symmetries

3.1 Introduction

The algebraic methods described in the previous chapter can be (and have been) used to derive a variety of results within the framework of the interacting boson–fermion model-1. It is convenient here to divide the exposition into three parts (Sects. 3.2, 3.3 and 3.4 respectively) by considering Bose–Fermi symmetries associated with each of the dynamic symmetries of the interacting boson model-1 discussed in Volume 1:

$$U^B(6) \quad
\begin{cases}
\nearrow & U^B(5) \supset O^B(5) \supset O^B(3) \supset O^B(2), \quad \text{I} \\
\rightarrow & SU^B(3) \supset O^B(3) \supset O^B(2), \quad \text{II} \\
\searrow & O^B(6) \supset O^B(5) \supset O^B(3) \supset O^B(2). \quad \text{III}
\end{cases}$$

We shall begin with the dynamic symmetries associated with symmetry III since this is the case that has been most extensively investigated and for which there are the best experimental examples. We then proceed with symmetries I and II.

The exposition in this chapter will, to a certain extent, be repetitive, since we shall follow the same logic scheme to study the 13 cases of practical interest. The scheme consists in the following steps.

(i) *Identifying the appropriate algebraic structure* (lattice of algebras). Here, the boson algebraic structure will be always taken to be $U^B(6)$, while the fermion algebraic structure will depend on the values of the angular momenta, j, taken into consideration.

(ii) *Constructing the corresponding basis.* This procedure provides the quantum numbers needed to classify the states uniquely.

(iii) *Writing the Hamiltonian in terms of Casimir operators and finding its eigenvalues.*

(iv) *Evaluating matrix elements of interest.* In the study of odd–even nuclei, to which this chapter is particularly addressed, the matrix elements of interest are usually those that allow a calculation of:

(i) electric quadrupole (E2) transitions and moments;

(ii) magnetic dipole (M1) transitions and moments;

(iii) electric monopole (E0) transitions and moments (nuclear radii);

(iv) one-nucleon transfer reaction intensities;

(v) two-nucleon transfer reaction intensities.

We will present many closed formulas for these quantities. These formulas are of particular interest for a direct comparison with the experimental data. We do this comparison explicitly for several nuclei throughout the periodic table. The agreement between theory and experiment ranges from excellent (as in $^{195}_{78}\text{Pt}_{117}$, $^{63}_{29}\text{Cu}_{34}$ and $^{103}_{45}\text{Rh}_{58}$) to good ($^{193}_{77}\text{Ir}_{116}$ and $^{197}_{79}\text{Au}_{118}$) to only qualitative ($^{185}_{74}\text{W}_{111}$ and $^{169}_{69}\text{Tm}_{100}$). The various classification schemes presented in this chapter should allow one to study, at least approximately, the structure of any odd–even nucleus with mass $A \geq 70$.

In addition, we present here explicit expressions for several isoscalar factors involving the algebras U(6), O(6), O(5) and O(3). These can be useful also in the context of other models which make use of group theoretic techniques.

3.2 Symmetries associated with O(6)

We begin our discussion with Bose–Fermi symmetries associated with $O^B(6)$. We consider here five cases, each denoted by the Bose–Fermi algebra that occurs at the earliest stage in the combination of the corresponding boson and fermion algebras. We label the symmetries with a roman numeral, I, II and III, as in Volume 1, and a subscript $1, 2, \ldots$, following the order of presentation.

3.2.1 $\text{Spin}^{BF}(6)$ *(III$_1$)*

3.2.1.1 Lattice of algebras This symmetry was the first to be investigated in detail (Iachello, 1980; Iachello and Kuyucak, 1981;

Kuyucak, 1982). It corresponds to bosons with O(6) symmetry and fermions occupying a state with $j = 3/2$. This symmetry exploits the isomorphisms of the algebras discussed in Chapter 2, $SU(4) \approx O(6) \approx Spin(6)$, $Sp(4) \approx O(5) \approx Spin(5)$ and $SU(2) \approx O(3) \approx Spin(3)$. The lattice of algebras is:

$$
\begin{array}{ccc}
U^B(6) & \otimes & U^F(4) \\
\downarrow & & \downarrow \\
O^B(6) & & SU^F(4) \\
\downarrow \quad \searrow^a & & \swarrow^a \downarrow \\
O^B(5) \quad Spin^{BF}(6) & & Sp^F(4) \\
\downarrow \quad \searrow^b & \downarrow & \swarrow^b \downarrow \\
O^B(3) \quad Spin^{BF}(5) & & SU^F(2) \\
\searrow^c & \downarrow & \swarrow^c \\
& Spin^{BF}(3) & \\
& \downarrow & \\
& Spin^{BF}(2) & .
\end{array}
\tag{3.1}
$$

We consider here in detail the route a. The combined algebra is $Spin^{BF}(6)$, whose generators can be written explicitly as

$$
\begin{aligned}
G^{(3)}_\mu &= [d^\dagger \times \tilde{d}]^{(3)}_\mu + \tfrac{1}{\sqrt{2}}[a^\dagger_{3/2} \times \tilde{a}_{3/2}]^{(3)}_\mu & 7 \\
G^{(2)}_\mu &= [s^\dagger \times \tilde{d} + d^\dagger \times \tilde{s}]^{(2)}_\mu + [a^\dagger_{3/2} \times \tilde{a}_{3/2}]^{(2)}_\mu & 5 \\
G^{(1)}_\mu &= [d^\dagger \times \tilde{d}]^{(1)}_\mu - \tfrac{1}{\sqrt{2}}[a^\dagger_{3/2} \times \tilde{a}_{3/2}]^{(1)}_\mu & 3 \\
& & \overline{} \\
& & 15.
\end{aligned}
$$
$$\tag{3.2}$$

Deleting $G^{(2)}_\mu$ gives the generators of $Spin^{BF}(5)$ and further deleting $G^{(3)}_\mu$ gives those of $Spin^{BF}(3)$.

3.2.1.2 Basis states We consider here only the case with one fermion, $N_F = 1$. The basis states are of the form

$$
\left.\begin{array}{l}
\begin{array}{ccccc}
U^B(6) & \otimes & U^F(4) & \supset O^B(6) \otimes SU^F(4) & \supset Spin^{BF}(6) \\
\downarrow & & \downarrow & \downarrow & \downarrow \\
[N_B = N] & & \{N_F = 1\} & \Sigma & (\sigma_1, \sigma_2, \sigma_3)
\end{array} \\[1em]
\begin{array}{ccccc}
\supset Spin^{BF}(5) & \supset Spin^{BF}(3) & \supset Spin^{BF}(2) \\
\downarrow & \downarrow & \downarrow \\
(\tau_1, \tau_2) & \nu_\Delta, J & M_J
\end{array}
\end{array}\right\}.
$$

$$(3.3)$$

Although, in general, three labels are needed to characterize the representations of $O(6)$, only one is non-zero here, $(\Sigma, 0, 0)$, since $O^B(6)$ describes bosons. The values of Σ are given by the rule discussed in Volume 1,

$$[N] : \Sigma = N, N - 2, \ldots, 1 \text{ or } 0; \qquad (N = \text{ odd or even}). \quad (3.4)$$

The next step involves the combination of $O^B(6)$ and $SU^F(4)$ to $Spin^{BF}(6)$. There are three labels that characterize the representations of $Spin(6)$. For these, one can use either Murnaghan's notation $(\sigma_1, \sigma_2, \sigma_3)$, (Murnaghan, 1938) or, by exploiting the isomorphism of $Spin(6)$ and $SU(4)$, one can use the $SU(4)$ notation (n_1, n_2, n_3), where n_1, n_2 and n_3 denote the number of boxes in the Young tableau. The labels (n_1, n_2, n_3) can be converted into the labels $(\sigma_1, \sigma_2, \sigma_3)$ using the relations

$$n_1 = \sigma_1 + \sigma_2; \qquad n_2 = \sigma_1 - \sigma_3; \qquad n_3 = \sigma_2 - \sigma_3. \quad (3.5)$$

The representations of $Spin^{BF}(6)$ can be obtained by taking the outer product of the representations of $O^B(6)$ and $SU^F(4)$. It is convenient to use here the $SU(4)$ labels since the multiplication rules for unitary groups are simpler. For $N_F = 1$ one must consider the product

$$
\overbrace{\begin{array}{c}\square\,\square\,\cdots\,\square \\ \square\,\square\,\cdots\,\square\end{array}}^{\Sigma} \otimes\, \square =
\overbrace{\begin{array}{c}\square\,\square\,\cdots\,\square\,\square \\ \square\,\square\,\cdots\,\square\end{array}}^{\Sigma} \oplus
\overbrace{\begin{array}{c}\square\,\square\,\cdots\,\square \\ \square\,\square\,\cdots\,\square \\ \square\end{array}}^{\Sigma}. \quad (3.6)
$$

Converting to the Spin(6) notation using (3.5), we obtain

$$(\Sigma, 0, 0) \otimes (\tfrac{1}{2}, \tfrac{1}{2}, \tfrac{1}{2}) = (\Sigma + \tfrac{1}{2}, \tfrac{1}{2}, \tfrac{1}{2}) \oplus (\Sigma - \tfrac{1}{2}, \tfrac{1}{2}, -\tfrac{1}{2}). \qquad (3.7)$$

With the values of Σ given by (3.4), the resulting values of $(\sigma_1, \sigma_2, \sigma_3)$ are

$$\sigma_1 = N + \tfrac{1}{2}, N - \tfrac{3}{2}, \ldots, \tfrac{1}{2} \text{ or } \tfrac{3}{2}; \quad \sigma_2 = \tfrac{1}{2}; \quad \sigma_3 = \tfrac{1}{2},$$
$$\sigma_1 = N - \tfrac{1}{2}, N - \tfrac{5}{2}, \ldots, \tfrac{1}{2} \text{ or } \tfrac{3}{2}; \quad \sigma_2 = \tfrac{1}{2}; \quad \sigma_3 = -\tfrac{1}{2}. \quad (3.8)$$

The representations $(\sigma_1, \tfrac{1}{2}, \tfrac{1}{2})$ and $(\sigma_1, \tfrac{1}{2}, -\tfrac{1}{2})$ are, as mentioned in Sect. 2.6, equivalent. Thus, it is sufficient to label them as $(\sigma_1, \sigma_2, |\sigma_3|)$, reducing (3.8) to

$$\sigma_1 = N + \tfrac{1}{2}, N - \tfrac{1}{2}, N - \tfrac{3}{2}, \ldots, \tfrac{1}{2}; \qquad \sigma_2 = |\sigma_3| = \tfrac{1}{2}. \quad (3.9)$$

The representations of $\mathrm{Spin}^{\mathrm{BF}}(5)$ are labelled by two quantum numbers (τ_1, τ_2). The values of (τ_1, τ_2) contained in each $\mathrm{Spin}^{\mathrm{BF}}(6)$ representation $(\sigma_1, \sigma_2 = \tfrac{1}{2}, |\sigma_3| = \tfrac{1}{2})$ are given by

$$(\sigma_1, \tfrac{1}{2}, \tfrac{1}{2}) : (\tau_1, \tau_2) = (\sigma_1, \tfrac{1}{2}), (\sigma_1 - 1, \tfrac{1}{2}), \ldots, (\tfrac{1}{2}, \tfrac{1}{2}). \qquad (3.10)$$

The representations of $\mathrm{Spin}^{\mathrm{BF}}(3)$ are labelled by one quantum number J. However, when going from Spin(5) to Spin(3), one encounters the problem of non-fully reducibility discussed in Volume 1, p. 27. One needs a further quantum number ν_Δ to characterize uniquely the states. For $N_{\mathrm{F}} = 1$ this quantum number takes on the values

$$\nu_\Delta = 0, \tfrac{1}{2}, 1, \tfrac{3}{2}, \ldots. \qquad (3.11)$$

The integer values apply to the representations with $\sigma_3 = \tfrac{1}{2}$, while the half-integer values apply to those with $\sigma_3 = -\tfrac{1}{2}$. When $N_{\mathrm{F}} = 1$, the angular momentum J is given by (Iachello, 1980)

$$(\tau_1, \tfrac{1}{2}) : J = 2\tau_1 - 6\nu_\Delta + \tfrac{1}{2}, 2\tau_1 - 6\nu_\Delta - \tfrac{1}{2}, \ldots,$$
$$\tau_1 - 3\nu_\Delta + 1 - \tfrac{1}{4}[1 - (-)^{2\nu_\Delta}]. \qquad (3.12)$$

The missing label for the Spin(5) $\supset$ Spin(3) can be defined in a different way, as discussed by Van der Jeugt (1985). The values of M_J are given as usual by the half-integers

Table 3-1 Angular momentum content of the lowest Spin(5) representations $(\tau_1, \tau_2 = \frac{1}{2})$

(τ_1, τ_2)	$\nu_\Delta = 0$	$\nu_\Delta = \frac{1}{2}$	$\nu_\Delta = 1$
$(\frac{1}{2}, \frac{1}{2})$	3/2		
$(\frac{3}{2}, \frac{1}{2})$	7/2,5/2	1/2	
$(\frac{5}{2}, \frac{1}{2})$	11/2,9/2,7/2	5/2,3/2	
$(\frac{7}{2}, \frac{1}{2})$	15/2,13/2,11/2,9/2	9/2,7/2,5/2	3/2

$$-J \leq M_J \leq +J. \tag{3.13}$$

In Table 3.1 we give the decomposition Spin(5) $\supset$ Spin(3) for the lowest Spin(5) representations. Summarizing, the basis states can be labelled by

$$|[N_B = N], \{N_F = 1\}, \Sigma, (\sigma_1, \sigma_2, \sigma_3), (\tau_1, \tau_2), \nu_\Delta, J, M_J\rangle. \tag{3.14}$$

3.2.1.3 Energy eigenvalues Energy eigenvalues can be obtained using the method discussed in Sect. 2.11. We write the Hamiltonian as

$$
\begin{aligned}
H^{(\mathrm{III}_{1a})} &= e_0 + e_1 \mathcal{C}_1(\mathrm{U}^\mathrm{B}6) + e_2 \mathcal{C}_2(\mathrm{U}^\mathrm{B}6) + e_3 \mathcal{C}_1(\mathrm{U}^\mathrm{F}4) + e_4 \mathcal{C}_2(\mathrm{U}^\mathrm{F}4) \\
&\quad + e_5 \mathcal{C}_1(\mathrm{U}^\mathrm{B}6)\mathcal{C}_1(\mathrm{U}^\mathrm{F}4) + \eta \mathcal{C}_2(\mathrm{O}^\mathrm{B}6) + \eta' \mathcal{C}_2(\mathrm{Spin}^\mathrm{BF}6) \\
&\quad + \beta \mathcal{C}_2(\mathrm{Spin}^\mathrm{BF}5) + \gamma \mathcal{C}_2(\mathrm{Spin}^\mathrm{BF}3). \tag{3.15}
\end{aligned}
$$

The Casimir invariant of $\mathrm{Spin}^\mathrm{BF}(2)$ is not included in (3.15), since we assume that the nucleus is not placed in an external electric or magnetic field. The Hamiltonian (3.15) is diagonal in the basis (3.3) with eigenvalues given in Table 2.8 of Volume 1 and Tables 2.8 and 2.9 of this volume. They are

$$
\begin{aligned}
E^{(\mathrm{III}_{1a})}&(N_B = N, N_F = 1, \Sigma, (\sigma_1, \sigma_2, \sigma_3), (\tau_1, \tau_2), \nu_\Delta, J, M_J) \\
&= E_{01} + 2\eta\Sigma(\Sigma + 4) + 2\eta'[\sigma_1(\sigma_1 + 4) + \sigma_2(\sigma_2 + 2) + \sigma_3^2] \\
&\quad + 2\beta[\tau_1(\tau_1 + 3) + \tau_2(\tau_2 + 1)] + 2\gamma J(J + 1), \tag{3.16}
\end{aligned}
$$

where

$$E_{01} = e_0 + e_1 N + e_2 N(N + 5) + e_3 + 4e_4 + e_5 N. \tag{3.17}$$

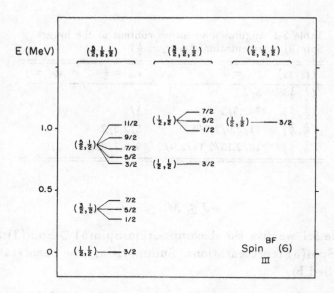

Fig. 3-1 A typical spectrum with $\text{Spin}^{\text{BF}}(6)$ (III) symmetry for $N_\text{B} = 2$, $N_\text{F} = 1$. The $\text{Spin}^{\text{BF}}(6)$ quantum numbers are shown on top, the $\text{Spin}^{\text{BF}}(5)$ quantum numbers to the left and the angular momentum J to the right of each level.

The quantity E_{01} does not contribute to the excitation energies but only to the binding energies. A spectrum of states corresponding to (3.16) is shown in Fig. 3.1.

For the calculation of the wave functions, it is of interest to write down explicitly the form of the Casimir invariants of the spinor algebras appearing in (3.15). They are given in terms of the operators (3.2) as follows:

$$\mathcal{C}_2(\text{Spin}^{\text{BF}}6) = 2G^{(2)} \cdot G^{(2)} + 4G^{(1)} \cdot G^{(1)} + 4G^{(3)} \cdot G^{(3)},$$

$$\mathcal{C}_2(\text{Spin}^{\text{BF}}5) = 4G^{(1)} \cdot G^{(1)} + 4G^{(3)} \cdot G^{(3)},$$

$$\mathcal{C}_2(\text{Spin}^{\text{BF}}3) = 20G^{(1)} \cdot G^{(1)}. \qquad (3.18)$$

3.2.1.4 Wave functions. Isoscalar factors In order to compute matrix elements of operators one needs wave functions obtained by coupling $O^\text{B}(6)$ and $SU^\text{F}(4)$ representations, since that is the level at which boson and fermion algebras are combined. For some cases of interest these wave functions have been explicitly constructed. Consider the $\mathrm{Spin}^{\mathrm{BF}}(6)$ wave function

$$|[N_{\mathrm{B}}=N],\{N_{\mathrm{F}}=1\},(\sigma_1=N+\tfrac{1}{2},\sigma_2=\tfrac{1}{2},\sigma_3=\tfrac{1}{2}),$$
$$(\tau_1,\tau_2=\tfrac{1}{2}),\nu_\Delta=0,J,M_J),$$

denoted in short by $|N+\tfrac{1}{2};\tau_1;J\rangle$. This wave function can be expanded into products of wave functions for $\mathrm{O}^{\mathrm{B}}(6)$,

$$|[N_{\mathrm{B}}=N],\{N_{\mathrm{F}}=0\},(\sigma_1=N,\sigma_2=0,\sigma_3=0),$$
$$(\tau_1=\tau',\tau_2=0),\nu_\Delta=0,L',M_L'),$$

denoted by $|N;\tau';L'\rangle$, and the wave function for $\mathrm{SU}^{\mathrm{F}}(4)$. Since $N_{\mathrm{F}}=1$, this wave function is just the fundamental representation

$$|\{N_{\mathrm{F}}=1\},(\sigma_1=\tfrac{1}{2},\sigma_2=\tfrac{1}{2},\sigma_3=\tfrac{1}{2}),(\tau_1=\tfrac{1}{2},\tau_2=\tfrac{1}{2}),$$
$$\nu_\Delta=0,J=3/2,M_J),$$

denoted by $|\tfrac{1}{2};\tfrac{1}{2};3/2\rangle$. The expansion is

$$|N+\tfrac{1}{2};\tau_1;J\rangle = \sum_{\tau',L'} \xi^{N,\tau',L'}_{N+\frac{1}{2},\tau_1,J}|N,\tau',L';\tfrac{1}{2},\tfrac{1}{2},3/2;J\rangle. \qquad (3.19)$$

Since we are interested in reduced matrix elements of operators, we have omitted magnetic quantum numbers from (3.19). The ξs are the expansion coefficients and the sum goes over $\tau'=\tau_1\pm\tfrac{1}{2}$, $L'=J\pm3/2, J\pm1/2$. The expansion coefficients can also be interpreted as isoscalar factors for the group chain $\mathrm{SU}(4)\approx\mathrm{Spin}(6)\supset\mathrm{Sp}(4)\approx\mathrm{Spin}(5)\supset\mathrm{SU}(2)\approx\mathrm{Spin}(3)$.

A technique which has been found useful for obtaining the isoscalar factors is that of considering matrix elements of operators which can be evaluated in a straightforward way. In the present case it is sufficient to consider the operator $G^{(2)}\cdot G^{(2)}$. On the one hand, one has:

$$G^{(2)}\cdot G^{(2)} = A^{(2)}\cdot A^{(2)} + B^{(2)}\cdot B^{(2)} + 2A^{(2)}\cdot B^{(2)}. \qquad (3.20)$$

On the other hand, by using (3.18), this operator can be written as

$$G^{(2)}\cdot G^{(2)} = \tfrac{1}{2}\left[\mathcal{C}_2(\mathrm{Spin}^{\mathrm{BF}}6) - \mathcal{C}_2(\mathrm{Spin}^{\mathrm{BF}}5)\right]. \qquad (3.21)$$

Since

$$A^{(2)} \cdot A^{(2)} = \tfrac{1}{2} \left[\mathcal{C}_2(\mathrm{SU^F}4) - \mathcal{C}_2(\mathrm{Sp^F}4) \right],$$

$$B^{(2)} \cdot B^{(2)} = \tfrac{1}{2} \left[\mathcal{C}_2(\mathrm{O^B}6) - \mathcal{C}_2(\mathrm{O^B}5) \right], \tag{3.22}$$

one obtains, by taking matrix elements of $G^{(2)} \cdot G^{(2)}$ between (3.19) and the state $|N, \tau_1 - \tfrac{1}{2}, L; \tfrac{1}{2}, \tfrac{1}{2}, 3/2; J\rangle$,

$$
\left[(N + \tfrac{1}{2})(N + \tfrac{9}{2}) + \tfrac{3}{2} - \tau_1(\tau_1 + 3) - \tfrac{3}{4} \right] \xi_{N+\frac{1}{2},\tau_1,J}^{N,\tau_1-\frac{1}{2},L}
$$

$$
= \left[N(N+4) - (\tau_1 - \tfrac{1}{2})(\tau_1 + \tfrac{5}{2}) + \tfrac{5}{4} \right] \xi_{N+\frac{1}{2},\tau_1,J}^{N,\tau_1-\frac{1}{2},L}
$$

$$
+ \sum_{\tau',L'} 2(-)^{J+L'+3/2} \left\{ \begin{matrix} L & 3/2 & J \\ 3/2 & L' & 2 \end{matrix} \right\}
$$

$$
\times \langle \tfrac{1}{2}; \tfrac{1}{2}; 3/2 \parallel A^{(2)} \parallel \tfrac{1}{2}; \tfrac{1}{2}; 3/2 \rangle
$$

$$
\times \langle N; \tau_1 - \tfrac{1}{2}; L \parallel B^{(2)} \parallel N; \tau'; L' \rangle \xi_{N+\frac{1}{2},\tau_1,J}^{N,\tau',L'}, \tag{3.23}
$$

where one has used the expression for the eigenvalues of the Casimir invariants given in the previous chapter. The double bar $\parallel$ in (3.23) denotes reduced matrix elements with respect to the rotation group (de-Shalit and Talmi, 1963) and its definition is given in Eq. (2.100) of Volume 1. Since

$$\langle \tfrac{1}{2}; \tfrac{1}{2}; 3/2 \parallel A^{(2)} \parallel \tfrac{1}{2}; \tfrac{1}{2}; 3/2 \rangle = -\sqrt{5}, \tag{3.24}$$

one finds

$$
\xi_{N+\frac{1}{2},\tau_1,J}^{N,\tau_1-\frac{1}{2},L} = \frac{-2\sqrt{5}}{N - \tau_1 + \frac{1}{2}} \sum_{L'} (-)^{J+L'+3/2} \left\{ \begin{matrix} L & 3/2 & J \\ 3/2 & L' & 2 \end{matrix} \right\}
$$

$$
\times \langle N; \tau_1 - \tfrac{1}{2}; L \parallel B^{(2)} \parallel N; \tau_1 + \tfrac{1}{2}; L' \rangle \xi_{N+\frac{1}{2},\tau_1,J}^{N,\tau_1+\frac{1}{2},L'}. \tag{3.25}
$$

Since the matrix elements of the operator $B^{(2)}$ between $\mathrm{O^B}(6)$ states are known (see Volume 1, p. 56), Eq. (3.25) gives a relation between the ξs. A second relation can be obtained by taking matrix elements of $G^{(2)} \cdot G^{(2)}$ between (3.19) and the state

$$|N, \tau_1 + \tfrac{1}{2}, L; \tfrac{1}{2}, \tfrac{1}{2}, 3/2; J\rangle.$$

This gives

$$\xi^{N,\tau_1+\frac{1}{2},L}_{N+\frac{1}{2},\tau_1,J} = \frac{-2\sqrt{5}}{N+\tau_1+\frac{7}{2}} \sum_{L'} (-)^{J+L'+3/2} \begin{Bmatrix} L & 3/2 & J \\ 3/2 & L' & 2 \end{Bmatrix}$$

$$\times \langle N;\tau_1+\tfrac{1}{2};L \parallel B^{(2)} \parallel N;\tau_1-\tfrac{1}{2};L'\rangle \xi^{N,\tau_1-\frac{1}{2},L'}_{N+\frac{1}{2},\tau_1,J}. \tag{3.26}$$

Equations (3.25)–(3.26) provide a set of eight linear homogeneous equations in the eight isoscalar factors, and, together with the normalization condition,

$$\sum_{\tau,L} \left(\xi^{N,\tau,L}_{N+\frac{1}{2},\tau_1,J} \right)^2 = 1, \tag{3.27}$$

completely determine the isoscalar factors. The system of eight equations can be further simplified by making use of Racah's factorization lemma, which allows one to write a ξ-coefficient as the product of two isoscalar factors associated with $SU(4) \supset Sp(4)$ and $Sp(4) \supset SU(2)$,

$$\xi^{N,\tau,L}_{N+\frac{1}{2},\tau_1,J} = \left\langle \begin{array}{cc|c} (N,0,0) & (\frac{1}{2},\frac{1}{2},\frac{1}{2}) & (N+\frac{1}{2},\frac{1}{2},\frac{1}{2}) \\ (\tau,0) & (\frac{1}{2},\frac{1}{2}) & (\tau_1,\frac{1}{2}) \end{array} \right\rangle$$

$$\times \left\langle \begin{array}{cc|c} (\tau,0) & (\frac{1}{2},\frac{1}{2}) & (\tau_1,\frac{1}{2}) \\ L & 3/2 & J \end{array} \right\rangle. \tag{3.28}$$

Both isoscalar factors separately satisfy orthonormalization conditions. For example, for the $SU(4) \supset Sp(4)$ isoscalar factors we have:

$$\sum_{\tau} \left\langle \begin{array}{cc|c} (N,0,0) & (\frac{1}{2},\frac{1}{2},\frac{1}{2}) & (\sigma_1,\frac{1}{2},\frac{1}{2}) \\ (\tau,0) & (\frac{1}{2},\frac{1}{2}) & (\tau_1,\frac{1}{2}) \end{array} \right\rangle$$

$$\times \left\langle \begin{array}{cc|c} (N,0,0) & (\frac{1}{2},\frac{1}{2},\frac{1}{2}) & (\sigma'_1,\frac{1}{2},\frac{1}{2}) \\ (\tau,0) & (\frac{1}{2},\frac{1}{2}) & (\tau_1,\frac{1}{2}) \end{array} \right\rangle = \delta_{\sigma_1,\sigma'_1}. \tag{3.29}$$

Another useful property of isoscalar factors, which follows from orthonormality, is that in the case of unique coupling (the *stretched* case), they reduce to unity. For example, if $J = 2\tau + \frac{3}{2}$ only one value is allowed for L, namely $L = 2\tau$, and consequently one finds

$$\left\langle \begin{array}{cc|c} (\tau,0) & (\frac{1}{2},\frac{1}{2}) & (\tau+\frac{1}{2},\frac{1}{2}) \\ 2\tau & 3/2 & 2\tau+\frac{3}{2} \end{array} \right\rangle^2 = +1. \tag{3.30}$$

Table 3-2 General expressions for the $SU(4) \supset Sp(4)$ isoscalar factors

$$\left\langle \begin{array}{cc} (N,0,0) & (\frac{1}{2},\frac{1}{2},\frac{1}{2}) \\ (\tau,0) & (\frac{1}{2},\frac{1}{2}) \end{array} \middle| \begin{array}{c} (\sigma_1,\frac{1}{2},\frac{1}{2}) \\ (\tau_1,\frac{1}{2}) \end{array} \right\rangle$$

	$\sigma_1 = N - \frac{1}{2}$	$\sigma_1 = N + \frac{1}{2}$
$\tau_1 = \tau - \frac{1}{2}$	$\left[\dfrac{N + \tau_1 + \frac{7}{2}}{2(N+2)}\right]^{\frac{1}{2}}$	$-\left[\dfrac{N - \tau_1 + \frac{1}{2}}{2(N+2)}\right]^{\frac{1}{2}}$
$\tau_1 = \tau + \frac{1}{2}$	$\left[\dfrac{N - \tau_1 + \frac{1}{2}}{2(N+2)}\right]^{\frac{1}{2}}$	$\left[\dfrac{N + \tau_1 + \frac{7}{2}}{2(N+2)}\right]^{\frac{1}{2}}$

Putting $J = 2\tau_1 + \frac{1}{2}$ and $L = 2\tau_1 \pm 1$ in (3.26) and making use of (3.30), one finds general expressions for the $SU(4) \supset Sp(4)$ isoscalar factors, shown in Table 3.2. Since the decomposition $Sp(4) \supset SU(2)$ is not fully reducible, no general expressions are available for the corresponding isoscalar factors. Cases of interest, however, can be computed on the basis of (3.25)–(3.26) and are given in Table 3.3.

3.2.1.5 Electromagnetic transitions and moments; E2 The knowledge of the isoscalar factors allows one to compute matrix elements of all operators between the basis states (3.3). We consider first the matrix elements of the E2 transition operator. Using the notation of Chapter 1 and Volume 1, we can write this operator as

$$T_\mu^{(E2)} = T_{B,\mu}^{(E2)} + T_{F,\mu}^{(E2)}, \tag{3.31}$$

with

$$\begin{aligned} T_{B,\mu}^{(E2)} &= \alpha_2 [s^\dagger \times \tilde{d} + d^\dagger \times \tilde{s}]_\mu^{(2)} + \beta_2 [d^\dagger \times \tilde{d}]_\mu^{(2)}, \\ T_{F,\mu}^{(E2)} &= f_2 [a^\dagger_{3/2} \times \tilde{a}_{3/2}]_\mu^{(2)}. \end{aligned} \tag{3.32}$$

In the calculation of matrix elements of operators, simplifications occur whenever the operator can be written in terms of generators of algebras in the chain. In the present case, this implies $\beta_2 = 0$, since then the operator $T_{B,\mu}^{(E2)}$ is a generator of $O^B(6)$ and thus

$$T_\mu^{(E2)} = \alpha_2 B_\mu^{(2)} + f_2 A_\mu^{(2)}, \tag{3.33}$$

Table 3-3 A selected number of $Sp(4) \supset SU(2)$ isoscalar factors

$$\left\langle \begin{array}{cc} (\tau,0) & (\frac{1}{2},\frac{1}{2}) \\ L & 3/2 \end{array} \right| \left. \begin{array}{c} (\tau_1,\frac{1}{2}) \\ J \end{array} \right\rangle$$

$(\tau_1,\frac{1}{2})$	J	L	Isoscalar factor
$(\tau + \frac{1}{2},\frac{1}{2})$	$2\tau + \frac{3}{2}$	2τ	1
$(\tau + \frac{1}{2},\frac{1}{2})$	$2\tau + \frac{1}{2}$	2τ	1
$(\tau + \frac{1}{2},\frac{1}{2})$	$2\tau - \frac{1}{2}$	2τ	$-\left[\dfrac{4(\tau - 1)}{(2\tau + 3)(4\tau - 1)}\right]^{\frac{1}{2}}$
$(\tau + \frac{1}{2},\frac{1}{2})$	$2\tau - \frac{1}{2}$	$2\tau - 2$	$\left[\dfrac{(2\tau + 1)(4\tau + 1)}{(2\tau + 3)(4\tau - 1)}\right]^{\frac{1}{2}}$
$(\tau - \frac{1}{2},\frac{1}{2})$	$2\tau - \frac{1}{2}$	2τ	$\left[\dfrac{(2\tau + 1)(4\tau + 1)}{(2\tau + 3)(4\tau - 1)}\right]^{\frac{1}{2}}$
$(\tau - \frac{1}{2},\frac{1}{2})$	$2\tau - \frac{1}{2}$	$2\tau - 2$	$\left[\dfrac{4(\tau - 1)}{(2\tau + 3)(4\tau - 1)}\right]^{\frac{1}{2}}$
$(\tau - \frac{1}{2},\frac{1}{2})$	$2\tau - \frac{3}{2}$	2τ	$\left[\dfrac{4(\tau - 1)\tau(4\tau + 1)}{(2\tau - 1)(2\tau + 3)(4\tau - 1)}\right]^{\frac{1}{2}}$
$(\tau - \frac{1}{2},\frac{1}{2})$	$2\tau - \frac{3}{2}$	$2\tau - 2$	$-\left[\dfrac{(2\tau + 1)^2}{(\tau - 1)(2\tau + 3)(4\tau - 1)}\right]^{\frac{1}{2}}$
$(\tau - \frac{1}{2},\frac{1}{2})$	$2\tau - \frac{3}{2}$	$2\tau - 3$	$\left[\dfrac{(\tau - 2)(4\tau - 1)}{(\tau - 1)(2\tau - 1)(2\tau + 3)}\right]^{\frac{1}{2}}$

contains only generators of $O^B(6)$ and $SU^F(4)$. Further simplification occurs when $\alpha_2 = f_2$, since then

$$T_\mu^{(E2)} = \alpha_2 G_\mu^{(2)}, \qquad (3.34)$$

is a generator of $\text{Spin}^{BF}(6)$.

When $T^{(E2)}$ can be written as in (3.34), it has selection rules $\Delta\sigma_1 = \Delta\sigma_2 = \Delta\sigma_3 = 0$, since the generator $G^{(2)}$ cannot connect different $\text{Spin}^{BF}(6)$ representations. An inspection of the tensorial character of $G^{(2)}$ with respect to $\text{Spin}^{BF}(5)$ also shows that this operator in the case $N_F = 1$ (i.e. for odd–even nuclei) satisfies the selection rules $\Delta\tau_1 = 0, \pm 1, \Delta\tau_2 = 0$. These selection rules should be contrasted with those discussed in Volume 1, p. 56, $\Delta\tau_1 = \pm 1, \Delta\tau_2 = 0$ (even–even nuclei, $N_F = 0$). A consequence of

Table 3-4 Some $B(E2)$ values and electric quadrupole moments in the $\mathrm{Spin^{BF}}(6)$ limit

---

$B(E2; N + \frac{1}{2}, \tau_1 + 1, 2\tau_1 + \frac{5}{2} \to N + \frac{1}{2}, \tau_1, 2\tau_1 + \frac{1}{2})$

$\quad = \alpha_2^2 (N - \tau_1 + \frac{1}{2})(N + \tau_1 + \frac{9}{2}) \dfrac{\tau_1 + \frac{1}{2}}{2\tau_1 + 4}$

$B(E2; N + \frac{1}{2}, \tau_1 + 1, 2\tau_1 + \frac{3}{2} \to N + \frac{1}{2}, \tau_1, 2\tau_1 + \frac{1}{2})$

$\quad = \alpha_2^2 (N - \tau_1 + \frac{1}{2})(N + \tau_1 + \frac{9}{2}) \dfrac{3}{(2\tau_1 + 4)(4\tau_1 + 1)}$

$B(E2; N + \frac{1}{2}, \tau_1 + 1, 2\tau_1 + \frac{3}{2} \to N + \frac{1}{2}, \tau_1, 2\tau_1 - \frac{1}{2})$

$\quad = \alpha_2^2 (N - \tau_1 + \frac{1}{2})(N + \tau_1 + \frac{9}{2}) \dfrac{(\tau_1 - \frac{1}{2})(4\tau_1 + 5)}{(2\tau_1 + 4)(4\tau_1 + 1)}$

$B(E2; N + \frac{1}{2}, \tau_1 + 1, 2\tau_1 + \frac{1}{2} \to N + \frac{1}{2}, \tau_1, 2\tau_1 + \frac{1}{2})$

$\quad = \alpha_2^2 (N - \tau_1 + \frac{1}{2})(N + \tau_1 + \frac{9}{2}) \dfrac{16(\tau_1 - \frac{1}{2})(\tau_1 + \frac{3}{2})^2}{(2\tau_1)(2\tau_1 + 4)^2(4\tau_1 + 1)}$

$B(E2; N + \frac{1}{2}, \tau_1 + 1, 2\tau_1 + \frac{1}{2} \to N + \frac{1}{2}, \tau_1, 2\tau_1 - \frac{1}{2})$

$\quad = \alpha_2^2 (N - \tau_1 + \frac{1}{2})(N + \tau_1 + \frac{9}{2}) \dfrac{50(2\tau_1 + 2)}{(2\tau_1 + 4)^2(4\tau_1 - 1)(4\tau_1 + 1)}$

$B(E2; N + \frac{1}{2}, \tau_1 + 1, 2\tau_1 + \frac{1}{2} \to N + \frac{1}{2}, \tau_1, 2\tau_1 - \frac{3}{2})$

$\quad = \alpha_2^2 (N - \tau_1 + \frac{1}{2})(N + \tau_1 + \frac{9}{2}) \dfrac{(\tau_1 - \frac{3}{2})(2\tau_1 + 2)^2(4\tau_1 + 3)}{(2\tau_1)(2\tau_1 + 4)^2(4\tau_1 - 1)}$

$B(E2; N + \frac{1}{2}, \tau_1, 2\tau_1 + \frac{1}{2} \to N + \frac{1}{2}, \tau_1, 2\tau_1 - \frac{1}{2})$

$\quad = \alpha_2^2 \left(\dfrac{2N + 5}{2\tau_1 + 4} \right)^2 \dfrac{2(\tau_1 - \frac{1}{2})(4\tau_1 + 3)}{(4\tau_1 - 1)(4\tau_1 + 1)}$

$B(E2; N + \frac{1}{2}, \tau_1, 2\tau_1 + \frac{1}{2} \to N + \frac{1}{2}, \tau_1, 2\tau_1 - \frac{3}{2})$

$\quad = \alpha_2^2 \left(\dfrac{2N + 5}{2\tau_1 + 4} \right)^2 \dfrac{4(\tau_1 - \frac{1}{2})(\tau_1 - \frac{3}{2})}{(2\tau_1)(2\tau_1 + 2)(4\tau_1 - 1)}$

$B(E2; N + \frac{1}{2}, \tau_1, 2\tau_1 - \frac{1}{2} \to N + \frac{1}{2}, \tau_1, 2\tau_1 - \frac{3}{2})$

$\quad = \alpha_2^2 \left(\dfrac{2N + 5}{2\tau_1 + 4} \right)^2 \dfrac{12(\tau_1 - \frac{3}{2})}{(2\tau_1)(2\tau_1 + 2)(4\tau_1 - 3)(4\tau_1 - 1)}$

$Q(N + \frac{1}{2}, \tau_1, 2\tau_1 + \frac{1}{2})$

$\quad = -\alpha_2 \left(\dfrac{2N + 5}{2\tau_1 + 4} \right) \left[\dfrac{(\tau_1 + \frac{1}{2})(2\tau_1 + 2)(4\tau_1 + 3)}{(2\tau_1)(4\tau_1 + 1)} \right]^{\frac{1}{2}}$

$Q(N + \frac{1}{2}, \tau_1, 2\tau_1 - \frac{1}{2})$

$\quad = +\alpha_2 \left(\dfrac{2N + 5}{2\tau_1 + 4} \right) \left[\dfrac{(\tau_1 + \frac{1}{2})(2\tau_1)(4\tau_1 - 5)^2}{(2\tau_1 - 1)(4\tau_1 - 1)(4\tau_1 + 1)} \right]^{\frac{1}{2}}$

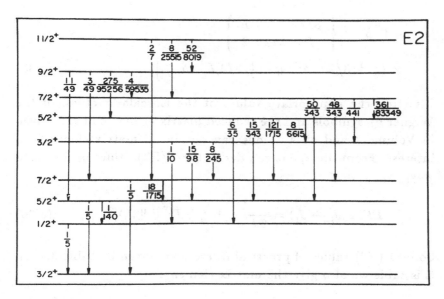

Fig. 3-2 $B(E2)$ values, in units of α_2^2, for the $\text{Spin}^{\text{BF}}(6)$ (III) symmetry. The $\Delta\tau_1 = 0$ transitions should be multiplied by $(2N+5)^2$ and $\Delta\tau_1 = 1$ transitions by $(N - \tau_1 + \frac{1}{2})(N + \tau_1 + \frac{9}{2})$, where τ_1 characterizes the final state.

these selection rules is that quadrupole moments vanish in even–even nuclei, since they correspond to $\Delta\tau_1 = 0$. However, they do not vanish any longer in odd–even nuclei, since $\Delta\tau_1 = 0$ matrix elements are allowed in the spinor representations of $\text{Spin}^{\text{BF}}(6)$.

Here we present the calculation of reduced matrix elements of $G^{(2)}$ between states belonging to the $\text{Spin}^{\text{BF}}(6)$ representation $(\sigma_1 = N + \frac{1}{2}, \sigma_2 = \frac{1}{2}, \sigma_3 = \frac{1}{2})$. These are obtained by expanding the wave functions as in the previous subsection,

$$\langle N + \tfrac{1}{2}; \tau_1; J \parallel G^{(2)} \parallel N + \tfrac{1}{2}; \tau_1'; J' \rangle$$
$$= \sqrt{(2J+1)(2J'+1)} \sum_{\tau,L} \sum_{\tau',L'} \xi_{N+\frac{1}{2},\tau_1,J}^{N,\tau,L} \xi_{N+\frac{1}{2},\tau_1',J'}^{N,\tau',L'} (-1)^{L+J'+3/2}$$
$$\times \left[\left\{ \begin{matrix} L & J & 3/2 \\ J' & L' & 2 \end{matrix} \right\} \langle N; \tau; L \parallel B^{(2)} \parallel N; \tau'; L' \rangle \right.$$

$$+ (-)^{J-J'} \left\{ \begin{matrix} 3/2 & J & L \\ J' & 3/2 & 2 \end{matrix} \right\}$$

$$\times \langle \tfrac{1}{2}; \tfrac{1}{2}; 3/2 \parallel A^{(2)} \parallel \tfrac{1}{2}; \tfrac{1}{2}; 3/2 \rangle \delta_{\tau,\tau'} \delta_{L,L'} \Big]. \tag{3.35}$$

Inserting the appropriate values of the isoscalar factors, of the Wigner $6j$-symbols and the reduced matrix elements of $B^{(2)}$ given in Volume 1 and of $A^{(2)}$, one can obtain all matrix elements of interest. From these, one can derive the $B(E2)$ values in the usual way,

$$B(E2; J_i \rightarrow J_f) = \frac{1}{2J_i + 1} |\langle J_f \parallel T^{(E2)} \parallel J_i \rangle|^2. \tag{3.36}$$

Some $B(E2)$ values of practical interest are given in Table 3.4. In this table we also give the matrix elements

$$Q(N + \tfrac{1}{2}, \tau_1, J) = \langle N + \tfrac{1}{2}; \tau_1; J \parallel T^{(E2)} \parallel N + \tfrac{1}{2}; \tau_1; J \rangle, \tag{3.37}$$

from which one can calculate quadrupole moments using

$$Q_J = \sqrt{\frac{16\pi}{5}} \sqrt{\frac{J(2J-1)}{(2J+1)(J+1)(2J+3)}} \langle J \parallel T^{(E2)} \parallel J \rangle. \tag{3.38}$$

A summary of $B(E2)$ values between low-lying states is shown in Fig. 3.2.

3.2.1.6 Electromagnetic transitions and moments; M1 In contrast to even–even nuclei ($N_F = 0$) where M1 transitions are largely retarded, magnetic dipole transitions occur in odd–even nuclei ($N_F = 1$) with considerable strength. The M1 operator can be written as

$$T_\mu^{(M1)} = T_{B,\mu}^{(M1)} + T_{F,\mu}^{(M1)}, \tag{3.39}$$

with

$$T_{B,\mu}^{(M1)} = \beta_1 [d^\dagger \times \tilde{d}]_\mu^{(1)},$$
$$T_{F,\mu}^{(M1)} = f_1 [a^\dagger{}_{3/2} \times \tilde{a}_{3/2}]_\mu^{(1)}. \tag{3.40}$$

Introducing the operators $B^{(1)}$ and $A^{(1)}$, one can rewrite (3.39) as

$$T_\mu^{(M1)} = \beta_1 B_\mu^{(1)} - \frac{t_1}{\sqrt{2}} A_\mu^{(1)}, \qquad (3.41)$$

where $f_1 = -t_1/\sqrt{2}$. If, in addition, $t_1 = \beta_1$, the M1 operator can be written as

$$T_\mu^{(M1)} = \beta_1 G_\mu^{(1)}. \qquad (3.42)$$

Since $G^{(1)}$ is proportional to the total angular momentum operator, the only non-zero matrix elements of the operator (3.42) are diagonal and all M1 transitions are forbidden. Furthermore, the equality $t_1 = \beta_1$ is a poor approximation compared to actual microscopic calculations. We shall therefore consider, in this case, the general form (3.41), with $\beta_1 \neq t_1$. The operator (3.41) has the selection rules $\Delta\sigma_1 = \Delta\sigma_2 = \Delta\sigma_3 = 0$, $\Delta\tau_1 = 0, \pm 1, \Delta\tau_2 = 0$. Its non-vanishing matrix elements can be obtained by expanding the wave functions as in Sect. 3.2.1.4,

$$\langle N + \tfrac{1}{2}; \tau_1; J \parallel T^{(M1)} \parallel N + \tfrac{1}{2}; \tau_1'; J' \rangle$$

$$= \sqrt{(2J+1)(2J'+1)} \sum_{\tau,L} \sum_{\tau',L'} \xi_{N+\frac{1}{2},\tau_1,J}^{N,\tau,L} \xi_{N+\frac{1}{2},\tau_1',J'}^{N,\tau',L'} (-1)^{L+J'+1/2}$$

$$\times \left[\left\{ \begin{matrix} L & J & 3/2 \\ J' & L' & 1 \end{matrix} \right\} \langle N; \tau; L \parallel \beta_1 B^{(1)} \parallel N; \tau'; L' \rangle \right.$$

$$- \frac{(-)^{J-J'}}{\sqrt{2}} \left\{ \begin{matrix} 3/2 & J & L \\ J' & 3/2 & 1 \end{matrix} \right\}$$

$$\left. \times \langle \tfrac{1}{2}; \tfrac{1}{2}; 3/2 \parallel t_1 A^{(1)} \parallel \tfrac{1}{2}; \tfrac{1}{2}; 3/2 \rangle \delta_{\tau,\tau'} \delta_{L,L'} \right]. \qquad (3.43)$$

Inserting the appropriate values of the isoscalar factors, of the Wigner 6j-symbols and the reduced matrix elements of the operators $B^{(1)}$ and $A^{(1)}$, one can obtain all matrix elements of interest. From these, one can calculate the $B(M1)$ values

$$B(M1; J_i \to J_f) = \frac{1}{2J_i + 1} |\langle J_f \parallel T^{(M1)} \parallel J_i \rangle|^2, \qquad (3.44)$$

given in Table 3.5. In this table we also give the E2/M1 mixing ratios defined by

$$\langle \phi, J \parallel \Delta^{(E2/M1)} \parallel \phi', J' \rangle = \frac{\langle \phi, J \parallel T^{(E2)} \parallel \phi', J' \rangle}{\langle \phi, J \parallel T^{(M1)} \parallel \phi', J' \rangle}, \qquad (3.45)$$

Table 3-5 Some $B(M1)$ values and magnetic dipole moments in the $\text{Spin}^{\text{BF}}(6)$ limit

$$B(M1; N+\tfrac{1}{2}, \tau_1+1, 2\tau_1+\tfrac{3}{2} \to N+\tfrac{1}{2}, \tau_1, 2\tau_1+\tfrac{1}{2})$$
$$= (\beta_1 - t_1)^2 \frac{(N-\tau_1+\tfrac{1}{2})(N+\tau_1+\tfrac{9}{2})}{(2N+4)^2} \frac{4\tau_1+5}{10(2\tau_1+4)}$$

$$B(M1; N+\tfrac{1}{2}, \tau_1+1, 2\tau_1+\tfrac{1}{2} \to N+\tfrac{1}{2}, \tau_1, 2\tau_1+\tfrac{1}{2})$$
$$= (\beta_1 - t_1)^2 \frac{(N-\tau_1+\tfrac{1}{2})(N+\tau_1+\tfrac{9}{2})}{(2N+4)^2} \frac{8(\tau_1-\tfrac{1}{2})(2\tau_1+2)}{5(2\tau_1+4)^2(4\tau_1+1)}$$

$$B(M1; N+\tfrac{1}{2}, \tau_1+1, 2\tau_1+\tfrac{1}{2} \to N+\tfrac{1}{2}, \tau_1, 2\tau_1-\tfrac{1}{2})$$
$$= (\beta_1 - t_1)^2 \frac{(N-\tau_1+\tfrac{1}{2})(N+\tau_1+\tfrac{9}{2})}{(2N+4)^2} \frac{3(2\tau_1+2)(4\tau_1+3)}{5(2\tau_1+4)^2(4\tau_1+1)}$$

$$B(M1; N+\tfrac{1}{2}, \tau_1, 2\tau_1+\tfrac{1}{2} \to N+\tfrac{1}{2}, \tau_1, 2\tau_1-\tfrac{1}{2})$$
$$= (\beta_1 - t_1)^2 \left(\frac{2N+5}{2N+4}\right)^2 \frac{12(\tau_1-\tfrac{1}{2})(\tau_1+\tfrac{3}{2})^2}{5(2\tau_1+4)^2(4\tau_1+1)}$$

$$B(M1; N+\tfrac{1}{2}, \tau_1, 2\tau_1-\tfrac{1}{2} \to N+\tfrac{1}{2}, \tau_1, 2\tau_1-\tfrac{3}{2})$$
$$= (\beta_1 - t_1)^2 \left(\frac{2N+5}{2N+4}\right)^2 \frac{8(\tau_1-\tfrac{3}{2})(\tau_1+\tfrac{3}{2})^2(4\tau_1+1)}{5(2\tau_1)(2\tau_1+2)(2\tau_1+4)^2(4\tau_1-1)}$$

$$\langle N+\tfrac{1}{2}, \tau_1+1, 2\tau_1+\tfrac{3}{2} \| \Delta^{(E2/M1)} \| N+\tfrac{1}{2}, \tau_1, 2\tau_1+\tfrac{1}{2} \rangle$$
$$= A(2N+4)\left[\frac{30}{(4\tau_1+1)(4\tau_1+5)}\right]^{\frac{1}{2}}$$

$$\langle N+\tfrac{1}{2}, \tau_1+1, 2\tau_1+\tfrac{1}{2} \| \Delta^{(E2/M1)} \| N+\tfrac{1}{2}, \tau_1, 2\tau_1+\tfrac{1}{2} \rangle$$
$$= A(2N+4)\left[\frac{10(\tau_1+\tfrac{3}{2})^2}{(2\tau_1)(2\tau_1+2)}\right]^{\frac{1}{2}}$$

$$\langle N+\tfrac{1}{2}, \tau_1+1, 2\tau_1+\tfrac{1}{2} \| \Delta^{(E2/M1)} \| N+\tfrac{1}{2}, \tau_1, 2\tau_1-\tfrac{1}{2} \rangle$$
$$= A(2N+4)\left[\frac{250}{3(4\tau_1-1)(4\tau_1+3)}\right]^{\frac{1}{2}}$$

$$\mu(N+\tfrac{1}{2}, \tau_1, 2\tau_1+\tfrac{1}{2}) = \left[(4\tau_1-2)\beta_1 + 3t_1 + 4\frac{(N-\tau_1+\tfrac{1}{2})(2\tau_1+2)}{(2N+4)(2\tau_1+4)}(\beta_1-t_1)\right]$$
$$\times \left[\frac{(\tau_1+\tfrac{1}{2})(4\tau_1+3)}{10(4\tau_1+1)}\right]^{\frac{1}{2}}$$

$$\mu(N+\tfrac{1}{2}, \tfrac{3}{2}, \tfrac{5}{2}) = \frac{1}{5}\sqrt{\frac{3}{7}}\left[11\beta_1 + \frac{13}{2}t_1\right] + \frac{48}{35}\sqrt{\frac{3}{7}}\frac{N-1}{2N+4}(\beta_1-t_1)$$

$$\mu(N+\tfrac{1}{2}, \tfrac{3}{2}, \tfrac{1}{2}) = \sqrt{\frac{3}{5}}(\beta_1 - \tfrac{1}{2}t_1)$$

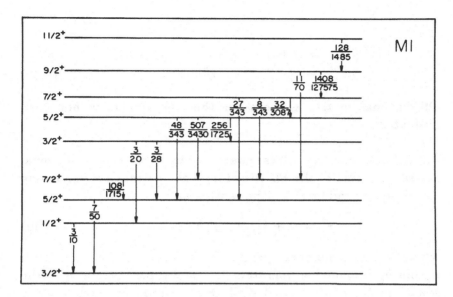

Fig. 3-3 $B(\text{M1})$ values, in units of $(\beta_1 - t_1)^2$, for the $\text{Spin}^{\text{BF}}(6)$ (III) symmetry. The $\Delta\tau_1 = 0$ transitions should be multiplied by $(2N+5)^2/(2N+4)^2$ and $\Delta\tau_1 = 1$ transitions by $(N - \tau_1 + \frac{1}{2})(N + \tau_1 + \frac{9}{2})/(2N+4)^2$, where τ_1 characterizes the final state.

in terms of the quantity $A = \alpha_2/(\beta_1 - t_1)$. In (3.45), ϕ and ϕ' denote all additional quantum numbers. Finally, we give the matrix elements

$$\mu(N + \tfrac{1}{2}, \tau_1, J) = \langle N + \tfrac{1}{2}; \tau_1; J \parallel T^{(\text{M1})} \parallel N + \tfrac{1}{2}; \tau_1; J \rangle, \quad (3.46)$$

from which the magnetic moments can be calculated using

$$\mu_J = \sqrt{\frac{4\pi}{3}} \sqrt{\frac{J}{(2J+1)(J+1)}} \langle J \parallel T^{(\text{M1})} \parallel J \rangle. \quad (3.47)$$

A summary of $B(\text{M1})$ values between low-lying states is shown in Fig. 3.3.

3.2.1.7 Electromagnetic transitions and moments; E0   The appropriate operator here is:

$$T_0^{(\text{E0})} = T_{\text{B},0}^{(\text{E0})} + T_{\text{F},0}^{(\text{E0})}, \quad (3.48)$$

with

$$T_{B,0}^{(E0)} = \gamma_0 + \alpha_0[s^\dagger \times \tilde{s}]_0^{(0)} + \beta_0[d^\dagger \times \tilde{d}]_0^{(0)},$$
$$T_{F,0}^{(E0)} = f_0[a^\dagger{}_{3/2} \times \tilde{a}_{3/2}]_0^{(0)}. \tag{3.49}$$

Off-diagonal matrix elements of this operator have not been evaluated explicitly.

3.2.1.8 Nuclear radii These quantities are related to the diagonal matrix elements of the E0 operator. The operator corresponding to the square radius r^2 can be written as

$$r^2 = r_c^2 + \tilde{\gamma}_0\hat{N}_B + \tilde{\beta}_0\hat{n}_d + f_0 A_0^{(0)}, \tag{3.50}$$

where $\hat{N}_B$ is the number operator for bosons, $\hat{N}_B = \hat{n}_s + \hat{n}_d$, with $\hat{n}_s$ and $\hat{n}_d$ the number operators for s and d bosons, and r_c^2 is the square radius of the closed shell. Square radii are given by the expectation values of (3.50) in the states (3.19), denoted here by

$$\langle r^2 \rangle_{\tau_1,J}^{(N+\frac{1}{2})} = r_c^2 + \tilde{\gamma}_0 N + \tilde{\beta}_0 \langle \hat{n}_d \rangle_{\tau_1,J}^{(N+\frac{1}{2})} + f_0 \langle A_0 \rangle_{\tau_1,J}^{(N+\frac{1}{2})}. \tag{3.51}$$

Inserting the appropriate values one finds

$$\langle r^2 \rangle_{\tau_1,J}^{(N+\frac{1}{2})} = r_c^2 + \tilde{\gamma}_0 N + \tilde{\beta}_0 \frac{N^2 + (\tau_1 - \frac{1}{2})(\tau_1 + \frac{7}{2})}{2(N+2)} - \frac{f_0}{2}. \tag{3.52}$$

From this it is possible to calculate isomer and isotope shifts. The isomer shift, $\delta\langle r^2 \rangle$, is the difference in r^2 between an excited state and the ground state. Since (3.52) is independent of J, we see that the isomer shift is identical for all states with the same τ_1. Using (3.51) we find

$$\delta\langle r^2 \rangle^{(N+\frac{1}{2})} \equiv \langle r^2 \rangle_{\tau_1=\frac{3}{2}}^{(N+\frac{1}{2})} - \langle r^2 \rangle_{\tau_1=\frac{1}{2}}^{(N+\frac{1}{2})}$$
$$= \tilde{\beta}_0 \left[\langle \hat{n}_d \rangle_{\tau_1=\frac{3}{2}}^{(N+\frac{1}{2})} - \langle \hat{n}_d \rangle_{\tau_1=\frac{1}{2}}^{(N+\frac{1}{2})} \right]$$
$$= \tilde{\beta}_0 \frac{5}{2(N+2)}. \tag{3.53}$$

This should be compared with the isomer shift in the corresponding even–even nucleus ($N_F = 0$), given by Eq. (2.217) of Volume

1, $2\tilde{\beta}_0/(N+1)$. Similarly, one can compute isotope shifts. These are defined as

$$\Delta\langle r^2\rangle^{(N+\frac{1}{2})} \equiv \langle r^2\rangle_{\tau_1=\frac{1}{2}}^{(N+\frac{3}{2})} - \langle r^2\rangle_{\tau_1=\frac{1}{2}}^{(N+\frac{1}{2})}$$

$$= \tilde{\gamma}_0 + \tilde{\beta}_0 \frac{N^2 + 5N + 2}{2(N+2)(N+3)}, \qquad (3.54)$$

to be compared with the corresponding expression when $N_F = 0$, given by Eq. (2.222) of Volume 1.

An interesting new quantity that appears here is the odd–even shift, defined as

$$\Delta'\langle r^2\rangle^{(N)} \equiv \langle r^2\rangle_{\tau_1=\frac{1}{2}}^{(N+\frac{1}{2})} - \langle r^2\rangle_{\tau_1=0}^{(N)}. \qquad (3.55)$$

This shift is given by

$$\Delta'\langle r^2\rangle^{(N)} = \tilde{\beta}_0 \frac{N}{(N+1)(N+2)} - \frac{f_0}{2}. \qquad (3.56)$$

3.2.1.9 One-nucleon transfer intensities In addition to electromagnetic transition rates, it is of interest to calculate intensities of transfer reactions. The form of the transfer operators when written in terms of bosons and fermions is, in general, very complex. We begin with a discussion of one-nucleon transfer reactions. There are two types of reactions possible here (see Sect. 1.4.5). The first is between an even–even nucleus with N bosons and an odd–even nucleus with N bosons and one fermion (and *vice versa*). The second is between an even–even nucleus with $N+1$ bosons and an odd–even nucleus with N bosons and one fermion (and *vice versa*). The transfer addition operator for the first type of reaction is, in lowest order, given by (1.37),

$$P_{+,m}^{(j)} = p_j a_{j,m}^{\dagger}. \qquad (3.57)$$

The removal operator $P_{-}^{(j)}$ is given by a similar expression with $a^{\dagger}$ replaced by $\tilde{a}$. Higher-order terms of the type (1.39) may be added if needed.

Using the isoscalar factors derived in the previous subsections, one can obtain the matrix elements of the operator (3.57) in closed form. In the present case, only one single-particle level $j = 3/2$ is included. The operator (3.57) transforms then as the $(\frac{1}{2}, \frac{1}{2}, \frac{1}{2})$ representation of $\mathrm{Spin}^{\mathrm{BF}}(6)$ and as the $(\frac{1}{2}, \frac{1}{2})$ representation of $\mathrm{Spin}^{\mathrm{BF}}(5)$. It has the selection rules $\Delta\sigma_1 = \pm\frac{1}{2}$, $\Delta\sigma_2 = \Delta|\sigma_3| = \frac{1}{2}$, $\Delta\tau_1 = \Delta\tau_2 = \pm\frac{1}{2}$. Explicit evaluation of the matrix elements gives

$$\langle [N_{\mathrm{B}} = N]; \{N_{\mathrm{F}} = 1\}; N + \tfrac{1}{2}; \tfrac{1}{2}; 3/2 \parallel a^{\dagger}_{3/2} \parallel [N_{\mathrm{B}} = N]; \{N_{\mathrm{F}} = 0\}; N; \tau; L\rangle$$
$$= 2\left[\left(\frac{N+4}{2(N+2)}\right)^{\frac{1}{2}} \delta_{\tau,0}\delta_{L,0} - \left(\frac{N}{2(N+2)}\right)^{\frac{1}{2}} \delta_{\tau,1}\delta_{L,2}\right], \quad (3.58)$$

and similar expressions for other matrix elements. It is customary to define, as a measure of the intensities of transfer reactions, the quantities

$$I([N_{\mathrm{B}}]; \{N_{\mathrm{F}}\}; \sigma_1; \tau_1; J \to [N'_{\mathrm{B}}]; \{N'_{\mathrm{F}}\}; \sigma'_1; \tau'_1; J')$$
$$= |\langle [N'_{\mathrm{B}}]; \{N'_{\mathrm{F}}\}; \sigma'_1; \tau'_1; J' \parallel P^{(j)}_{\pm} \parallel [N_{\mathrm{B}}]; \{N_{\mathrm{F}}\}; \sigma_1; \tau_1; J\rangle|^2. \quad (3.59)$$

Using (3.58) and similar expressions one obtains:
(i) even–even to odd–even,

$$I([N]; \{0\}; N; 0; 0 \to [N]; \{1\}; N + \tfrac{1}{2}; \tfrac{1}{2}; 3/2) = 4\zeta^2 \left(\frac{N+4}{2(N+2)}\right),$$
$$I([N]; \{0\}; N; 0; 0 \to [N]; \{1\}; N - \tfrac{1}{2}; \tfrac{1}{2}; 3/2) = 4\zeta^2 \left(\frac{N}{2(N+2)}\right);$$
$$(3.60)$$

(ii) odd–even to even–even,

$$I([N]; \{1\}; N + \tfrac{1}{2}; \tfrac{1}{2}; 3/2 \to [N]; \{0\}; N; 0; 0) = 4\zeta^2 \left(\frac{N+4}{2(N+2)}\right),$$
$$I([N]; \{1\}; N + \tfrac{1}{2}; \tfrac{1}{2}; 3/2 \to [N]; \{0\}; N; 1; 2) = 4\zeta^2 \left(\frac{N}{2(N+2)}\right),$$
$$(3.61)$$

where $\zeta = p_{3/2}$. These formulas describe intensities for reactions of the first type, $N_{\mathrm{B}} = N, N_{\mathrm{F}} = 0 \rightleftharpoons N_{\mathrm{B}} = N, N_{\mathrm{F}} = 1$.

For reactions of the second type, $N_{\mathrm{B}} = N + 1, N_{\mathrm{F}} = 0 \rightleftharpoons N_{\mathrm{B}} = N$, $N_{\mathrm{F}} = 1$, the appropriate transfer operator is, in lowest order,

$$P'^{(j)}_{+,m} = p'^{(j)}_{0j}[s^\dagger \times \tilde{a}_j]^{(j)}_m + \sum_{j'} p'^{(j)}_{2j'}[d^\dagger \times \tilde{a}_{j'}]^{(j)}_m. \qquad (3.62)$$

In the present case, there is only one term $j' = j = 3/2$ in the summation. Furthermore, if the coefficient $p'^{(3/2)}_{2,3/2}$ is equal to $-\sqrt{5}p'^{(3/2)}_{0,3/2}$, the operator $P'^{(j)}_+$ transforms as the representation $(\frac{1}{2},\frac{1}{2},\frac{1}{2})$ of Spin$^{\mathrm{BF}}(6)$ and $(\frac{1}{2},\frac{1}{2})$ of Spin$^{\mathrm{BF}}(5)$. It has thus the same selection rules as before, $\Delta\sigma_1 = \pm\frac{1}{2}$, $\Delta\sigma_2 = \Delta|\sigma_3| = \frac{1}{2}$, $\Delta\tau_1 = \Delta\tau_2 = \pm\frac{1}{2}$. Using the isoscalar factors of the previous subsections, one obtains:

(i) even–even to odd–even,

$$I([N + 1]; \{0\}; N + 1; 0; 0 \rightarrow [N]; \{1\}; N + \tfrac{1}{2}; \tfrac{1}{2}; 3/2) = 4\theta^2(N + 1),$$
$$I([N + 1]; \{0\}; N + 1; 0; 0 \rightarrow [N]; \{1\}; N - \tfrac{1}{2}; \tfrac{1}{2}; 3/2) = 0; \qquad (3.63)$$

(ii) odd–even to even–even,

$$I([N]; \{1\}; N + \tfrac{1}{2}; \tfrac{1}{2}; 3/2 \rightarrow [N + 1]; \{0\}; N + 1; 0; 0) = 4\theta^2(N + 1),$$
$$I([N]; \{1\}; N + \tfrac{1}{2}; \tfrac{1}{2}; 3/2 \rightarrow [N + 1]; \{0\}; N + 1; 1; 2) = 4\theta^2(N + 5),$$
$$(3.64)$$

where $\theta = p'^{(3/2)}_{0,3/2}$. The situation for transfer to the same odd–even nucleus is summarized in Fig. 3.4.

3.2.1.10 Two-nucleon transfer intensities Two-nucleon transfer reactions from an even–odd nucleus to another even–odd nucleus can be calculated in the same way as for even–even nuclei. The $L = 0$ transfer operators can be written as in Eq. (2.228) of Volume 1,

$$P^{(0)}_{+,\pi,0} = \alpha_\pi s^\dagger \left(\frac{N_\pi + 1}{N + 1}\right)^{\frac{1}{2}} \left(\Omega_\pi - N_\pi - \frac{N_\pi}{N}\hat{n}_{\mathrm{d}}\right)^{\frac{1}{2}},$$

$$P^{(0)}_{+,\nu,0} = \alpha_\nu s^\dagger \left(\frac{N_\nu + 1}{N + 1}\right)^{\frac{1}{2}} \left(\Omega_\nu - N_\nu - \frac{N_\nu}{N}\hat{n}_{\mathrm{d}}\right)^{\frac{1}{2}}, \qquad (3.65)$$

and their Hermitian conjugates. In order to calculate intensities of two-nucleon transfer reactions, one needs therefore to evaluate

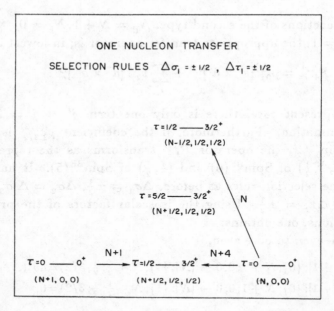

Fig. 3-4 Schematic illustration of the selection rules for the one-nucleon transfer operator in the Spin^{BF}(6) (III) symmetry.

matrix elements of the operators $s^\dagger$ and $\tilde{s}$. These operators have the selection rules $\Delta\sigma_1 = \pm 1$, $\Delta\sigma_2 = \Delta\sigma_3 = 0$ and $\Delta\tau_1 = \Delta\tau_2 = 0$. The same selection rules hold approximately for the operators $P^{(0)}_{+,\pi}$ and $P^{(0)}_{+,\nu}$, since they are only weakly broken by the presence of the square-root factors in (3.65). For transitions between ground states of odd–even nuclei, corresponding to two-neutron transfer, one obtains:

$$I([N]; \{1\}; N + \tfrac{1}{2}; \tfrac{1}{2}; 3/2 \to [N+1]; \{1\}; N + \tfrac{3}{2}; \tfrac{1}{2}; 3/2)$$
$$= \alpha_\nu^2 (N_\nu + 1) \frac{N+5}{2(N+3)} \left(\Omega_\nu - N_\nu - \frac{N}{2(N+2)} N_\nu \right). \quad (3.66)$$

This should be compared with Eq. (2.233) of Volume 1 which, written in the present notation, reads:

$$I([N]; \{0\}; N; 0; 0 \to [N+1]; \{0\}; N+1; 0; 0)$$
$$= \alpha_\nu^2 (N_\nu + 1) \frac{N+4}{2(N+2)} \left(\Omega_\nu - N_\nu - \frac{N-1}{2(N+1)} N_\nu \right). \quad (3.67)$$

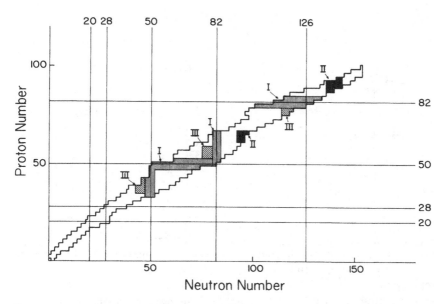

Fig. 3-5   Regions of the periodic table where examples of boson dynamic symmetries have been found: (I) U(5); (II) SU(3); (III) O(6).

Expressions for two-proton transfer are obtained from (3.66) and (3.67) by replacing the index ν by π.

3.2.1.11 Examples of spectra with $\text{Spin}^{\text{BF}}(6)$ *symmetry*   In order to have spectra with $\text{Spin}^{\text{BF}}(6)$ symmetry, two conditions must be met: (a) the adjacent even–even nucleus has $O^{\text{B}}(6)$ symmetry; (b) the odd nucleon occupies a single-particle level with $j = 3/2$. As mentioned in Volume 1, there are, at present, three regions where the $O^{\text{B}}(6)$ symmetry seems to be appropriate, the Os–Pt region (Arima and Iachello, 1979), the Xe–Ba region (Casten and von Brentano, 1985) and the Kr–Sr region (Kaup and Gelberg, 1979). These three regions are shown in Fig. 3.5.   Inspection of the single-particle level structure then shows that a $\text{Spin}^{\text{BF}}(6)$ symmetry could occur for: (a) odd-proton nuclei in the Os–Pt region with the odd proton occupying the $2d_{3/2}$ level; (b) odd-neutron nuclei in the Xe–Ba region with the odd neutron occupying the

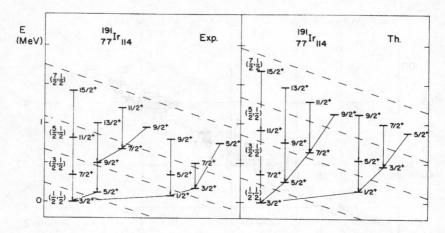

Fig. 3-6   An example of a spectrum with $\text{Spin}^{BF}(6)$ (III) symmetry: $^{191}_{77}\text{Ir}_{114}$ ($N_B = 8$, $N_F = 1$). The theoretical spectrum is calculated using (3.16) with $2\beta = 40$ KeV and $2\gamma = 10$ KeV. All states in the figure belong to the $\text{Spin}^{BF}(6)$ representation $\left(\frac{17}{2}, \frac{1}{2}, \frac{1}{2}\right)$ and thus the other terms in (3.16) do not contribute to the excitation energies. The $\text{Spin}^{BF}(5)$ quantum numbers are shown in parentheses to the left of the levels.

$2\text{d}_{3/2}$ level; (c) odd-proton nuclei in the Kr–Sr region with the odd proton occupying the $2\text{p}_{3/2}$ level.

A difficulty with a simple treatment of odd–even nuclei in terms of the $\text{Spin}^{BF}(6)$ symmetry is that the single-particle levels with $j = 3/2$ are close to other single-particle levels with which they mix. In particular, in the regions (a), (b) and (c) the single-particle levels $3\text{s}_{1/2}$ (a,b) and $2\text{p}_{1/2}$ (c) play an important role. The $\text{Spin}^{BF}(6)$ symmetry is thus an approximate symmetry and should only be viewed as providing a guide for more detailed numerical calculations.

An example of a nuclear spectrum with $\text{Spin}^{BF}(6)$ symmetry, $^{191}_{77}\text{Ir}_{114}$, is shown in Fig. 3.6. A similar situation occurs for $^{193}_{77}\text{Ir}_{116}$, $^{191}_{79}\text{Au}_{112}$, $^{193}_{79}\text{Au}_{114}$, and $^{197}_{79}\text{Au}_{118}$ (Wood, 1981a; Vervier, 1981). In some of these nuclei also states belonging to the $\sigma_1 = N - \frac{1}{2}$ multiplet have been observed (Cizewski *et al.*, 1987). The identification of the quantum numbers corresponding to these levels is more difficult and, therefore, they have been omitted from Fig. 3.6.

Table 3-6 Comparison between experimental and calculated $B(E2)$ values in $^{193}_{77}\text{Ir}_{116}$ and $^{197}_{79}\text{Au}_{118}$

E_i (KeV)	$(\sigma_1,\tau_1,J)_i$	$\to$	E_f (KeV)	$(\sigma_1,\tau_1,J)_f$	$B(E2)\ (e^2b^2)$ Exp[a]	Th[b]
			$^{193}_{77}\text{Ir}_{116}$			
73	$\frac{15}{2},\frac{3}{2},1/2$		0	$\frac{15}{2},\frac{1}{2},3/2$	0.220(30)	0.306
139	$\frac{15}{2},\frac{3}{2},5/2$		0	$\frac{15}{2},\frac{1}{2},3/2$	0.445(44)	0.306
358	$\frac{15}{2},\frac{3}{2},7/2$		0	$\frac{15}{2},\frac{1}{2},3/2$	0.258(5)	0.306
180	$\frac{15}{2},\frac{5}{2},3/2$		0	$\frac{15}{2},\frac{1}{2},3/2$	0.079(12)	0
362	$\frac{15}{2},\frac{5}{2},5/2$		0	$\frac{15}{2},\frac{1}{2},3/2$	0.0106(3)	0
621	$\frac{15}{2},\frac{5}{2},7/2$		0	$\frac{15}{2},\frac{1}{2},3/2$	0.059(3)	0
362	$\frac{15}{2},\frac{5}{2},5/2$		73	$\frac{15}{2},\frac{3}{2},1/2$	0.450(70)	0.244
358	$\frac{15}{2},\frac{3}{2},7/2$		139	$\frac{15}{2},\frac{3}{2},5/2$	0.190(40)	0.069
362	$\frac{15}{2},\frac{5}{2},5/2$		139	$\frac{15}{2},\frac{3}{2},5/2$	≤ 0.1	0.062
621	$\frac{15}{2},\frac{5}{2},7/2$		139	$\frac{15}{2},\frac{3}{2},5/2$	0.134(30)	0.207
621	$\frac{15}{2},\frac{5}{2},7/2$		358	$\frac{15}{2},\frac{3}{2},7/2$	0.059(31)	0.199
362	$\frac{15}{2},\frac{5}{2},5/2$		180	$\frac{15}{2},\frac{5}{2},3/2$	0.170(30)	0.008
621	$\frac{15}{2},\frac{5}{2},7/2$		180	$\frac{15}{2},\frac{5}{2},3/2$	≤ 0.003	0.015
			$^{197}_{79}\text{Au}_{118}$			
77	$\frac{11}{2},\frac{3}{2},1/2$		0	$\frac{11}{2},\frac{1}{2},3/2$	0.260(14)	0.231
279	$\frac{11}{2},\frac{3}{2},5/2$		0	$\frac{11}{2},\frac{1}{2},3/2$	0.209(5)	0.231
547	$\frac{11}{2},\frac{3}{2},7/2$		0	$\frac{11}{2},\frac{1}{2},3/2$	0.226(9)	0.231
269	$\frac{11}{2},\frac{5}{2},3/2$		0	$\frac{11}{2},\frac{1}{2},3/2$	0.083(6)	0
503	$\frac{11}{2},\frac{5}{2},5/2$		0	$\frac{11}{2},\frac{1}{2},3/2$	≤ 0.003	0
737	$\frac{11}{2},\frac{5}{2},7/2$		0	$\frac{11}{2},\frac{1}{2},3/2$	≤ 0.004	0
855	$\frac{11}{2},\frac{5}{2},9/2$		279	$\frac{11}{2},\frac{3}{2},5/2$	0.258(38)	0.228
1231	$\frac{11}{2},\frac{5}{2},11/2$		547	$\frac{11}{2},\frac{3}{2},7/2$	0.269(60)	0.290

[a] From Mundy *et al.* (1984) and Vervier (1987).
[b] With $\alpha_2 = 0.135\ eb$ for $^{193}_{77}\text{Ir}_{116}$ and $\alpha_2 = 0.152\ eb$ for $^{197}_{79}\text{Au}_{116}$.

Table 3-7 Comparison between experimental and calculated $B(M1)$ values in $^{193}_{77}\text{Ir}_{116}$ and $^{197}_{79}\text{Au}_{118}$

E_i (KeV)	$(\sigma_1, \tau_1, J)_i$	$\rightarrow$	E_f (KeV)	$(\sigma_1, \tau_1, J)_f$	$B(M1)$ (μ_N^2) Exp[a]	Th[b]
			$^{193}_{77}\text{Ir}_{116}$			
73	$\frac{15}{2}, \frac{3}{2}, 1/2$		0	$\frac{15}{2}, \frac{1}{2}, 3/2$	0.0026(4)	0.024
139	$\frac{15}{2}, \frac{3}{2}, 5/2$		0	$\frac{15}{2}, \frac{1}{2}, 3/2$	0.058(7)	0.011
180	$\frac{15}{2}, \frac{5}{2}, 3/2$		0	$\frac{15}{2}, \frac{1}{2}, 3/2$	0.0048(29)	0
362	$\frac{15}{2}, \frac{5}{2}, 5/2$		0	$\frac{15}{2}, \frac{1}{2}, 3/2$	0.0284(10)	0
180	$\frac{15}{2}, \frac{5}{2}, 3/2$		73	$\frac{15}{2}, \frac{3}{2}, 1/2$	0.113(68)	0.011
358	$\frac{15}{2}, \frac{3}{2}, 7/2$		139	$\frac{15}{2}, \frac{3}{2}, 5/2$	0.054(20)	0.021
362	$\frac{15}{2}, \frac{5}{2}, 5/2$		139	$\frac{15}{2}, \frac{3}{2}, 5/2$	≤ 0.004	0.010
621	$\frac{15}{2}, \frac{5}{2}, 7/2$		139	$\frac{15}{2}, \frac{3}{2}, 5/2$	0.021(4)	0.006
621	$\frac{15}{2}, \frac{5}{2}, 7/2$		358	$\frac{15}{2}, \frac{3}{2}, 7/2$	0.010(4)	0.002
362	$\frac{15}{2}, \frac{5}{2}, 5/2$		180	$\frac{15}{2}, \frac{5}{2}, 3/2$	0.109(13)	0.018
			$^{197}_{79}\text{Au}_{118}$			
77	$\frac{11}{2}, \frac{3}{2}, 1/2$		0	$\frac{11}{2}, \frac{1}{2}, 3/2$	0.0078(2)	0.019
279	$\frac{11}{2}, \frac{3}{2}, 5/2$		0	$\frac{11}{2}, \frac{1}{2}, 3/2$	0.061(11)	0.009
269	$\frac{11}{2}, \frac{5}{2}, 3/2$		0	$\frac{11}{2}, \frac{1}{2}, 3/2$	≤ 0.0004	0
269	$\frac{11}{2}, \frac{5}{2}, 3/2$		77	$\frac{11}{2}, \frac{3}{2}, 1/2$	0.160(26)	0.008
547	$\frac{11}{2}, \frac{3}{2}, 7/2$		279	$\frac{11}{2}, \frac{3}{2}, 5/2$	0.0152(62)	0.018

[a] From Mundy et al. (1984) and Bolotin et al. (1979).
[b] With $\beta_1 = 0.545 \, \mu_N$ and $t_1 = -0.008 \, \mu_N$ in $^{193}_{77}\text{Ir}_{116}$ and $\beta_1 = 0.504 \, \mu_N$ and $t_1 = 0.008 \, \mu_N$ in $^{197}_{79}\text{Au}_{118}$.

One can investigate also to what extent electromagnetic transition rates can be described by the $\text{Spin}^{\text{BF}}(6)$ symmetry. A comparison between experimental and calculated $B(E2)$ values in $^{193}_{77}\text{Ir}_{116}$ and $^{197}_{79}\text{Au}_{118}$ is shown in Table 3.6. In each nucleus, the parameter in the E2 transition operator (3.34), α_2, is fitted to the $B(E2)$ values for transitions from the lowest $(\tau_1, \tau_2) = (\frac{3}{2}, \frac{1}{2})$ multiplet to the ground state. The agreement appears to be reasonable. For the calculation of $B(M1)$ values, the parameters in the transition operator (3.41), β_1 and t_1, can be obtained from the

Table 3-8 Comparison between experimental and calculated electric quadrupole and magnetic dipole moments in $^{193}_{77}\text{Ir}_{116}$ and $^{197}_{79}\text{Au}_{118}$

E (KeV)	σ_1, τ_1, J	$Q\,(eb)$ Exp[a]	Th[c]	$\mu\,(\mu_N)$ Exp[b]	Th[d]
			$^{193}_{77}\text{Ir}_{116}$		
0	$\frac{15}{2},\frac{1}{2},3/2$	0.751(9)	0.813	0.1591(6)	0.159
73	$\frac{15}{2},\frac{3}{2},1/2$			0.504(3)	0.355
139	$\frac{15}{2},\frac{3}{2},5/2$			0.528(30)	0.666
358	$\frac{15}{2},\frac{3}{2},7/2$			1.645(245)	0.868
180	$\frac{15}{2},\frac{5}{2},3/2$			1.020(375)	0.461
621	$\frac{15}{2},\frac{5}{2},7/2$			0.525(385)	0.851
522	$\frac{15}{2},\frac{5}{2},9/2$			3.780(1.125)	1.427
			$^{197}_{79}\text{Au}_{118}$		
0	$\frac{11}{2},\frac{1}{2},3/2$	0.594(10)	0.723	0.145746	0.145
77	$\frac{11}{2},\frac{3}{2},1/2$			0.420(4)	0.324
279	$\frac{11}{2},\frac{3}{2},5/2$			≤ 0.11	0.607

[a] From Tanaka *et al.* (1983) and Harmatz (1981).
[b] From Kölbl *et al.* (1986) and Harmatz (1981).
[c] With parameters as given in Table 3.6.
[d] With parameters as given in Table 3.7.

magnetic dipole moments of the ground state of the odd–even nucleus and of the first-excited 2^+ state of a neighboring even–even nucleus. If calculated in this way, one finds that major discrepancies occur, Table 3.7. This situation is similar to that encountered in even–even nuclei and is compounded by the fact that M1 transitions in odd–even nuclei are dominated, to a large extent, by the single-particle part of the M1 operator and thus are very sensitive to admixtures of other single-particle levels, such as the $3s_{1/2}$ level. This situation persists for static moments, as shown in Table 3.8.

Similar results are obtained from the analysis of electromagnetic transitions and moments of the nuclei $^{191}_{77}\text{Ir}_{114}$, $^{191}_{79}\text{Au}_{112}$, $^{193}_{79}\text{Au}_{114}$ and $^{195}_{79}\text{Au}_{116}$.

Intensities of transfer reactions can also be analyzed using the $\text{Spin}^{\text{BF}}(6)$ symmetry and are reported in Table 3.9. The general

Table 3-9 Comparison between experimental and calculated intensities of one-proton transfer reactions to and from $^{193}_{77}\text{Ir}_{116}$

		(i) $^{192}_{76}\text{Os}_{116} \rightarrow$ $^{193}_{77}\text{Ir}_{116}$			
$(\sigma_1,\tau_1,J)_i$	$\rightarrow$ $(\sigma_1,\tau_1,J)_f$	E_f (KeV)	$(\alpha,t)^{a,c}$	$(^3\text{He},d)^{b,c}$	Th^c
8,0,0	$\frac{15}{2},\frac{1}{2},3/2$	0	1.00	1.00	1.00
	$\frac{15}{2},\frac{5}{2},3/2$	180	0.07	0.06	0
	$\frac{13}{2},\frac{1}{2},3/2$	460	< 0.01	not seen	0

	(ii) $^{193}_{77}\text{Ir}_{116} \rightarrow$ $^{192}_{76}\text{Os}_{116}$				
$(\sigma_1,\tau_1,J)_i$	$\rightarrow$ $(\sigma_1,\tau_1,J)_f$	E_f (KeV)	$(t,\alpha)^{a,c}$	$(d,\,^3\text{He})^{b,c}$	Th^c
$\frac{15}{2},\frac{1}{2},3/2$	8,0,0	0	1.00	1.00	1.00
	8,1,2	205	—	1.12	1.50
	8,2,2	489	—	0.38	0
	8,3,0	956	≤ 0.01	not seen	0

	(iii) $^{194}_{78}\text{Pt}_{116} \rightarrow$ $^{193}_{77}\text{Ir}_{116}$				
$(\sigma_1,\tau_1,J)_i$	$\rightarrow$ $(\sigma_1,\tau_1,J)_f$	E_f (KeV)	$(t,\alpha)^{a,c}$	$(d,\,^3\text{He})^{b,c}$	Th^c
7,0,0	$\frac{15}{2},\frac{1}{2},3/2$	0	1.00	1.00	1.00
	$\frac{15}{2},\frac{5}{2},3/2$	180	0.07	0.08	0
	$\frac{13}{2},\frac{1}{2},3/2$	460	0.69	0.74	0.64

	(iv) $^{193}_{77}\text{Ir}_{116} \rightarrow$ $^{194}_{78}\text{Pt}_{116}$				
$(\sigma_1,\tau_1,J)_i$	$\rightarrow$ $(\sigma_1,\tau_1,J)_f$	E_f (KeV)	$(\alpha,t)^{a,c}$	$(^3\text{He},d)^{b,c}$	Th^c
$\frac{15}{2},\frac{1}{2},3/2$	7,0,0	0	1.00	1.00	1.00
	7,1,2	328	0.56	0.58	0.64
	7,2,2	622	0.70	0.78	0
	7,3,0	1267	0.33	0.36	0

^{a} From Cizewski et $al.$ (1983) and Vergnes et $al.$ (1981).

^{b} From Price et $al.$ (1971), Blasi et $al.$ (1982) and Iwasaki et $al.$ (1981).

^{c} Intensity to ground state is normalized to 1.00.

features of the data appear to be reproduced by the calculations, with the exception of the violation of the $\Delta\tau_1 = \pm\frac{1}{2}$ selection rule in the fourth reaction, $^{193}_{77}\text{Ir}_{116} \rightarrow$ $^{194}_{78}\text{Pt}_{116}$. The situation for two-nucleon transfer reactions is summarized in Table 3.10.

Table 3-10 Comparison between experimental and calculated intensities of two-nucleon
transfer reactions in Pt and Ir nuclei

Reaction	I_{exp}[a]	I_{th}[b]
$^{192}_{78}\mathrm{Pt}_{114}(t,p)\,^{194}_{78}\mathrm{Pt}_{116}$	0.97(13)	1.02
$^{191}_{77}\mathrm{Ir}_{114}(t,p)\,^{193}_{77}\mathrm{Ir}_{116}$	1.00(10)	1.00

[a] From Cizewski *et al.* (1981).
[b] Normalized to the Ir reaction.

3.2.2 $\mathrm{Spin}^{BF}(3)$ *(III$_2$)*

3.2.2.1 Lattice of algebras This is a somewhat trivial case
(Iachello and Kuyucak, 1981) which corresponds to bosons with
O(6) symmetry and fermions occupying a state with $j = 1/2$. The
lattice of algebras is:

$$
\begin{array}{ccc}
U^B(6) & \otimes & U^F(2) \\
\downarrow & & | \\
O^B(6) & & | \\
\downarrow & & | \\
O^B(5) & & | \\
\downarrow & & \downarrow \\
O^B(3) & & SU^F(2) \\
& \searrow \quad \swarrow & \\
& \mathrm{Spin}^{BF}(3) & \\
& \downarrow & \\
& \mathrm{Spin}^{BF}(2) &
\end{array}
\qquad (3.68)
$$

We only discuss here basis states and energy eigenvalues.

3.2.2.2 Basis states The basis states for $N_F = 1$ are of the form

$$
\left|
\begin{array}{ccccc}
U^B(6) & \otimes & U^F(2) & \supset O^B(6) \otimes SU^F(2) & \supset O^B(5) \otimes SU^F(2) \\
\downarrow & & \downarrow & \downarrow & \downarrow \\
[N_B = N] & & \{N_F = 1\} & \sigma & \tau \\[2mm]
\multicolumn{5}{l}{\supset O^B(3) \otimes SU^F(2) \supset \mathrm{Spin}^{BF}(3) \supset \mathrm{Spin}^{BF}(2)} \\
\downarrow & & \downarrow & \downarrow & \\
\nu_\Delta, L & & J & M_J &
\end{array}
\right\rangle .
$$

$$ (3.69) $$

The values of the quantum numbers $\sigma, \tau, \nu_\Delta, L$ are the same as those described in Volume 1 for chain III. The quantum number J is given by angular momentum coupling,

$$J = L \pm 1/2; \qquad L \neq 0,$$
$$J = 1/2; \qquad L = 0, \qquad (3.70)$$

and $-J \leq M_J \leq +J$ as usual.

3.2.2.3 Energy eigenvalues   By writing the Hamiltonian in terms of Casimir invariants

$$
\begin{aligned}
H^{(\mathrm{III}_2)} &= e_0 + e_1 \mathcal{C}_1(\mathrm{U}^\mathrm{B}6) + e_2 \mathcal{C}_2(\mathrm{U}^\mathrm{B}6) + e_3 \mathcal{C}_1(\mathrm{U}^\mathrm{F}2) + e_4 \mathcal{C}_2(\mathrm{U}^\mathrm{F}2) \\
&\quad + e_5 \mathcal{C}_1(\mathrm{U}^\mathrm{B}6)\mathcal{C}_1(\mathrm{U}^\mathrm{F}2) + \eta \mathcal{C}_1(\mathrm{O}^\mathrm{B}6) + \beta \mathcal{C}_2(\mathrm{O}^\mathrm{B}5) \\
&\quad + \gamma \mathcal{C}_2(\mathrm{O}^\mathrm{B}3) + \gamma' \mathcal{C}_2(\mathrm{Spin}^\mathrm{BF}3), \qquad (3.71)
\end{aligned}
$$

one obtains the energy eigenvalues

$$
\begin{aligned}
E^{(\mathrm{III}_2)}(N_\mathrm{B} &= N, N_\mathrm{F} = 1, \sigma, \tau, \nu_\Delta, L, J, M_J) \\
&= E_{02} + 2\eta\sigma(\sigma + 4) + 2\beta\tau(\tau + 3) + 2\gamma L(L + 1) \\
&\quad + 2\gamma' J(J + 1), \qquad (3.72)
\end{aligned}
$$

where

$$E_{02} = e_0 + e_1 N + e_2 N(N + 5) + e_3 + 2e_4 + e_5 N. \qquad (3.73)$$

The energy spectrum associated with (3.72) is shown in Fig. 3.7.

3.2.3 $\mathrm{U}^\mathrm{BF}(6) \otimes \mathrm{U}^\mathrm{F}_s(2)$ *(III$_3$)*

In the previous subsections we have discussed two examples of spinor symmetries. We consider now three examples of pseudo-spin symmetries.

3.2.3.1 Lattice of algebras   This symmetry has been extensively investigated (Bijker and Iachello, 1985). It corresponds to bosons

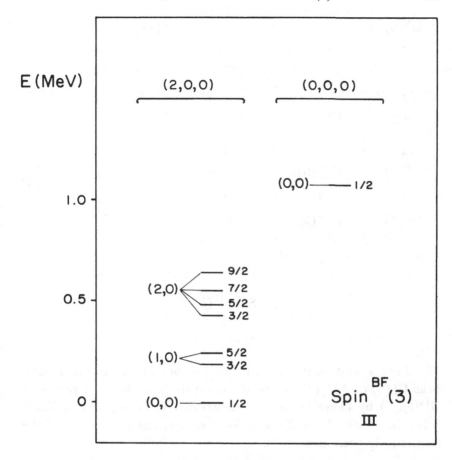

Fig. 3-7   A typical spectrum with $\text{Spin}^{BF}(3)$ (III) symmetry for $N_B = 2, N_F = 1$. The $O^B(6)$ quantum numbers are shown on top, the $O^B(5)$ quantum numbers to the left and the angular momentum J to the right of each level.

with O(6) symmetry and fermions with $j = 1/2$, $3/2, 5/2$. This symmetry exploits the breaking of the fermion algebra into a pseudo-orbital and pseudo-spin part discussed in Sect. 2.5.2. The pseudo-orbital part has $k = 0, 2$ and the pseudo-spin part has $s = 1/2$. The lattice of algebras is:

$$
\begin{array}{ccccccc}
\mathrm{U^B(6)} & & & \otimes & & \mathrm{U^F(12)} & \\
\downarrow & \searrow^a & & & & \downarrow & \\
& & & \mathrm{U_k^F(6)} & \otimes & \mathrm{U_s^F(2)} & \\
\downarrow & & \nearrow^a & \downarrow & & & \\
\mathrm{O^B(6)} & \mathrm{U^{BF}(6)} & & \mathrm{O_k^F(6)} & & & \\
\downarrow & \searrow^b \quad \downarrow \quad \nearrow^b & & \downarrow & & & \\
\mathrm{O^B(5)} & \mathrm{O^{BF}(6)} & & \mathrm{O_k^F(5)} & & & \\
\downarrow & \searrow^c \quad \downarrow \quad \nearrow^c & & \downarrow & & & \\
\mathrm{O^B(3)} & \mathrm{O^{BF}(5)} & & \mathrm{O_k^F(3)} & & & \\
& \searrow^d \quad \downarrow \quad \nearrow^d & & & & & \\
& \mathrm{O^{BF}(3)} & & & & \mathrm{SU_s^F(2)} & \\
& & \searrow & & \nearrow & & \\
& & & \mathrm{Spin^{BF}(3)} & & & \\
& & & \downarrow & & & \\
& & & \mathrm{Spin^{BF}(2)} & & &
\end{array}
\qquad (3.74)
$$

We discuss first route a. The generators of the fermion algebras $\mathrm{U}_k^F(6)$ and $\mathrm{U}_s^F(2)$ can be obtained from the bilinear products $A^{(\lambda)}(j,j')$ by using (2.25). Their explicit form is given in Table 3.11. In terms of the K- and S-operators, the generators of the fermion algebras are:

$$
\begin{aligned}
\{G\}_{\mathrm{U}_k^F(6)} &= \{K_\mu^{(\lambda)}(l,l'); \lambda = 0,\dots,4; l,l' = 0,2\}, \\
\{G\}_{\mathrm{O}_k^F(6)} &= \{K_\mu^{(1)}(2,2); K_\mu^{(2)}(0,2) + K_\mu^{(2)}(2,0); K_\mu^{(3)}(2,2)\}, \\
\{G\}_{\mathrm{O}_k^F(5)} &= \{K_\mu^{(\lambda)}(2,2); \lambda = 1,3\}, \\
\{G\}_{\mathrm{O}_k^F(3)} &= \{K_\mu^{(1)}(2,2)\}, \\
\{G\}_{\mathrm{U}_s^F(2)} &= \{S_\mu^{(\lambda)}(\tfrac{1}{2},\tfrac{1}{2}); \lambda = 0,1\}, \\
\{G\}_{\mathrm{SU}_s^F(2)} &= \{S_\mu^{(1)}(\tfrac{1}{2},\tfrac{1}{2})\}.
\end{aligned}
\qquad (3.75)
$$

The generators of the Bose–Fermi algebras are obtained by adding the fermion and boson generators,

$$
G_\mu^{(\lambda)}(l,l') = B_\mu^{(\lambda)}(l,l') + K_\mu^{(\lambda)}(l,l'), \qquad (3.76)
$$

Table 3-11 Decomposition of pseudo-orbital ($k = 0, 2$) and pseudo-spin ($s = 1/2$) operators in terms of bilinear products of fermion operators $A_\mu^{(\lambda)}(j, j')$

$(j,j') =$	$(\tfrac{1}{2},\tfrac{1}{2})$	$(\tfrac{1}{2},\tfrac{3}{2})$	$(\tfrac{1}{2},\tfrac{5}{2})$	$(\tfrac{3}{2},\tfrac{1}{2})$	$(\tfrac{3}{2},\tfrac{3}{2})$	$(\tfrac{3}{2},\tfrac{5}{2})$	$(\tfrac{5}{2},\tfrac{1}{2})$	$(\tfrac{5}{2},\tfrac{3}{2})$	$(\tfrac{5}{2},\tfrac{5}{2})$
$K_0^{(0)}(0,0)$	$-\sqrt{2}$								
$K_0^{(0)}(2,2)$					$-\sqrt{\tfrac{4}{5}}$				$-\sqrt{\tfrac{6}{5}}$
$K_\mu^{(1)}(2,2)$					$-\sqrt{\tfrac{18}{25}}$	$\sqrt{\tfrac{2}{25}}$		$-\sqrt{\tfrac{2}{25}}$	$-\sqrt{\tfrac{28}{25}}$
$K_\mu^{(2)}(0,2)$		$-\sqrt{\tfrac{4}{5}}$	$-\sqrt{\tfrac{6}{5}}$						
$K_\mu^{(2)}(2,0)$				$\sqrt{\tfrac{4}{5}}$			$-\sqrt{\tfrac{6}{5}}$		
$K_\mu^{(2)}(2,2)$					$-\sqrt{\tfrac{14}{25}}$	$\sqrt{\tfrac{6}{25}}$		$-\sqrt{\tfrac{6}{25}}$	$-\sqrt{\tfrac{24}{25}}$
$K_\mu^{(3)}(2,2)$					$-\sqrt{\tfrac{8}{25}}$	$\sqrt{\tfrac{12}{25}}$		$-\sqrt{\tfrac{12}{25}}$	$-\sqrt{\tfrac{18}{25}}$
$K_\mu^{(4)}(2,2)$						$\sqrt{\tfrac{4}{5}}$		$-\sqrt{\tfrac{4}{5}}$	$-\sqrt{\tfrac{2}{5}}$
$S_0^{(0)}(\tfrac{1}{2},\tfrac{1}{2})$	1				$\sqrt{2}$				$\sqrt{3}$
$S_\mu^{(1)}(\tfrac{1}{2},\tfrac{1}{2})$	1				$-\sqrt{\tfrac{2}{5}}$	$\sqrt{\tfrac{8}{5}}$		$-\sqrt{\tfrac{8}{5}}$	$\sqrt{\tfrac{7}{5}}$

and are given by

$$\{G\}_{\mathrm{UBF}(6)} = \{G_\mu^{(\lambda)}(l,l'); \lambda = 0, \ldots, 4; l, l' = 0, 2\},$$

$$\{G\}_{\mathrm{OBF}(6)} = \{G_\mu^{(1)}(2,2); G_\mu^{(2)}(0,2) + G_\mu^{(2)}(2,0); G_\mu^{(3)}(2,2)\},$$

$$\{G\}_{\mathrm{OBF}(5)} = \{G_\mu^{(\lambda)}(2,2); \lambda = 1, 3\},$$

$$\{G\}_{\mathrm{OBF}(3)} = \{G_\mu^{(1)}(2,2)\}. \tag{3.77}$$

Finally, the generators of the spinor groups, $\mathrm{Spin}^{\mathrm{BF}}(3)$ and $\mathrm{Spin}^{\mathrm{BF}}(2)$, can be obtained by adding the pseudo-orbital and pseudo-spin generators,

$$\{G\}_{\mathrm{Spin}^{\mathrm{BF}}(3)} = \{G_\mu^{(1)}(2,2) - \tfrac{1}{\sqrt{20}}S_\mu^{(1)}(\tfrac{1}{2},\tfrac{1}{2})\},$$

$$\{G\}_{\mathrm{Spin}^{\mathrm{BF}}(2)} = \{G_0^{(1)}(2,2) - \tfrac{1}{\sqrt{20}}S_0^{(1)}(\tfrac{1}{2},\tfrac{1}{2})\}. \tag{3.78}$$

3.2.3.2 Basis states. Route a   The basis states for this symmetry can be written as

$$
\left| \begin{array}{l}
U^B(6) \otimes U^F(12) \supset U^B(6) \otimes \quad U^F_k(6) \quad \otimes U^F_s(2) \\
\quad \downarrow \qquad \downarrow \qquad\qquad \downarrow \qquad\qquad\qquad \downarrow \\
\ [N_B] \qquad \{N_F\} \qquad\qquad\qquad [N'_1,\ldots,N'_6] \\[4pt]
\supset \quad U^{BF}(6) \quad \otimes U^F_s(2) \supset \quad O^{BF}(6) \quad \otimes U^F_s(2) \supset O^{BF}(5) \otimes U^F_s(2) \\
\qquad \downarrow \qquad\qquad\qquad\qquad \downarrow \qquad\qquad\qquad\qquad \downarrow \\
\ [N_1,\ldots,N_6] \qquad\qquad (\sigma_1,\sigma_2,\sigma_3) \qquad\qquad (\tau_1,\tau_2) \\[4pt]
\supset O^{BF}(3) \otimes SU^F_s(2) \supset \text{Spin}^{BF}(3) \supset \text{Spin}^{BF}(2) \\
\qquad \downarrow \qquad\qquad \downarrow \qquad\qquad \downarrow \qquad\qquad \downarrow \\
\qquad \nu_\Delta, L \qquad\quad S \qquad\qquad J \qquad\qquad M_J
\end{array} \right\rangle .
$$

$$(3.79)$$

We consider the case $N_B = N$ and $N_F = 1$, for which $N'_1 = 1$ and $N'_2 = \cdots = N'_6 = 0$. The values of $N_1,\ldots,N_6$ can be obtained from the multiplication rule for U(6),

$$[N] \otimes [1] = [N+1,0,0,0,0,0] \oplus [N,1,0,0,0,0]. \qquad (3.80)$$

To simplify notation, we shall not write the last four zeros and thus denote the representations by $[N_1, N_2]$. When $N_2 = 0$, the decomposition of representations $[N+1,0]$ of $U^{BF}(6)$ into representations $(\sigma_1, \sigma_2, \sigma_3)$ of $O^{BF}(6)$ is:

$$
[N+1,0] : (\sigma_1,\sigma_2,\sigma_3) = (N+1,0,0), (N-1,0,0),
$$
$$
\ldots, \begin{cases} (0,0,0), & N = \text{odd}, \\ (1,0,0), & N = \text{even}, \end{cases} \qquad (3.81)
$$

When $N_2 = 1$, the decomposition is:

$$
[N,1] : (\sigma_1,\sigma_2,\sigma_3) = (N,1,0), (N-1,0,0), (N-2,1,0),
$$
$$
\ldots, \begin{cases} (1,0,0), & N = \text{even}, \\ (1,1,0), & N = \text{odd}. \end{cases} \qquad (3.82)
$$

Next, we need the decomposition of representations of $O^{BF}(6)$ into those of $O^{BF}(5)$. For the $(\sigma_1,0,0)$ representations the result is given in Volume 1,

$$(\sigma_1,0,0) : (\tau_1,\tau_2) = (\sigma_1,0), (\sigma_1-1,0),\ldots,(0,0). \qquad (3.83)$$

Table 3-12 Angular momentum content of the lowest O(5) representations $(\tau_1, \tau_2 = 1)$

(τ_1, τ_2)	$\nu_\Delta = 0$	$\nu_\Delta = \frac{1}{3}$	$\nu_\Delta = \frac{2}{3}$	$\nu_\Delta = 1$	$\nu_\Delta = \frac{4}{3}$	$\nu_\Delta = \frac{5}{3}$
(1,1)	3	1				
(2,1)	5,4	3,2	1			
(3,1)	7,6,5	5,4,3	3,2	1		
(4,1)	9,8,7,6	7,6,5,4	5,4,3	3,2	1	
(5,1)	11,10,9,8,7	9,8,7,6,5	7,6,5,4	5,4,3	3,2	1

When $\sigma_2 = 1$, the decomposition is (Balantekin *et al.*, 1983):

$$(\sigma_1, 1, 0) : (\tau_1, \tau_2) = (\sigma_1, 0), (\sigma_1 - 1, 0), \dots, (1, 0),$$
$$(\sigma_1, 1), (\sigma_1 - 1, 1), \dots, (1, 1). \quad (3.84)$$

The values of L contained in a representation $(\tau_1, \tau_2 = 0)$ of $O^{BF}(5)$ are given in Volume 1,

$$(\tau_1, 0) : L = 2\tau_1 - 6\nu_\Delta, 2\tau_1 - 6\nu_\Delta - 2, \dots, \tau_1 - 3\nu_\Delta + 1, \tau_1 - 3\nu_\Delta, \quad (3.85)$$

where the extra label ν_Δ had been introduced to label the states uniquely $(\nu_\Delta = 0, 1, 2, \dots; 3\nu_\Delta \le \tau_1)$. For $(\tau_1, \tau_2 = 1)$ the values of L are given by

$$(\tau_1, 1) : L = 2\tau_1 - 6\nu_\Delta + 1, 2\tau_1 - 6\nu_\Delta, \dots, \tau_1 - 3\nu_\Delta + 2 - (1 - \delta_{\nu_\Delta, 0}), \quad (3.86)$$

with $\nu_\Delta = 0, \frac{1}{3}, \frac{2}{3}, \dots$. The allowed L-values are shown in Table 3.12 for $(\tau_1 \le 5, \tau_2 = 1)$. Finally, since $S = 1/2$ in this case, the total angular momentum J is given by $J = L \pm 1/2$ $(L \ne 0)$ or $J = 1/2$ $(L = 0)$, and $M_J = -J, -J + 1, \dots, +J$. Omitting the nonessential quantum numbers $N_1', \dots, N_6', N_3, \dots, N_6$ and S, the basis states can be labelled by

$$|[N_B = N], \{N_F = 1\}, [N + 1 - i, i], (\sigma_1, \sigma_2, \sigma_3), (\tau_1, \tau_2), \nu_\Delta, L, J, M_J\rangle. \quad (3.87)$$

3.2.3.3 Energy eigenvalues. Route a These are obtained in the usual way by expanding the Hamiltonian in terms of Casimir invariants,

$$
\begin{aligned}
H^{(\mathrm{III}_{3a})} &= e_0 + e_1 \mathcal{C}_1(\mathrm{U^B}6) + e_2 \mathcal{C}_2(\mathrm{U^B}6) + e_3 \mathcal{C}_1(\mathrm{U^F}12) + e_4 \mathcal{C}_2(\mathrm{U^F}12) \\
&\quad + e_5 \mathcal{C}_1(\mathrm{U^B}6)\mathcal{C}_1(\mathrm{U^F}12) + \eta \mathcal{C}_2(\mathrm{U^{BF}}6) + \eta' \mathcal{C}_2(\mathrm{O^{BF}}6) \\
&\quad + \beta \mathcal{C}_2(\mathrm{O^{BF}}5) + \gamma \mathcal{C}_2(\mathrm{O^{BF}}3) + \gamma' \mathcal{C}_2(\mathrm{SU}_s^\mathrm{F}2) \\
&\quad + \gamma'' \mathcal{C}_2(\mathrm{Spin^{BF}}3).
\end{aligned}
\tag{3.88}
$$

The Casimir invariants of the coupled algebras can be written in terms of generators as

$$
\begin{aligned}
\mathcal{C}_2(\mathrm{U^{BF}}6) &= G^{(0)}(0,0) \cdot G^{(0)}(0,0) + \sum_{\lambda=0}^{4} G^{(\lambda)}(2,2) \cdot G^{(\lambda)}(2,2) \\
&\quad + G^{(2)}(0,2) \cdot G^{(2)}(2,0) + G^{(2)}(2,0) \cdot G^{(2)}(0,2), \\
\mathcal{C}_2(\mathrm{O^{BF}}6) &= 2\left(G^{(2)}(0,2) + G^{(2)}(2,0)\right) \cdot \left(G^{(2)}(0,2) + G^{(2)}(2,0)\right) \\
&\quad + 4 \sum_{\lambda=1,3} G^{(\lambda)}(2,2) \cdot G^{(\lambda)}(2,2), \\
\mathcal{C}_2(\mathrm{O^{BF}}5) &= 4 \sum_{\lambda=1,3} G^{(\lambda)}(2,2) \cdot G^{(\lambda)}(2,2), \\
\mathcal{C}_2(\mathrm{O^{BF}}3) &= 20 G^{(1)}(2,2) \cdot G^{(1)}(2,2), \\
\mathcal{C}_2(\mathrm{Spin^{BF}}3) &= 2 J^{(1)} \cdot J^{(1)}, \quad J_\mu^{(1)} = \left(\sqrt{10}\, G_\mu^{(1)}(2,2) - \tfrac{1}{\sqrt{2}} S_\mu^{(1)}(\tfrac{1}{2},\tfrac{1}{2})\right),
\end{aligned}
\tag{3.89}
$$

From (3.88) one can obtain the energy eigenvalues

$$
\begin{aligned}
E^{(\mathrm{III}_{3a})}&(N_\mathrm{B} = N, N_\mathrm{F} = 1, [N+1-i, i], (\sigma_1, \sigma_2, \sigma_3), (\tau_1, \tau_2), \nu_\Delta, J, M_J) \\
&= E_{03} + \eta[(N+1-i)(N+6-i) + i(i+3)] \\
&\quad + 2\eta'[\sigma_1(\sigma_1+4) + \sigma_2(\sigma_2+2) + \sigma_3^2] \\
&\quad + 2\beta[\tau_1(\tau_1+3) + \tau_2(\tau_2+1)] \\
&\quad + 2\gamma L(L+1) + 2\gamma'' J(J+1),
\end{aligned}
\tag{3.90}
$$

where

$$
E_{03} = e_0 + e_1 N + e_2 N(N+5) + e_3 + 12 e_4 + e_5 N + \tfrac{3}{2}\gamma'. \tag{3.91}
$$

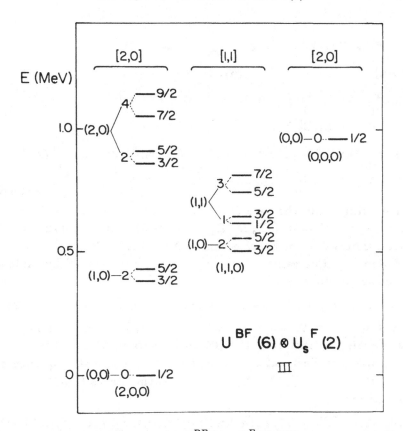

Fig. 3-8   A typical spectrum with $U^{BF}(6) \otimes U_s^F(2)$ (III) symmetry for $N_B = 1, N_F = 1$. The $U^{BF}(6)$ quantum numbers are shown on top (square brackets), the $O^{BF}(6)$ quantum numbers at the bottom, the $O^{BF}(5)$ and $O^{BF}(3)$ quantum numbers to the left and the angular momentum J to the right of each level.

The excitation spectrum associated with (3.90) is shown in Fig. 3.8.

3.2.3.4 Basis states. Route b   It is also of interest to study route b of the lattice (3.74). The basis states for this route can be written as

$$
\begin{aligned}
&\left| \; U^B(6) \otimes U^F(12) \supset U^B(6) \otimes \quad U_k^F(6) \quad \otimes U_s^F(2) \right. \\
&\quad\;\; \downarrow \qquad\qquad\qquad\quad \downarrow \qquad\qquad\qquad \downarrow \\
&\quad\;\; [N_B] \qquad\quad \{N_F\} \qquad\qquad\qquad [N_1',\ldots,N_6'] \\[4pt]
&\supset O^B(6) \otimes \quad O_k^F(6) \quad \otimes U_s^F(2) \supset O^{BF}(6) \otimes U_s^F(2) \\
&\quad\;\; \downarrow \qquad\qquad \downarrow \qquad\qquad\qquad\qquad \downarrow \\
&\quad\;\; \Sigma \qquad\quad (\Sigma_1',\Sigma_2',\Sigma_3') \qquad\qquad (\sigma_1,\sigma_2,\sigma_3) \\[4pt]
&\supset O^{BF}(5) \otimes U_s^F(2) \supset O^{BF}(3) \otimes SU_s^F(2) \supset Spin^{BF}(3) \supset Spin^{BF}(2) \\
&\quad\;\; \downarrow \qquad\qquad\quad \downarrow \qquad\qquad \downarrow \qquad\qquad \downarrow \qquad\qquad\quad \downarrow \\
&\left. \quad\;\; (\tau_1,\tau_2) \qquad\qquad \nu_\Delta, L \qquad\quad S \qquad\qquad J \qquad\qquad M_J \right\rangle .
\end{aligned}
$$

$$(3.92)$$

Considering again the case $N_B = N$ and $N_F = 1$, one has $N_1' = 1$, $N_2' = \cdots = N_6' = 0$, $(\Sigma_1',\Sigma_2',\Sigma_3') = (1,0,0)$ and $S = 1/2$ for all states. The values of Σ are given by $\Sigma = N, N-2, \ldots, 1$ or 0, as in Volume 1. The values of $(\sigma_1,\sigma_2,\sigma_3)$ can be obtained by taking the outer product

$$(\Sigma,0,0) \otimes (1,0,0) = (\Sigma+1,0,0) \oplus (\Sigma,1,0) \oplus (\Sigma-1,0,0), \quad (3.93)$$

except for $\Sigma = 0$, when $(\sigma_1,\sigma_2,\sigma_3) = (1,0,0)$ only. The further decomposition from $O^{BF}(6)$ to its subalgebras is the same as discussed in Sect. 3.2.3.2. Omitting the nonessential quantum numbers $N_1',\ldots,N_6'$, $(\Sigma_1',\Sigma_2',\Sigma_3')$ and S, the states (3.92) can be labelled by

$$|[N_B = N], \{N_F = 1\}, \Sigma, (\sigma_1,\sigma_2,\sigma_3), (\tau_1,\tau_2), \nu_\Delta, L, J, M_J\rangle. \quad (3.94)$$

3.2.3.5 Energy eigenvalues. Route b

Expanding the Hamiltonian in Casimir invariants of the algebras in b,

$$
\begin{aligned}
H^{(III3b)} &= e_0 + e_1 C_1(U^B 6) + e_2 C_2(U^B 6) + e_3 C_1(U^F 12) + e_4 C_2(U^F 12) \\
&\quad + e_5 C_1(U^B 6) C_1(U^F 12) + \eta'' C_2(O^B 6) + \eta' C_2(O^{BF} 6) \\
&\quad + \beta C_2(O^{BF} 5) + \gamma C_2(O^{BF} 3) + \gamma' C_2(SU_s^F 2) + \gamma'' C_2(Spin^{BF} 3),
\end{aligned}
$$

$$(3.95)$$

one obtains the energy eigenvalues

$$
\begin{aligned}
E^{(III3b)}&(N_B = N, N_F = 1, \Sigma, (\sigma_1,\sigma_2,\sigma_3), (\tau_1,\tau_2), \nu_\Delta, J, M_J) \\
&= E_{03} + 2\eta'' \Sigma(\Sigma+4) + 2\eta'[\sigma_1(\sigma_1+4) + \sigma_2(\sigma_2+2) + \sigma_3^2] \\
&\quad + 2\beta[\tau_1(\tau_1+3) + \tau_2(\tau_2+1)] + 2\gamma L(L+1) + 2\gamma'' J(J+1).
\end{aligned}
$$

$$(3.96)$$

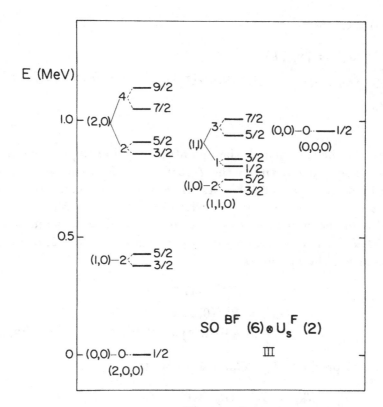

Fig. 3-9   A typical spectrum with $O^{BF}(6) \otimes U_s^F(2)$ (III) symmetry for $N_B = 1, N_F = 1$. The $O^{BF}(6)$ quantum numbers are shown at the bottom, the $O^{BF}(5)$ and $O^{BF}(3)$ quantum numbers to the left and the angular momentum J to the right of each level.

The excitation spectrum associated with (3.96) is shown in Fig. 3.9.

3.2.3.6 Wave functions. Isoscalar factors   It is convenient to discuss the wave functions simultaneously for routes a and b. We begin this time with route b, where the coupling of bosons and fermions occurs at the level O(6). The wave function can be obtained by expanding $O^{BF}(6)$ states into states of the product

$$O^B(6) \otimes O_k^F(6),$$

$$
|(\sigma_1, \sigma_2, \sigma_3), (\tau_1, \tau_2), L\rangle
$$
$$
= \sum_{\substack{\tau', L' \\ \tau'', L''}} \xi^{(\Sigma,0,0),(\tau',0),L'}_{(\sigma_1,\sigma_2,\sigma_3),(\tau_1,\tau_2),L} |(\Sigma,0,0),(\tau',0),L';(1,0,0),(\tau'',0),L'';L\rangle,
$$

$$(3.97)$$

where we have omitted the quantum number ν_Δ. The procedure used to determine the ξ-coefficients is identical to that described in Sect. 3.2.1.4. As an example, consider the case $|(\sigma_1, \sigma_2, \sigma_3), (\tau_1, \tau_2), L\rangle = |(\Sigma + 1, 0, 0), (\tau, 0), L\rangle$. When $\tau'' = 0, L'' = 0$, the sum of τ', L' goes over $\tau' = \tau, L' = L$, while when $\tau'' = 1, L'' = 2$, the sum goes over $\tau' = \tau \pm 1, L' = L \pm 2, L \pm 1, L$. Introducing the operators

$$
\begin{aligned}
G^{(2)}_{\mathrm{BF},\mu} &= G^{(2)}_{\mathrm{B},\mu} + G^{(2)}_{\mathrm{F},\mu}, \\
G^{(2)}_{\mathrm{B},\mu} &= B^{(2)}_\mu(0,2) + B^{(2)}_\mu(2,0), \\
G^{(2)}_{\mathrm{F},\mu} &= K^{(2)}_\mu(0,2) + K^{(2)}_\mu(2,0),
\end{aligned}
\qquad (3.98)
$$

the scalar product $G^{(2)}_{\mathrm{BF}} \cdot G^{(2)}_{\mathrm{BF}}$ can be written as

$$
\begin{aligned}
G^{(2)}_{\mathrm{BF}} \cdot G^{(2)}_{\mathrm{BF}} &= G^{(2)}_{\mathrm{B}} \cdot G^{(2)}_{\mathrm{B}} + G^{(2)}_{\mathrm{F}} \cdot G^{(2)}_{\mathrm{F}} + 2G^{(2)}_{\mathrm{B}} \cdot G^{(2)}_{\mathrm{F}} \\
&= \tfrac{1}{2}\left[\mathcal{C}_2(O^{\mathrm{BF}}6) - \mathcal{C}_2(O^{\mathrm{BF}}5) \right].
\end{aligned}
\qquad (3.99)
$$

One then takes matrix elements of (3.99) between $|(\Sigma + 1, 0, 0), (\tau, 0), L\rangle$ and the state $|(\Sigma, 0, 0), (\tau + 1, 0), L_1; (1, 0, 0), (1, 0), 2; L\rangle$ and obtains

$$
[(\Sigma + 1)(\Sigma + 5) - \tau(\tau + 3)]\xi^{(\Sigma,0,0),(\tau+1,0),L_1}_{(\Sigma+1,0,0),(\tau,0),L}
$$
$$
= [\Sigma(\Sigma + 4) - (\tau + 1)(\tau + 4) + 1]\xi^{(\Sigma,0,0),(\tau+1,0),L_1}_{(\Sigma+1,0,0),(\tau,0),L}
$$
$$
+ \sum_{\substack{\tau', L' \\ \tau'', L''}} 2(-)^{L+L'} \left\{ \begin{array}{ccc} L_1 & 2 & L \\ L'' & L' & 2 \end{array} \right\}
$$
$$
\times \langle (\Sigma, 0, 0), (\tau + 1, 0), L_1 \| G^{(2)}_{\mathrm{B}} \| (\Sigma, 0, 0), (\tau', 0), L'\rangle
$$
$$
\times \langle (1, 0, 0), (1, 0), 2 \| G^{(2)}_{\mathrm{F}} \| (1, 0, 0), (\tau'', 0), L''\rangle \xi^{(\Sigma,0,0),(\tau',0),L'}_{(\Sigma+1,0,0),(\tau,0),L}.
$$

$$(3.100)$$

Because of the selection rules for $G_{\mathrm{F}}^{(2)}$ one has $\tau'' = 0, L'' = 0$ and inserting the appropriate values for the $6j$-symbol one obtains

$$\xi_{(\Sigma+1,0,0),(\tau,0),L}^{(\Sigma,0,0),(\tau+1,0),L_1} = \frac{(-)^{L+L_1}}{(\Sigma+\tau+4)\sqrt{2L+1}}\xi_{(\Sigma+1,0,0),(\tau,0),L}^{(\Sigma,0,0),(\tau,0),L}$$

$$\times \langle (\Sigma,0,0),(\tau+1,0),L_1 \parallel G_{\mathrm{B}}^{(2)} \parallel (\Sigma,0,0),(\tau,0),L \rangle.$$

$$(3.101)$$

Two more relations can be obtained by taking matrix elements of (3.99) between the state $|(\Sigma+1,0,0),(\tau,0),L\rangle$ and the states

$$|(\Sigma,0,0),(\tau-1,0),L_1;(1,0,0),(1,0),2;L\rangle$$

and

$$|(\Sigma,0,0),(\tau,0),L;(1,0,0),(0,0),0;L\rangle.$$

This gives

$$\xi_{(\Sigma+1,0,0),(\tau,0),L}^{(\Sigma,0,0),(\tau-1,0),L_1} = \frac{(-)^{L+L_1}}{(\Sigma-\tau+1)\sqrt{2L+1}}\xi_{(\Sigma+1,0,0),(\tau,0),L}^{(\Sigma,0,0),(\tau,0),L}$$

$$\times \langle (\Sigma,0,0),(\tau-1,0),L_1 \parallel G_{\mathrm{B}}^{(2)} \parallel (\Sigma,0,0),(\tau,0),L \rangle,$$

$$(3.102)$$

and

$$\xi_{(\Sigma+1,0,0),(\tau,0),L}^{(\Sigma,0,0),(\tau,0),L} = \frac{1}{\Sigma\sqrt{2L+1}} \sum_{\tau',L'} \xi_{(\Sigma+1,0,0),(\tau,0),L}^{(\Sigma,0,0),(\tau',0),L'}$$

$$\times \langle (\Sigma,0,0),(\tau,0),L \parallel G_{\mathrm{B}}^{(2)} \parallel (\Sigma,0,0),(\tau',0),L' \rangle.$$

$$(3.103)$$

Since the matrix elements of the operator $G_{\mathrm{B}}^{(2)}$ are known, Volume 1, the set of Eqs. (3.101)–(3.103), together with the normalization condition,

$$\sum_{\tau',L'} \left(\xi_{(\Sigma+1,0,0),(\tau,0),L}^{(\Sigma,0,0),(\tau',0),L'} \right)^2 = 1,$$

$$(3.104)$$

determines the expansion coefficients. Similar equations can be derived for $(\sigma_1,\sigma_2,\sigma_3) = (\Sigma-1,0,0)$ or $(\Sigma,1,0)$. These equations can again be simplified using Racah's factorization lemma, which states that the ξ-coefficients can be written as

$$\xi_{(\sigma_1,\sigma_2,\sigma_3),(\tau_1,\tau_2),L}^{(\Sigma,0,0),(\tau',0),L'} = \left\langle \begin{array}{cc|c} (\Sigma,0,0) & (1,0,0) & (\sigma_1,\sigma_2,\sigma_3) \\ (\tau',0) & (0,0) & (\tau_1,\tau_2) \end{array} \right\rangle$$

$$\times \left\langle \begin{array}{cc|c} (\tau',0) & (0,0) & (\tau_1,\tau_2) \\ L' & 0 & L \end{array} \right\rangle, \quad (3.105)$$

for $\tau'' = 0, L'' = 0$ and

$$\xi_{(\sigma_1,\sigma_2,\sigma_3),(\tau_1,\tau_2),L}^{(\Sigma,0,0),(\tau',0),L'} = \left\langle \begin{array}{cc|c} (\Sigma,0,0) & (1,0,0) & (\sigma_1,\sigma_2,\sigma_3) \\ (\tau',0) & (1,0) & (\tau_1,\tau_2) \end{array} \right\rangle$$

$$\times \left\langle \begin{array}{cc|c} (\tau',0) & (1,0) & (\tau_1,\tau_2) \\ L' & 2 & L \end{array} \right\rangle, \quad (3.106)$$

for $\tau'' = 1, L'' = 2$. The last isoscalar factor in (3.105), associated with $O(5) \supset O(3)$, is trivially equal to 1 for $(\tau_1,\tau_2)L = (\tau',0)L'$ and 0 otherwise, since it involves multiplication with the $O(5)$-scalar representation $(0,0)$. The $O(5) \supset O(3)$ isoscalar factors with $(\tau_1,\tau_2) = (\tau'+1,0)$ in (3.106) can be related to the reduced matrix elements of $d^\dagger$, discussed in Sect. 2.6.1 of Volume 1,

$$\left\langle \begin{array}{cc|c} (\tau',0) & (1,0) & (\tau'+1,0) \\ L' & 2 & L \end{array} \right\rangle$$

$$= \frac{1}{\sqrt{(\tau'+1)(2L+1)}}$$

$$\times \langle n_d = \tau'+1, v = \tau'+1, L \parallel d^\dagger \parallel n_d = \tau', v = \tau', L' \rangle. \quad (3.107)$$

The $O(5) \supset O(3)$ isoscalar factors with $(\tau_1,\tau_2) = (\tau'-1,0)$ are related to (3.107) by (Van Isacker *et al.*, 1984)

$$\left\langle \begin{array}{cc|c} (\tau',0) & (1,0) & (\tau'-1,0) \\ L' & 2 & L \end{array} \right\rangle$$

$$= (-)^{L+L'} \left[\frac{\tau'(2\tau'+1)(2L'+1)}{(\tau'+2)(2\tau'+3)(2L+1)} \right]^{\frac{1}{2}} \quad (3.108)$$

$$\times \left\langle \begin{array}{cc|c} (\tau'-1,0) & (1,0) & (\tau',0) \\ L & 2 & L' \end{array} \right\rangle,$$

Table 3-13 A selected number of O(5) $\supset$ O(3) isoscalar factors

$$\left\langle \begin{array}{cc} (\tau',0) & (1,0) \\ L' & 2 \end{array} \middle| \begin{array}{c} (\tau_1,\tau_2) \\ L \end{array} \right\rangle$$

$(\tau_1,\tau_2)^a$	L	L'	Isoscalar factor
$(\tau'+1,0)$	$2\tau'+2$	$2\tau'$	1
$(\tau'+1,0)$	$2\tau'$	$2\tau'$	$\left[\dfrac{2(2\tau'+1)}{(\tau'+1)(4\tau'-1)}\right]^{\frac{1}{2}}$
$(\tau'+1,0)$	$2\tau'$	$2\tau'-2$	$\left[\dfrac{(\tau'-1)(4\tau'+3)}{(\tau'+1)(4\tau'-1)}\right]^{\frac{1}{2}}$
$(\tau'+1,0)$	$2\tau'-1$	$2\tau'$	$\left[\dfrac{2(\tau'-1)(4\tau'+1)}{(\tau'+1)(2\tau'-1)(4\tau'-1)}\right]^{\frac{1}{2}}$
$(\tau'+1,0)$	$2\tau'-1$	$2\tau'-2$	$-\left[\dfrac{3(2\tau'+1)}{(\tau'-1)(\tau'+1)(4\tau'-1)}\right]^{\frac{1}{2}}$
$(\tau'+1,0)$	$2\tau'-1$	$2\tau'-3$	$\left[\dfrac{(\tau'-2)\tau'(2\tau'+1)}{(\tau'-1)(\tau'+1)(2\tau'-1)}\right]^{\frac{1}{2}}$
$(\tau'+1,0)$	$2\tau'-2$	$2\tau'$	$\left[\dfrac{32(\tau'-2)(\tau'-1)\tau'}{(\tau'+1)(2\tau'-1)(2\tau'+3)(4\tau'-3)(4\tau'-1)}\right]^{\frac{1}{2}}$
$(\tau',1)$	$2\tau'+1$	$2\tau'$	1
$(\tau',1)$	$2\tau'$	$2\tau'$	$\left[\dfrac{(\tau'-1)(4\tau'+3)}{(\tau'+1)(4\tau'-1)}\right]^{\frac{1}{2}}$
$(\tau',1)$	$2\tau'$	$2\tau'-2$	$-\left[\dfrac{2(2\tau'+1)}{(\tau'+1)(4\tau'-1)}\right]^{\frac{1}{2}}$

^{a} Isoscalar factors with $(\tau_1,\tau_2) = (\tau'-1,0)$ can be obtained from Eq. (3.108).

and those with $(\tau_1,\tau_2) = (\tau',1)$ can, in simple cases, be obtained from orthonormality. From the expressions for the reduced matrix elements in Table 2.9 of Volume 1, one finds the results quoted in Table 3.13. Inserting the appropriate values for the O(5) $\supset$ O(3) isoscalar factors in Eqs. (3.101) and (3.102) with $L = 2\tau$ and $L_1 = 2\tau \pm 2$, one derives the general expressions for the O(6) $\supset$ O(5) isoscalar factors shown in Table 3.14. Identical expressions (up to a phase factor) have been obtained with a different technique by Hecht (1987).

Table 3-14 General expressions for the O(6) ⊃ O(5) isoscalar factors

$$\left\langle \begin{array}{cc|c} (\Sigma,0,0) & (1,0,0) & (\sigma_1,\sigma_2,0) \\ (\tau',0) & (\tau'',0) & (\tau_1,\tau_2) \end{array} \right\rangle$$

(σ_1,σ_2)	(τ_1,τ_2)	τ'	τ''	Isoscalar factor
$(\Sigma-1,0)$	$(\tau,0)$	τ	0	$-\left[\dfrac{(\Sigma-\tau)(\Sigma+\tau+3)}{2(\Sigma+2)(\Sigma+3)}\right]^{\frac{1}{2}}$
$(\Sigma-1,0)$	$(\tau,0)$	$\tau-1$	1	$\left[\dfrac{\tau(\Sigma-\tau)(\Sigma-\tau+1)}{2(2\tau+3)(\Sigma+2)(\Sigma+3)}\right]^{\frac{1}{2}}$
$(\Sigma-1,0)$	$(\tau,0)$	$\tau+1$	1	$\left[\dfrac{(\tau+3)(\Sigma+\tau+3)(\Sigma+\tau+4)}{2(2\tau+3)(\Sigma+2)(\Sigma+3)}\right]^{\frac{1}{2}}$
$(\Sigma+1,0)$	$(\tau,0)$	τ	0	$\left[\dfrac{(\Sigma-\tau+1)(\Sigma+\tau+4)}{2(\Sigma+1)(\Sigma+2)}\right]^{\frac{1}{2}}$
$(\Sigma+1,0)$	$(\tau,0)$	$\tau-1$	1	$\left[\dfrac{\tau(\Sigma+\tau+3)(\Sigma+\tau+4)}{2(2\tau+3)(\Sigma+1)(\Sigma+2)}\right]^{\frac{1}{2}}$
$(\Sigma+1,0)$	$(\tau,0)$	$\tau+1$	1	$\left[\dfrac{(\tau+3)(\Sigma-\tau)(\Sigma-\tau+1)}{2(2\tau+3)(\Sigma+1)(\Sigma+2)}\right]^{\frac{1}{2}}$
$(\Sigma,1)$	$(\tau,0)$	τ	0	$-\left[\dfrac{\tau(\tau+3)}{(\Sigma+1)(\Sigma+3)}\right]^{\frac{1}{2}}$
$(\Sigma,1)$	$(\tau,0)$	$\tau-1$	1	$\left[\dfrac{(\tau+3)(\Sigma-\tau+1)(\Sigma+\tau+3)}{(2\tau+3)(\Sigma+1)(\Sigma+3)}\right]^{\frac{1}{2}}$
$(\Sigma,1)$	$(\tau,0)$	$\tau+1$	1	$-\left[\dfrac{\tau(\Sigma-\tau)(\Sigma+\tau+4)}{(2\tau+3)(\Sigma+1)(\Sigma+3)}\right]^{\frac{1}{2}}$
$(\Sigma,1)$	$(\tau,1)$	τ	1	1

We now come to a discussion of isoscalar factors for route a. Here bosons and fermions are coupled at the level of U(6). The wave functions of the basis states (3.87) can be expanded into those of the product group $U^{\mathrm{B}}(6) \otimes U^{\mathrm{F}}_k(6)$,

$$|[N_1,N_2],(\sigma_1,\sigma_2,\sigma_3),(\tau_1,\tau_2),L\rangle$$
$$= \sum_{\substack{\Sigma,\tau',L' \\ \tau'',L''}} \xi^{[N],(\Sigma,0,0),(\tau',0),L'}_{[N_1,N_2],(\sigma_1,\sigma_2,\sigma_3),(\tau_1,\tau_2),L}$$
$$\times |[N],(\Sigma,0,0),(\tau',0),L';[1],(1,0,0),(\tau'',0),L'';L\rangle,$$
$$\tag{3.109}$$

where we have deleted again the label ν_Δ. Racah's factorization lemma is particularly useful here, since it allows to split the expansion coefficients in (3.109) into the product of coefficients for the group reduction $U(6) \supset O(6)$ and $O(6) \supset O(5) \supset O(3)$,

$$\xi^{[N],(\Sigma,0,0),(\tau',0),L'}_{[N_1,N_2],(\sigma_1,\sigma_2,\sigma_3),(\tau_1,\tau_2),L} = \xi^{[N],(\Sigma,0,0)}_{[N_1,N_2],(\sigma_1,\sigma_2,\sigma_3)} \; \xi^{(\Sigma,0,0),(\tau',0),L'}_{(\sigma_1,\sigma_2,\sigma_3),(\tau_1,\tau_2),L}.$$
(3.110)

The second factor in the product has been determined previously while the first one is a short-hand notation for the $U(6) \supset O(6)$ isoscalar factor,

$$\xi^{[N],(\Sigma,0,0)}_{[N_1,N_2],(\sigma_1,\sigma_2,\sigma_3)} = \left\langle \begin{array}{cc} [N] & [1] \\ (\Sigma,0,0) & (1,0,0) \end{array} \right| \left. \begin{array}{c} [N_1,N_2] \\ (\sigma_1,\sigma_2,\sigma_3) \end{array} \right\rangle.$$
(3.111)

This transformation bracket can be evaluated by calculating the reduced matrix elements of the operator $s^\dagger$ between states with different $U(6)$ quantum numbers. As an example, consider

$$\langle [N+1,0],(\Sigma+1,0,0),(\tau,0)L \, \| \, s^\dagger \, \| \, [N],(\Sigma,0,0),(\tau,0),L \rangle$$
$$= \sqrt{(N+1)(2L+1)} \; \xi^{[N],(\Sigma,0,0)}_{[N+1,0],(\Sigma+1,0,0)} \; \xi^{(\Sigma,0,0),(\tau,0),L}_{(\Sigma+1,0,0),(\tau,0),L}. \quad (3.112)$$

The last factor in (3.112) is known from the previous discussion and one finds

$$\langle [N+1,0],(\Sigma+1,0,0),(\tau,0)L \, \| \, s^\dagger \, \| \, [N],(\Sigma,0,0),(\tau,0),L \rangle$$
$$= \left[\frac{(2L+1)(N+1)(\Sigma-\tau+1)(\Sigma+\tau+4)}{2(\Sigma+1)(\Sigma+2)} \right]^{\frac{1}{2}} \xi^{[N],(\Sigma,0,0)}_{[N+1,0],(\Sigma+1,0,0)}.$$
(3.113)

On the other hand, this matrix element can be evaluated directly by expanding the wave functions into a spherical basis. The result becomes simple for $\tau = \Sigma$,

$$\langle [N+1,0],(\Sigma+1,0,0),(\Sigma,0)L \, \| \, s^\dagger \, \| \, [N],(\Sigma,0,0),(\Sigma,0),L \rangle$$
$$= \sum_{n_d} \sqrt{(2L+1)(N-n_d+1)} \; \zeta^{\Sigma+1}_{n_d,\tau=\Sigma} \; \zeta^{\Sigma}_{n_d,\tau=\Sigma}$$
$$= \left[\frac{(2L+1)(N+\Sigma+6)}{2(\Sigma+3)} \right]^{\frac{1}{2}}, \quad (3.114)$$

Table 3-15 General expressions for the U(6) ⊃ O(6) isoscalar factors

$$\left\langle \begin{array}{cc|c} [N] & [1] & [N_1, N_2] \\ (\Sigma,0,0) & (1,0,0) & (\sigma_1,\sigma_2,0) \end{array} \right\rangle$$

	$[N_1, N_2] = [N+1, 0]$	$[N_1, N_2] = [N, 1]$
$(\sigma_1,\sigma_2)=(\Sigma-1,0)$	$\left[\dfrac{(\sigma_1+4)(N-\sigma_1+1)}{2(\sigma_1+2)(N+1)}\right]^{\frac{1}{2}}$	$-\left[\dfrac{\sigma_1(N+\sigma_1+5)}{2(\sigma_1+2)(N+1)}\right]^{\frac{1}{2}}$
$(\sigma_1,\sigma_2)=(\Sigma+1,0)$	$\left[\dfrac{\sigma_1(N+\sigma_1+5)}{2(\sigma_1+2)(N+1)}\right]^{\frac{1}{2}}$	$\left[\dfrac{(\sigma_1+4)(N-\sigma_1+1)}{2(\sigma_1+2)(N+1)}\right]^{\frac{1}{2}}$
$(\sigma_1,\sigma_2)=(\Sigma,1)$		1

where the expansion coefficients ζ are defined in Eq. (2.57) of
Volume 1 and general expressions for them are given by Castaños
et al. (1979). Comparison between (3.113) for $\tau = \Sigma$ and (3.114)
gives in this case

$$\zeta^{[N],(\Sigma,0,0)}_{[N+1,0],(\Sigma+1,0,0)} = \left[\frac{(\Sigma+1)(N+\Sigma+6)}{2(\Sigma+3)(N+1)}\right]^{\frac{1}{2}}. \tag{3.115}$$

Repeating this procedure for other cases, the coefficients shown in
Table 3.15 are obtained.

The total wave function (3.87) can now be obtained from the
expansion (3.109) and a recoupling of angular momenta from
$(L', L'')L, 1/2, J$ to $L', (L'', 1/2)j, J$,

$$|[N_1, N_2], (\sigma_1, \sigma_2, \sigma_3), (\tau_1, \tau_2), L, J\rangle$$
$$= \sum_{\substack{\Sigma, \tau', L' \\ \tau'', L''}} \zeta^{[N],(\Sigma,0,0),(\tau',0),L'}_{[N_1, N_2],(\sigma_1\sigma_2\sigma_3),(\tau_1,\tau_2),L} \sum_j (-)^{L'+L''+1/2+J}$$

$$\times \sqrt{(2L+1)(2j+1)} \left\{ \begin{array}{ccc} L' & L'' & L \\ 1/2 & J & j \end{array} \right\} |[N], \Sigma, \tau', L'; j; J\rangle.$$
$$\tag{3.116}$$

3.2.3.7 Electromagnetic transitions and moments; E2 The most
general E2 transition operator can be written as
$$T^{(\mathrm{E}2)}_\mu = \alpha_2 [s^\dagger \times \tilde{d} + d^\dagger \times \tilde{s}]^{(2)}_\mu + \beta_2 [d^\dagger \times \tilde{d}]^{(2)}_\mu + \sum_{jj'} f^{(2)}_{jj'} [a^\dagger_j \times \tilde{a}_{j'}]^{(2)}_\mu.$$
$$\tag{3.117}$$

The coefficients $f_{jj'}^{(2)}$ can be related to the matrix elements of the single-particle quadrupole operator $r^2 Y^{(2)}$,

$$f_{jj'}^{(2)} = -\frac{e_{\mathrm{F}}}{\sqrt{5}} \langle j \parallel r^2 Y^{(2)} \parallel j' \rangle, \qquad (3.118)$$

where e_{F} denotes an overall effective charge. In general, these matrix elements must be evaluated numerically. However, as in the corresponding cases discussed in Volume 1, they can be obtained in closed form whenever they can be written in terms of generators of the algebras appearing in a particular chain. It turns out that, for E2 operators, it is a good approximation to write

$$T_\mu^{(\mathrm{E2})} = \alpha_2 \left(B_\mu^{(2)}(0,2) + B_\mu^{(2)}(2,0) \right) + f_2 \left(K_\mu^{(2)}(0,2) + K_\mu^{(2)}(2,0) \right)$$
$$+ \beta_2 B_\mu^{(2)}(2,2) + f_2' K_\mu^{(2)}(2,2). \qquad (3.119)$$

Further simplifications occur when $T^{(\mathrm{E2})}$ is written in terms only of generators of the combined algebras. Introducing the operators $G_{\mathrm{BF}}^{(2)}$ of (3.98) and

$$G_{\mathrm{BF},\mu}'^{(2)} = G_{\mathrm{B},\mu}'^{(2)} + G_{\mathrm{F},\mu}'^{(2)},$$
$$G_{\mathrm{B},\mu}'^{(2)} = B_\mu^{(2)}(2,2),$$
$$G_{\mathrm{F},\mu}'^{(2)} = K_\mu^{(2)}(2,2), \qquad (3.120)$$

one has, for $f_2 = \alpha_2$ and $f_2' = \beta_2$,

$$T_\mu^{(\mathrm{E2})} = \alpha_2 G_{\mathrm{BF},\mu}^{(2)} + \beta_2 G_{\mathrm{BF},\mu}'^{(2)}. \qquad (3.121)$$

Matrix elements of both terms in (3.121) can now be calculated easily. We consider, for simplicity, the case $\beta_2 = 0$. In this case, $T^{(\mathrm{E2})}$ is a generator of $\mathrm{O}^{\mathrm{BF}}(6)$ and thus has selection rules $\Delta\sigma_1 = \Delta\sigma_2 = \Delta\sigma_3 = 0$. The selection rules with respect to $\mathrm{O}^{\mathrm{BF}}(5)$ are derived from the tensorial character of the operator under $\mathrm{O}^{\mathrm{BF}}(5)$ transformations and are $\Delta\tau_1 = 0, \Delta\tau_2 = \pm 1$ or $\Delta\tau_1 = \pm 1, \Delta\tau_2 = 0$. A consequence of this selection rule is that, when $\beta_2 = 0$, all quadrupole moments vanish, since the quadrupole moments are related to the diagonal matrix elements of $T^{(\mathrm{E2})}$.

To evaluate the matrix elements of the E2 operator explicitly, one makes use of the isoscalar factors previously determined. It turns out that, if one restricts oneself to matrix elements between states belonging to the $O^{BF}(6)$ representations $(N + 1, 0, 0)$ and $(N, 1, 0)$ the results are identical for routes a and b, since the $U(6) \supset O(6)$ expansion coefficients are, in these cases, equal to 1. The matrix elements for the representation $(N + 1, 0, 0)$ can be easily obtained. Separating the pseudo-spin part, one has:

$$
\langle (N + 1, 0, 0), (\tau, 0), L, J \parallel G_{BF}^{(2)} \parallel (N + 1, 0, 0), (\tau', 0), L', J' \rangle
$$

$$
= (-)^{L+1/2+J'} \sqrt{(2J + 1)(2J' + 1)} \left\{ \begin{array}{ccc} L & J & 1/2 \\ J' & L' & 2 \end{array} \right\}
$$

$$
\times \langle (N+1, 0, 0), (\tau, 0), L \parallel G_{BF}^{(2)} \parallel (N+1, 0, 0), (\tau', 0), L' \rangle,
$$

$$(3.122)$$

where we have omitted ν_Δ and other redundant quantum numbers from the states (3.87) or (3.94). Since the reduced matrix elements of the $L = 2$ generator of $O(6)$ between states belonging to the symmetric representation $(\sigma, 0, 0)$ have been calculated (see Volume 1), Eq. (3.122) gives directly those of $G_{BF}^{(2)}$. For the representation $(N, 1, 0)$, the situation is more complex. One needs to separate first the pseudo-spin part, as in (3.122),

$$
\langle (N, 1, 0), (\tau_1, \tau_2), L, J \parallel G_{BF}^{(2)} \parallel (N, 1, 0), (\tau_1', \tau_2'), L', J' \rangle
$$

$$
= (-)^{L+1/2+J'} \sqrt{(2J + 1)(2J' + 1)} \left\{ \begin{array}{ccc} L & J & 1/2 \\ J' & L' & 2 \end{array} \right\}
$$

$$
\times \langle (N, 1, 0), (\tau_1, \tau_2), L \parallel G_{BF}^{(2)} \parallel (N, 1, 0), (\tau_1', \tau_2'), L' \rangle,
$$

$$(3.123)$$

and then expand the last factor in (3.123), obtaining

$$
\langle (N, 1, 0), (\tau_1, \tau_2), L \parallel G_{BF}^{(2)} \parallel (N, 1, 0), (\tau_1', \tau_2'), L' \rangle
$$

$$
= \sqrt{(2L + 1)(2L' + 1)} \sum_{\substack{\tau_B, L_B, \tau_F, L_F \\ \tau_B', L_B', \tau_F', L_F'}} \xi_{(N,1,0),(\tau_1,\tau_2),L}^{(N,0,0),(\tau_B,0),L_B} \xi_{(N,1,0),(\tau_1',\tau_2'),L'}^{(N,0,0),(\tau_B',0),L_B'}
$$

$$
\times \left[(-)^{L_B+L_F+L'} \left\{ \begin{array}{ccc} L_B & L & L_F \\ L' & L_B' & 2 \end{array} \right\} \right.
$$

$$
\times \langle (N, 0, 0), (\tau_B, 0), L_B \parallel G_B^{(2)} \parallel (N, 0, 0), (\tau_B', 0), L_B' \rangle \delta_{\tau_F, \tau_F'} \delta_{L_F, L_F'}
$$

$$+ (-)^{L_B + L_F' + L} \left\{ \begin{matrix} L_F & L & L_B \\ L' & L_F' & 2 \end{matrix} \right\}$$

$$\times \langle (1,0,0), (\tau_F, 0), L_F \parallel G_F^{(2)} \parallel (1,0,0), (\tau_F', 0), L_F' \rangle \delta_{\tau_B, \tau_B'} \delta_{L_B, L_B'} \Big].$$

$$(3.124)$$

Equation (3.124) now only contains the isoscalar factors ξ and the matrix elements of the O(6) generators between symmetric states, both of which have been previously determined. B(E2) values can thus be calculated using (3.124). Matrix elements of other parts of the E2 operator (3.119) can be computed with similar techniques. In Table 3.16 we list some B(E2) values between states of the representations $(N + 1, 0, 0)$ and $(N, 1, 0)$, calculated with an E2 operator with arbitrary α_2 and f_2 and $\beta_2 = f_2' = 0$.

3.2.3.8 Electromagnetic transitions and moments; M1 The most general form of the M1 transition operator is given in Chapter 1 as

$$T_\mu^{(M1)} = \beta_1 [d^\dagger \times \tilde{d}]_\mu^{(1)} + \sum_{jj'} f_{jj'}^{(1)} [a_j^\dagger \times \tilde{a}_{j'}]_\mu^{(1)}. \qquad (3.125)$$

Realistic values of the coefficients $f_{jj'}^{(1)}$ can be obtained by relating them to the matrix elements of the single-particle operators,

$$f_{jj'}^{(1)} = -\frac{f_1}{\sqrt{3}} \langle j \parallel g_l \vec{l} + g_s \vec{s} \parallel j' \rangle, \qquad (3.126)$$

where $\vec{l}$ denotes the orbital angular momentum of the odd nucleon, $\vec{s}$ its spin angular momentum, g_l and g_s are the orbital and spin g-factors and the coefficient f_1 denotes an overall strength. Contrary to the case of E2 transitions, the fermion part of the M1 operator cannot, to a good approximation, be written in terms only of the generators of $O^{BF}(3)$ and $Spin^{BF}(3)$. As a result, the evaluation of the matrix elements of the M1 operator in closed form is rather complex and will be omitted. Numerical calculations of M1 transition rates are possible with the computer program ODDA (Scholten, 1979) and will be discussed in Chapter 5.

3.2.3.9 One-nucleon transfer intensities In addition to E2 and M1 transition rates, an important tool in studying odd–even nuclei

Table 3-16 Some B(E2) values in the $U^{BF}(6) \otimes U_s^F(2)$ limit

$(N+1,0,0),(\tau+1,0),L',J' \to (N+1,0,0),(\tau,0),L,J$					
L'	J'	$\to$	L	J	$\dfrac{B(E2;J'\to J)}{(N-\tau+1)(N+\tau+5)}$
$2\tau+2$	$2\tau+\frac{5}{2}$		2τ	$2\tau+\frac{1}{2}$	$\left(\dfrac{N\alpha_2+f_2}{N+1}\right)^2 \dfrac{\tau+1}{2\tau+5}$
$2\tau+2$	$2\tau+\frac{3}{2}$		2τ	$2\tau+\frac{1}{2}$	$\left(\dfrac{N\alpha_2+f_2}{N+1}\right)^2 \dfrac{1}{(2\tau+5)(4\tau+1)}$
$2\tau+2$	$2\tau+\frac{3}{2}$		2τ	$2\tau-\frac{1}{2}$	$\left(\dfrac{N\alpha_2+f_2}{N+1}\right)^2 \dfrac{\tau(4\tau+5)}{(2\tau+5)(4\tau+1)}$
2τ	$2\tau+\frac{1}{2}$		2τ	$2\tau+\frac{1}{2}$	$\left(\dfrac{N\alpha_2+f_2}{N+1}\right)^2 \dfrac{4(\tau+1)}{(2\tau+5)(4\tau+1)}$
2τ	$2\tau+\frac{1}{2}$		2τ	$2\tau-\frac{1}{2}$	$\left(\dfrac{N\alpha_2+f_2}{N+1}\right)^2 \dfrac{6}{(2\tau+5)(4\tau-1)(4\tau+1)}$
2τ	$2\tau-\frac{1}{2}$		2τ	$2\tau-\frac{1}{2}$	$\left(\dfrac{N\alpha_2+f_2}{N+1}\right)^2 \dfrac{(2\tau-1)(2\tau+1)(4\tau+3)}{\tau(2\tau+5)(4\tau-1)(4\tau+1)}$
$(N,1,0),(\tau_1',\tau_2'),L',J' \to (N+1,0,0),(\tau_1,\tau_2),L,J$					
(τ_1',τ_2')	$\to$	(τ_1,τ_2)	$B(E2;J'\to J)^a$		
$(1,0)$		$(0,0)$	$(\alpha_2-f_2)^2 \dfrac{2N(N+3)}{5(N+1)(N+2)}$		
$(2,0)$		$(1,0)$	$(\alpha_2-f_2)^2 \dfrac{5(N-1)N(N+3)}{14(N+1)^2(N+2)}\mathcal{F}^2$		
$(1,1)$		$(1,0)$	$(\alpha_2-f_2)^2 \dfrac{N(N+5)}{2(N+1)(N+2)}\mathcal{F}^2$		
$(N+1,0,0),(\tau_1',\tau_2'),L',J' \to (N,1,0),(\tau_1,\tau_2),L,J$					
$(2,0)$		$(1,0)$	$(\alpha_2-f_2)^2 \dfrac{(N+3)(N+5)(N+6)}{28(N+1)^2(N+2)}\mathcal{F}^2$		
$(N,1,0),(\tau_1',\tau_2'),L',J' \to (N,1,0),(\tau_1,\tau_2),L,J$					
$(2,0)$		$(1,0)$	$\left((N^2+4N+5)\alpha_2-2f_2\right)^2 \dfrac{5(N-1)(N+5)}{28(N+1)^2(N+3)^2}\mathcal{F}^2$		
$(1,1)$		$(1,0)$	$\left((N^2+4N-1)\alpha_2+4f_2\right)^2 \dfrac{1}{4(N+1)(N+3)}\mathcal{F}^2$		

^{a} With $\mathcal{F}^2 \equiv \mathcal{F}(L',J',L,J)^2 = (2J+1)(2L'+1)\left\{\begin{array}{ccc} L & J & 1/2 \\ J' & L' & 2 \end{array}\right\}^2.$

Table 3-17 Tensorial character of one-nucleon transfer operators with respect to the groups appearing in (3.74)

Operator	$O^B(6)$ $(\Sigma,0,0)$	$U^{BF}(6)$ $[N_1,\ldots,N_6]$	$O^{BF}(6)$ $(\sigma_1,\sigma_2,\sigma_3)$	$O^{BF}(5)$ (τ_1,τ_2)	$O^{BF}(3)$ L
	(i) Reaction $N_B = N, N_F = 0 \rightleftharpoons N_B = N, N_F = 1$				
$P_+^{(1/2)}$	$(0,0,0)$	$[1,0,0,0,0,0]$	$(1,0,0)$	$(0,0)$	0
$P_-^{(1/2)}$	$(0,0,0)$	$[1,1,1,1,1,0]$	$(1,0,0)$	$(0,0)$	0
$P_+^{(j)}, j = 3/2, 5/2$	$(0,0,0)$	$[1,0,0,0,0,0]$	$(1,0,0)$	$(1,0)$	2
$P_-^{(j)}, j = 3/2, 5/2$	$(0,0,0)$	$[1,1,1,1,1,0]$	$(1,0,0)$	$(1,0)$	2
	(ii) Reaction $N_B = N + 1, N_F = 0 \rightleftharpoons N_B = N, N_F = 1$				
$P_\pm^{(1/2)}$	$(1,0,0)$	$[1,1,1,1,1,1]$	$(0,0,0)$	$(0,0)$	0
$P_{\pm,1}^{(j)}, j = 3/2, 5/2$	$(1,0,0)$	$[2,1,1,1,1,0]$	$(2,0,0)$	$(1,0)$	2
$P_{\pm,2}^{(j)}, j = 3/2, 5/2$	$(1,0,0)$	$[2,1,1,1,1,0]$	$(2,0,0)$	$(2,0)$	2

is provided by one-nucleon transfer reactions. An appropriate form of the transfer operators for transitions of the type $N_B = N$, $N_F = 0 \rightleftharpoons N_B = N, N_F = 1$ is:

$$P_{+,m}^{(j)} = p_j a_{j,m}^\dagger. \qquad (3.127)$$

The removal operator $P_-^{(j)}$ is given by a similar expression with $a^\dagger$ replaced by $\tilde{a}$. The tensorial character of $P_\pm^{(j)}$ with respect to the groups in the chains discussed in this section is specified in Table 3.17. The reduced matrix elements of $P_\pm^{(j)}$ can be calculated by expanding the wave functions as in Sect. 3.2.3.6 and inserting the appropriate expansion coefficients. The results for routes a and b are summarized in Table 3.18. From the matrix elements one can compute intensities as in (3.59). Sometimes, instead of the intensities, the spectroscopic strengths

$$S(J_i \rightarrow J_f) = \frac{1}{2J_i + 1} I(J_i \rightarrow J_f) \qquad (3.128)$$

are reported.

Table 3-18 One-nucleon transfer reaction intensities in the $U^{BF}(6) \otimes U_s^F(2)$ limit (routes a and b)

Final state	Transf j	Intensity
(i)^a Reaction $N_B = N, N_F = 0 \rightarrow N_B = N, N_F = 1$		
$\|(N+1,0,0),(0,0),0,1/2\rangle$	1/2	$\dfrac{N+4}{N+2}p_{1/2}^2$
$\|[N+1],(N-1,0,0),(0,0),0,1/2\rangle$	1/2	$\dfrac{N(N+3)}{(N+1)^2(N+2)}p_{1/2}^2$
$\|\Sigma=N,(N-1,0,0),(0,0),0,1/2\rangle$	1/2	$\dfrac{N}{N+2}p_{1/2}^2$
$\|(N+1,0,0),(1,0),2,j\rangle$	3/2,5/2	$\dfrac{(2j+1)(N+4)(N+5)}{10(N+1)(N+2)}p_j^2$
$\|(N,1,0),(1,0),2,j\rangle$	3/2,5/2	$\dfrac{4(2j+1)N(N+4)}{5(N+1)(N+3)}p_j^2$
$\|[N+1],(N-1,0,0),(1,0),2,j\rangle$	3/2,5/2	$\dfrac{(2j+1)(N-1)N}{10(N+1)^2(N+2)}p_j^2$
$\|\Sigma=N,(N-1,0,0),(1,0),2,j\rangle$	3/2,5/2	$\dfrac{(2j+1)(N-1)N}{10(N+2)(N+3)}p_j^2$
(ii)^b Reaction $N_B = N, N_F = 1 \rightarrow N_B = N, N_F = 0$		
$\|[N],(N,0,0),(0,0),0\rangle$	1/2	$\dfrac{N+4}{N+2}p_{1/2}^2$
$\|[N],(N,0,0),(1,0),2\rangle$	3/2,5/2	$\dfrac{(2j+1)N}{10(N+2)}p_j^2$

^{a} Initial state is $\|[N],(N,0,0),(0,0),0\rangle$.
^{b} Initial state is $\|(N+1,0,0),(0,0),0,1/2\rangle$.

For transfer reactions of the type $N_B = N+1, N_F = 0 \rightleftharpoons N_B = N, N_F = 1$, the transfer operator is given by

$$P_{+,m}^{\prime(j)} = p_{0j}^{\prime(j)}[s^\dagger \times \tilde{a}_j]_m^{(j)} + \sum_{j'} p_{2j'}^{\prime(j)}[d^\dagger \times \tilde{a}_{j'}]_m^{(j)}. \qquad (3.129)$$

For arbitrary $p_{0j}^{\prime(j)}$ and $p_{2j'}^{\prime(j)}$, the tensorial character of the transfer operator (3.129) with respect to $O^{BF}(6)$ can be obtained from the branching rule

$$(1,0,0)_B \otimes (1,0,0)_F = (2,0,0) \oplus (1,1,0) \oplus (0,0,0). \qquad (3.130)$$

By decomposing the representations of $O^{BF}(6)$ on the right-hand side of (3.130) into those of its subgroups $O^{BF}(5)$ and $O^{BF}(3)$, one finds that the transfer operator (3.129) has tensorial character $(2,0)$, $(1,1)$, $(1,0)$ and $(0,0)$ with respect to $O^{BF}(5)$ and $L = 4, 3, 2, 1, 0$ with respect to $O^{BF}(3)$. According to the splitting of the fermion angular momenta $j = 1/2, 3/2, 5/2$ into a pseudo-orbital $k = 0, 2$ part and a pseudo-spin $s = 1/2$ part, one assumes that the operator is a spherical tensor of rank $L = 0$ under $O^{BF}(3)$ for $j = 1/2$ transfer and a spherical tensor of rank $L = 2$ for $j = 3/2, 5/2$ transfer. For $j = 1/2$, the transfer operator becomes

$$P_{+,m}^{\prime(1/2)} = p_{1/2}' [s^\dagger \times \tilde{a}_{1/2}]_m^{(1/2)}$$
$$+ q_{1/2}' \left(\sqrt{2} [d^\dagger \times \tilde{a}_{3/2}]_m^{(1/2)} - \sqrt{3} [d^\dagger \times \tilde{a}_{5/2}]_m^{(1/2)} \right), \quad (3.131)$$

where

$$p_{1/2}' = p_{0,1/2}^{\prime(1/2)},$$
$$q_{1/2}' = \frac{1}{\sqrt{2}} p_{2,3/2}^{\prime(1/2)} = -\frac{1}{\sqrt{3}} p_{2,5/2}^{\prime(1/2)}. \quad (3.132)$$

If, in addition, $q_{1/2}' = p_{1/2}'$, the transfer operator further simplifies and transforms as $(0,0,0)$ under $O^{BF}(6)$. Under similar approximations, the operators for $j = 3/2, 5/2$ transfer can be written as

$$P_{+,m}^{\prime(j)} = p_j' P_{+,1,m}^{(j)} + q_j' P_{+,2,m}^{(j)}, \qquad j = 3/2, 5/2, \quad (3.133)$$

and transform as $(2,0,0)$ under $O^{BF}(6)$. In (3.133) we have

$$P_{+,1,m}^{(3/2)} = [s^\dagger \times \tilde{a}_{3/2}]_m^{(3/2)} + [d^\dagger \times \tilde{a}_{1/2}]_m^{(3/2)},$$
$$P_{+,2,m}^{(3/2)} = \sqrt{\tfrac{7}{2}} [d^\dagger \times \tilde{a}_{3/2}]_m^{(3/2)} + \sqrt{\tfrac{3}{2}} [d^\dagger \times \tilde{a}_{5/2}]_m^{(3/2)},$$
$$P_{+,1,m}^{(5/2)} = [s^\dagger \times \tilde{a}_{5/2}]_m^{(5/2)} + [d^\dagger \times \tilde{a}_{1/2}]_m^{(5/2)},$$
$$P_{+,2,m}^{(5/2)} = -[d^\dagger \times \tilde{a}_{3/2}]_m^{(5/2)} + 2[d^\dagger \times \tilde{a}_{5/2}]_m^{(5/2)}. \quad (3.134)$$

The selection rules of these transfer operators are determined by the transformation properties under the groups in (3.74) and are

summarized in Table 3.17. The reduced matrix elements can be
calculated as usual and are reported in Table 3.19.

3.2.3.10 Two-nucleon transfer intensities These intensities can
be calculated as in Sect. 3.2.1.10. We give here a selection of the
most important results for $L = 0$ transfer:
(i) Route a

$$
\begin{aligned}
I&([N+1,0],(N+1,0,0),(0,0),0,1/2 \\
&\rightleftharpoons [N+2,0],(N+2,0,0),(0,0),0,1/2) \\
&= \alpha_\rho^2(N_\rho+1)\frac{N+5}{N+3}\left(\Omega_\rho - N_\rho - \frac{N}{2(N+2)}N_\rho\right),
\end{aligned}
$$

$$
\begin{aligned}
I&([N+1,0],(N+1,0,0),(0,0),0,1/2 \\
&\rightarrow [N+2,0],(N,0,0),(0,0),0,1/2) \\
&= \alpha_\rho^2(N_\rho+1)\frac{(N+1)(N+4)}{(N+2)^2(N+3)}\left(\Omega_\rho - N_\rho - \frac{N}{2(N+2)}N_\rho\right),
\end{aligned}
$$

$$
\begin{aligned}
I&([N+1,0],(N+1,0,0),(0,0),0,1/2 \\
&\rightarrow [N+1,1],(N,0,0),(0,0),0,1/2) \\
&= \alpha_\rho^2(N_\rho+1)\frac{N}{(N+1)(N+2)^2}\left(\Omega_\rho - N_\rho - \frac{N}{2(N+2)}N_\rho\right);
\end{aligned}
$$

$$\tag{3.135}$$

(ii) Route b

$$
\begin{aligned}
I&(\Sigma = N,(N+1,0,0),(0,0),0,1/2 \\
&\rightleftharpoons \Sigma = N+1,(N+2,0,0),(0,0),0,1/2) \\
&= \alpha_\rho^2(N_\rho+1)\frac{N+5}{N+3}\left(\Omega_\rho - N_\rho - \frac{N}{2(N+2)}N_\rho\right),
\end{aligned}
$$

$$
\begin{aligned}
I&(\Sigma = N,(N+1,0,0),(0,0),0,1/2 \\
&\rightarrow \Sigma = N+1,(N,0,0),(0,0),0,1/2) \\
&= \alpha_\rho^2(N_\rho+1)\frac{4}{(N+1)(N+2)^2(N+3)} \\
&\quad \times \left(\Omega_\rho - N_\rho - \frac{N}{2(N+2)}N_\rho\right),
\end{aligned}
$$

$$
I(\Sigma = N,(N+1,0,0),(0,0),0,1/2
$$

Table 3-19 One-nucleon transfer reaction intensities in the $U^{BF}(6) \otimes U_s^F(2)$ limit (routes a and b)

Final state	Transf j	Intensity
$(i)^a$ Reaction $N_B = N+1, N_F = 0 \rightarrow N_B = N, N_F = 1$		
$\lvert (N+1,0,0),(0,0),0,1/2 \rangle$	$1/2$	$2(N+1)p_{1/2}'^2$
$\lvert (N+1,0,0),(1,0),2,j \rangle$	$3/2,5/2$	$\dfrac{4(2j+1)(N+5)}{5(N+2)^2}p_j'^2$
$\lvert (N,1,0),(1,0),2,j \rangle$	$3/2,5/2$	$\dfrac{2(2j+1)N(N+5)^2}{5(N+2)(N+3)}p_j'^2$
$\lvert [N+1],(N-1,0,0),(1,0),2,j \rangle$	$3/2,5/2$	$\dfrac{(2j+1)(N-1)N(N+4)}{5(N+1)(N+2)^2}p_j'^2$
$\lvert \Sigma = N,(N-1,0,0),(1,0),2,j \rangle$	$3/2,5/2$	$\dfrac{(2j+1)(N-1)N(N+1)(N+4)}{5(N+2)^2(N+3)}p_j'^2$
$\lvert (N+1,0,0),(2,0),2,j \rangle$	$3/2,5/2$	$\dfrac{(2j+1)N(N+5)(N+6)}{14(N+2)^2}q_j'^2$
$\lvert (N,1,0),(2,0),2,j \rangle$	$3/2,5/2$	$\dfrac{5(2j+1)(N-1)N(N+5)}{14(N+2)(N+3)}q_j'^2$
$\lvert [N+1],(N-1,0,0),(2,0),2,j \rangle$	$3/2,5/2$	$\dfrac{(2j+1)(N-2)(N-1)N}{14(N+1)(N+2)^2}q_j'^2$
$\lvert \Sigma = N,(N-1,0,0),(2,0),2,j \rangle$	$3/2,5/2$	$\dfrac{(2j+1)(N-2)(N-1)N(N+1)}{14(N+2)^2(N+3)}q_j'^2$
$(ii)^b$ Reaction $N_B = N, N_F = 1 \rightarrow N_B = N+1, N_F = 0$		
$\lvert [N+1],(N+1,0,0),(0,0),0 \rangle$	$1/2$	$2(N+1)p_{1/2}'^2$
$\lvert [N+1],(N+1,0,0),(1,0),2 \rangle$	$3/2,5/2$	$\dfrac{4(2j+1)(N+5)}{5(N+2)^2}p_j'^2$
$\lvert [N+1],(N-1,0,0),(1,0),2 \rangle$	$3/2,5/2$	$\dfrac{(2j+1)(N-1)N(N+4)}{5(N+1)(N+2)^2}p_j'^2$
$\lvert [N+1],(N+1,0,0),(2,0),2 \rangle$	$3/2,5/2$	$\dfrac{(2j+1)N(N+5)(N+6)}{14(N+2)^2}q_j'^2$
$\lvert [N+1],(N-1,0,0),(2,0),2 \rangle$	$3/2,5/2$	$\dfrac{(2j+1)(N-2)(N-1)N}{14(N+1)(N+2)^2}q_j'^2$

^{a} Initial state is $\lvert [N+1],(N+1,0,0),(0,0),0 \rangle$.
^{b} Initial state is $\lvert (N+1,0,0),(0,0),0,1/2 \rangle$.

$$\to \Sigma = N-1, (N,0,0), (0,0), 0, 1/2)$$

$$= \alpha_\rho^2 (N_\rho + 1) \frac{N(N+4)}{(N+1)(N+2)^2} \left(\Omega_\rho - N_\rho - \frac{N}{2(N+2)} N_\rho \right),$$

$$(3.136)$$

where $\rho = \pi(\nu)$ for proton (neutron) transfer.

3.2.3.11 Examples of spectra with $U^{\mathrm{BF}}(6) \otimes U_s^{\mathrm{F}}(2)$ *symmetry* Conditions to be met in order to have this type of symmetry are that: (a) the adjacent even–even nucleus has $O^{\mathrm{B}}(6)$ symmetry; (b) the odd nucleon occupies single-particle levels with angular momenta $j = 1/2, 3/2, 5/2$. In addition, the energies of these single-particle levels must satisfy certain constraints that relate them to the energies of the s and d bosons of the even–even nucleus (Bijker, 1984). Regions of the periodic table where the $O^{\mathrm{B}}(6)$ symmetry has been found to occur in even–even nuclei are shown by the shaded areas in Fig. 3.5. Inspection of the single-particle levels in these regions shows that symmetries of the $U^{\mathrm{BF}}(6) \otimes U_s^{\mathrm{F}}(2)$ type could occur for: (a) odd-neutron nuclei in the Os–Pt region, where the odd neutron occupies the $3p_{1/2}$, $3p_{3/2}$ and $2f_{5/2}$ orbits and (b) odd-neutron and odd-proton nuclei in the Kr–Sr region, where the odd particle occupies the $2p_{1/2}$, $2p_{3/2}$ and $1f_{5/2}$ orbits.

The nucleus $^{195}_{78}\mathrm{Pt}_{117}$ was shown to be an example of $U^{\mathrm{BF}}(6) \otimes U_s^{\mathrm{F}}(2)$ by Balantekin *et al.* (1983). As more experimental information became available about this nucleus (Warner *et al.*, 1982; Bruce *et al.*, 1985; Mauthofer *et al.*, 1986), this classification of levels in $^{195}_{78}\mathrm{Pt}_{117}$ was confirmed, but occasionally a reassignment of quantum numbers to the levels was needed on the basis of newly measured transition rates. In what follows we adopt the labelling of states as proposed in Mauthofer *et al.* (1986). The energy spectrum of $^{195}_{78}\mathrm{Pt}_{117}$ is better described with the eigenvalue expression (3.90) (Sun *et al.*, 1983; Bijker and Iachello, 1985) than with (3.96) (Balantekin *et al.*, 1983), Fig. 3.10. The assignment of quantum numbers to the levels is based on the experimental information on electromagnetic transition rates, in particular on E2 rates. The available data on $B(\mathrm{E2})$ values are shown in Table 3.20 and compared to a calculation with the transition operator (3.119) with $\beta_2 = f_2' = 0$

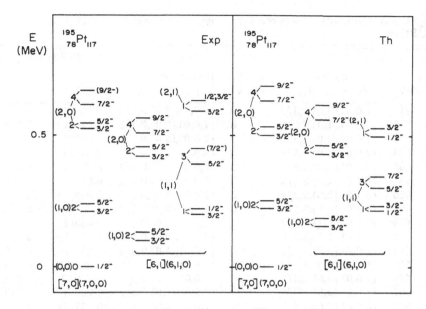

Fig. 3-10 An example of a spectrum with $U^{BF}(6) \otimes U_s^F(2)$ (III) symmetry: $^{195}_{78}Pt_{117}$ ($N_B = 6, N_F = 1$). The theoretical spectrum is calculated using (3.90) with $\eta + 2\eta' = 5$ KeV, $2\beta = 46$ KeV, $2\gamma = 4$ KeV and $2\gamma'' = 6$ KeV.

and $\alpha_2 = -f_2$. The latter equality is expected from microscopic considerations. The predictions are identical for both routes *a* and *b*. Three different types of E2 transitions are observed: $\Delta\sigma_1 = \Delta\sigma_2 = 0, \Delta(\tau_1 + \tau_2) = 1$ transitions, which are allowed and thus expected to be strong; $\Delta\sigma_1 \neq 0, \Delta\sigma_2 \neq 0, \Delta(\tau_1 + \tau_2) = 1$ transitions, which are only allowed for $\alpha_2 \neq f_2$ and thus expected to be weaker; and $\Delta\sigma_1 \neq 0, \Delta\sigma_2 \neq 0, \Delta(\tau_1 + \tau_2) \neq 1$ transitions, which are forbidden. With one exception (the 420 → 99 KeV transition), these selection rules, resulting from the use of a simple transition operator, are in good agreement with the $B(E2)$ values measured in $^{195}_{78}Pt_{117}$. Almost no information is available in these nuclei on M1 transition rates and moments.

One-neutron transfer reactions to and from $^{195}_{78}Pt_{117}$ have also been studied experimentally. In Table 3.21 we summarize the situation for the reactions leading to the odd–even nucleus $^{195}_{78}Pt_{117}$ (Vergnes *et al.*, 1983; Vergnes *et al.*, 1987). Again, the predictions

Table 3-20 Comparison between experimental and calculated $B(E2)$ values in $^{195}_{78}\text{Pt}_{117}$

E_i (KeV)	$((\sigma_1,\sigma_2,\sigma_3),(\tau_1,\tau_2),L,J)_i$	E_f (KeV)	$((\sigma_1,\sigma_2,\sigma_3),(\tau_1,\tau_2),L,J)_f$	$B(E2)(e^2b^2)$ Exp[a]	Th[b]
211	$(7,0,0),(1,0),2,3/2$	0	$(7,0,0),(0,0),0,1/2$	0.190(10)	0.179
239	$(7,0,0),(1,0),2,5/2$	0	$(7,0,0),(0,0),0,1/2$	0.170(10)	0.179
525	$(7,0,0),(2,0),2,3/2$	0	$(7,0,0),(0,0),0,1/2$	0.017(1)	0
544	$(7,0,0),(2,0),2,5/2$	0	$(7,0,0),(0,0),0,1/2$	0.008(4)	0
99	$(6,1,0),(1,0),2,3/2$	0	$(7,0,0),(0,0),0,1/2$	0.038(6)	0.035
130	$(6,1,0),(1,0),2,5/2$	0	$(7,0,0),(0,0),0,1/2$	0.066(4)	0.035
420	$(6,1,0),(2,0),2,3/2$	0	$(7,0,0),(0,0),0,1/2$	0.015(1)	0
455	$(6,1,0),(2,0),2,5/2$	0	$(7,0,0),(0,0),0,1/2$	≤ 0.00004	0
199	$(6,1,0),(1,1),1,3/2$	0	$(7,0,0),(0,0),0,1/2$	0.025(2)	0
389	$(6,1,0),(1,1),3,5/2$	0	$(7,0,0),(0,0),0,1/2$	0.007(1)	0
613	$(7,0,0),(2,0),4,7/2$	211	$(7,0,0),(1,0),2,3/2$	0.170(70)	0.215
508	$(6,1,0),(2,0),4,7/2$	211	$(7,0,0),(1,0),2,3/2$	0.055(17)	0.020
525	$(7,0,0),(2,0),2,3/2$	239	$(7,0,0),(1,0),2,5/2$	≤ 0.019	0.072
667	$(7,0,0),(2,0),4,9/2$	239	$(7,0,0),(1,0),2,5/2$	0.200(40)	0.239
563	$(6,1,0),(2,0),4,9/2$	239	$(7,0,0),(1,0),2,5/2$	0.091(22)	0.022
239	$(7,0,0),(1,0),2,5/2$	99	$(6,1,0),(1,0),2,3/2$	0.060(20)	0
525	$(7,0,0),(2,0),2,3/2$	99	$(6,1,0),(1,0),2,3/2$	≤ 0.033	0.007
613	$(7,0,0),(2,0),4,7/2$	99	$(6,1,0),(1,0),2,3/2$	0.005(3)	0.009
420	$(6,1,0),(2,0),2,3/2$	99	$(6,1,0),(1,0),2,3/2$	0.005(4)	0.177
508	$(6,1,0),(2,0),4,7/2$	99	$(6,1,0),(1,0),2,3/2$	0.240(50)	0.228
389	$(6,1,0),(1,1),3,5/2$	99	$(6,1,0),(1,0),2,3/2$	0.200(70)	0.219
525	$(7,0,0),(2,0),2,3/2$	130	$(6,1,0),(1,0),2,5/2$	0.009(5)	0.003
667	$(7,0,0),(2,0),4,9/2$	130	$(6,1,0),(1,0),2,5/2$	0.012(3)	0.010
563	$(6,1,0),(2,0),4,9/2$	130	$(6,1,0),(1,0),2,5/2$	0.240(40)	0.253
389	$(6,1,0),(1,1),3,5/2$	130	$(6,1,0),(1,0),2,5/2$	≤ 0.014	0.055

[a] From Bruce *et al.* (1985) and Mauthofer *et al.* (1986).
[b] With $\alpha_2 = -f_2 = 0.151\ eb$.

are identical for routes a and b, with the exception of the state at 927 KeV, which is excited differently for routes a and b in the $^{196}_{78}\text{Pt}_{118} \rightarrow\ ^{195}_{78}\text{Pt}_{117}$ reaction. The value $S_{th} = 0.12$ quoted in Table 3.21 is valid for route a. Overall, the agreement is quite good. The allowed transitions are strong and the forbidden ones are either weak or not seen. The coefficients p_j, p'_j and q'_j are fitted to the data and given in Table 3.21. One can also calculate the intensities of the inverse reactions $^{195}_{78}\text{Pt}_{117} \rightarrow\ ^{194}_{78}\text{Pt}_{116}$ and $^{195}_{78}\text{Pt}_{117} \rightarrow\ ^{196}_{78}\text{Pt}_{118}$. The calculated strengths (Table 3.22) are

Table 3-21 Comparison between experimental and calculated spectroscopic intensities for one-neutron transfer reactions leading to $^{195}_{78}\text{Pt}_{117}$

E_f	Final state	$^{196}_{78}\text{Pt}_{118} \rightarrow {}^{195}_{78}\text{Pt}_{117}$		$^{194}_{78}\text{Pt}_{116} \rightarrow {}^{195}_{78}\text{Pt}_{117}$	
(KeV)	$(\sigma_1,\sigma_2,\sigma_3),(\tau_1,\tau_2),$ L,J	$S_{\text{exp}}{}^a$	$S_{\text{th}}{}^b$	$S_{\text{exp}}{}^a$	$S_{\text{th}}{}^b$
0	(7,0,0),(0,0),0,1/2	1.08	1.08	0.54	0.56
99	(6,1,0),(1,0),2,3/2	1.21	1.17	0.68	0.78
130	(6,1,0),(1,0),2,5/2	2.27	2.05	1.52	1.51
199	(6,1,0),(1,1),1,3/2	0.20	0	0.04	0
211	(7,0,0),(1,0),2,3/2	0.30	0.30	0.18	0.03
222	(6,1,0),(1,1),1,1/2	not seen	0	≤ 0.06	0
239	(7,0,0),(1,0),2,5/2	0.46	0.53	0.08	0.05
389	(6,1,0),(1,1),3,5/2	not seen	0	not seen	0
420	(6,1,0),(2,0),2,3/2	not seen	0	≤ 0.03	0.03
455	(6,1,0),(2,0),2,5/2	≤ 0.07	0	not seen	0.00
525	(7,0,0),(2,0),2,3/2	not seen	0	0.04	0.02
544	(7,0,0),(2,0),2,5/2	0.06	0	≤ 0.09	0.00
927	(5,0,0),(0,0),0,1/2	0.12	0.12	0.06	0

[a] From Vergnes *et al.* (1987).
[b] With $p_{1/2} = 0.93$, $p_{3/2} = 0.62$, $p_{5/2} = 0.67$, $p'_{1/2} = 0.20$, $p'_{3/2} = 0.22$, $p'_{5/2} = 0.25$, $q'_{3/2} = 0.07$ and $q'_{5/2} = 0.00$.

in good agreement with experiment for the reaction $^{195}_{78}\text{Pt}_{117} \rightarrow {}^{194}_{78}\text{Pt}_{116}$, but show a sizeable breaking for the reaction $^{195}_{78}\text{Pt}_{117} \rightarrow {}^{196}_{78}\text{Pt}_{118}$ (Vergnes *et al.*, 1984). This situation is similar to that encountered in the study of the Spin^{BF}(6) symmetry.

Another possible example of this symmetry, $^{79}_{36}\text{Kr}_{43}$, has been discussed by Gelberg (1984).

3.2.4 $\overline{\text{U}}^{\text{BF}}(6) \otimes \overline{\text{U}}^{\text{F}}_s(2)$ (III$_4$)

The automorphism induced by the particle–hole conjugation, discussed in Sect. 2.9, allows one to construct quite often conjugate symmetries to a given Bose–Fermi symmetry. We discuss here one such example.

3.2.4.1 Lattice of algebras   This symmetry is the conjugate symmetry to that discussed in Sect. 3.2.3 (Bijker and Iachello, 1985). It corresponds to bosons with O(6) symmetry and fermion holes with $j = 1/2, 3/2, 5/2$. The lattice of algebras is:

Table 3-22 Comparison between experimental and
calculated spectroscopic intensities for one-neutron
transfer reactions starting from $^{195}_{78}\text{Pt}_{117}$

E_{f} (KeV)	Final state Σ, τ, L	Transferred j	$S_{\text{exp}}{}^{a}$	$S_{\text{th}}{}^{b}$
\multicolumn{5}{c}{(i) $^{195}_{78}\text{Pt}_{117} \rightarrow {}^{196}_{78}\text{Pt}_{118}$}
0	6,0,0	1/2	0.54	0.54
356	6,1,2	3/2	0.09	0.06
356	6,1,2	5/2	0.09	0.10
689	6,2,2	3/2	0.10	0
1135	6,3,0	1/2	0.11	0
1361	6,4,2	3/2,5/2	not seen	0
1402	4,0,0	1/2	0.19	0
1604	4,1,2	3/2,5/2	not seen	0
1677	6,5,2	3/2	0.22	0
\multicolumn{5}{c}{(ii) $^{195}_{78}\text{Pt}_{117} \rightarrow {}^{194}_{78}\text{Pt}_{116}$}
0	7,0,0	1/2	0.27	0.28
328	7,1,2	3/2	0.03	0.01
622	7,2,2	3/2	0.08	0.02
1267	7,3,0	1/2	0.02	0

^{a} From Yamazaki *et al.* (1978).
^{b} With parameters as given in Table 3.21.

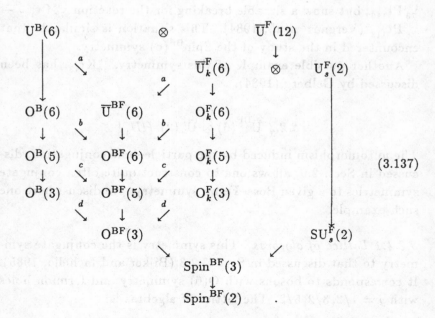

$$(3.137)$$

The generators of the conjugate algebras can be obtained from those discussed in Sect. 3.2.3 by using (2.46), and then combining the generators of $\overline{U}_k^{BF}(6)$ with those of $O^B(6)$ into

$$\bar{G}_\mu^{(\lambda)}(l,l') = B_\mu^{(\lambda)}(l,l') + \bar{K}_\mu^{(\lambda)}(l,l'). \tag{3.138}$$

As discussed in Sect. 2.9, particle–hole conjugation has no effect on orthogonal algebras and hence, for all algebras below $O^{BF}(6)$, the situation is similar to that of Sect. 3.2.3.

3.2.4.2 Basis states The main difference between the barred and the unbarred symmetries is in the basis states, which can be written as

$$
\left|
\begin{array}{l}
U^B(6) \otimes \overline{U}^F(12) \supset U^B(6) \otimes \quad \overline{U}_k^F(6) \quad \otimes \overline{U}_s^F(2) \\
\quad\downarrow \qquad\quad \downarrow \qquad\qquad\quad \downarrow \\
\;[N_B] \quad \{\bar{N}_F\} \qquad\qquad\quad [N_1',\dots,N_6'] \\[4pt]
\supset \overline{U}^{BF}(6) \otimes \overline{U}_s^F(2) \supset O^{BF}(6) \otimes \overline{U}_s^F(2) \supset O^{BF}(5) \otimes \overline{U}_s^F(2) \\
\qquad \downarrow \qquad\qquad\qquad\qquad \downarrow \qquad\qquad\qquad \downarrow \\
\;[N_1,\dots,N_6] \qquad\qquad (\sigma_1,\sigma_2,\sigma_3) \qquad\qquad (\tau_1,\tau_2) \\[4pt]
\supset O^{BF}(3) \otimes SU_s^F(2) \supset \mathrm{Spin}^{BF}(3) \supset \mathrm{Spin}^{BF}(2) \\
\quad\downarrow \qquad\quad \downarrow \qquad\qquad \downarrow \qquad\qquad \downarrow \\
\;\nu_\Delta,L \qquad S \qquad\qquad J \qquad\qquad M_J
\end{array}
\right\rangle. \tag{3.139}
$$

The representations of $\overline{U}^F(6)$ are now labelled by the number of holes, $\bar{N}_F = 12 - N_F$,

$$
\{\bar{N}_F\} = \quad
\left.
\begin{array}{c}
N_F \left\{ \begin{array}{c} \square \\ \square \\ \vdots \\ \square \end{array} \right. \\
\bar{N}_F \left\{ \begin{array}{c} \odot \\ \odot \end{array} \right.
\end{array}
\right\} 12, \tag{3.140}
$$

where we have introduced the symbol $\odot$ to denote holes in the Young tableau. The ones of interest here have $\bar{N}_F = 1$, $N_F = 11$. A similar situation occurs for $\overline{U}_k^F(6)$ where the representations of interest have 5 boxes. To determine which representations of $\overline{U}_k^{BF}(6)$ occur, one must consider the Kronecker product

$$[N] \otimes [1,1,1,1,1,0] = [N+1,1,1,1,1,0] \oplus [N,1,1,1,1,1]. \quad (3.141)$$

The reduction from $\overline{U}_k^{BF}(6)$ to $O_k^{BF}(6)$ is then given by the rules

$$[N,1,1,1,1,1] \equiv [N-1,0,0,0,0,0]:$$

$$(\sigma_1, \sigma_2, \sigma_3) = (N-1,0,0), (N-3,0,0), \ldots, \begin{cases} (0,0,0), & N = \text{odd}, \\ (1,0,0), & N = \text{even}, \end{cases}$$

$$(3.142)$$

and

$$[N+1,1,1,1,1,0]:(\sigma_1, \sigma_2, \sigma_3)$$

$$= (N+1,0,0), (N,1,0), (N-1,0,0), \ldots, \begin{cases} (1,0,0), & N = \text{even}, \\ (1,1,0), & N = \text{odd}. \end{cases}$$

$$(3.143)$$

The further decomposition of representations of $O^{BF}(6)$ into those of its subgroups is the same as in Sect. 3.2.3.2. Omitting the nonessential quantum numbers $N_1', \ldots, N_6'$ and $S = 1/2$, the basis states can be characterized by

$$|[N_B = N], \{\bar{N}_F = 1\}, [N+1-i,1,1,1,1,i],$$
$$(\sigma_1, \sigma_2, \sigma_3), (\tau_1, \tau_2), \nu_\Delta, L, J, M_J\rangle. \quad (3.144)$$

The basis states for route b in the chain (3.137) are labelled

$$\left| \begin{array}{l} U^B(6) \otimes \overline{U}^F(12) \supset U^B(6) \otimes \quad \overline{U}_k^F(6) \quad \otimes \overline{U}_s^F(2) \\ \quad \downarrow \qquad \downarrow \qquad \qquad \downarrow \qquad \qquad \downarrow \\ \quad [N_B] \quad \{\bar{N}_F\} \qquad \qquad [N_1', \ldots, N_6'] \\[6pt] \supset O^B(6) \otimes \quad O_k^F(6) \quad \otimes \overline{U}_s^F(2) \supset \quad O^{BF}(6) \quad \otimes U_s^F(2) \\ \quad \downarrow \qquad \quad \downarrow \qquad \qquad \downarrow \qquad \qquad \qquad \downarrow \\ \quad \Sigma \qquad (\Sigma_1', \Sigma_2', \Sigma_3') \qquad \qquad \qquad (\sigma_1, \sigma_2, \sigma_3) \\[6pt] \supset O^{BF}(5) \otimes \overline{U}_s^F(2) \supset O^{BF}(3) \otimes SU_s^F(2) \supset \text{Spin}^{BF}(3) \supset \text{Spin}^{BF}(2) \\ \quad \downarrow \qquad \qquad \quad \downarrow \qquad \quad \downarrow \qquad \quad \downarrow \qquad \qquad \downarrow \\ \quad (\tau_1, \tau_2) \qquad \qquad \nu_\Delta, L \qquad S \qquad J \qquad \qquad M_J \end{array} \right\rangle.$$

$$(3.145)$$

For $\bar{N}_F = 1$, the representations of $\overline{U}^F(12)$ and $\overline{U}_k^F(6)$ have 11 and 5 boxes, respectively. However, since one still has $(\Sigma_1', \Sigma_2', \Sigma_3') =$

$(1, 0, 0)$ and $\Sigma = N, N - 2, \ldots, 1$ or 0 (N odd or even), the further reduction to $O^{BF}(6)$, etc., is identical to that of Sect. 3.2.3.4. Omitting the nonessential labels, the states (3.145) can be written as

$$|[N_B = N], \{\bar{N}_F = 1\}, \Sigma, (\sigma_1, \sigma_2, \sigma_3), (\tau_1, \tau_2), \nu_\Delta, L, J, M_J). \tag{3.146}$$

3.2.4.3 Energy eigenvalues The Hamiltonian diagonal in the basis (3.139) is again

$$
\begin{aligned}
H^{(\mathrm{III}_{4a})} &= e_0 + e_1 C_1(U^B 6) + e_2 C_2(U^B 6) + e_3 C_1(\overline{U}^F 12) \\
&\quad + e_4 C_2(\overline{U}^F 12) + e_5 C_1(U^B 6) C_1(\overline{U}^F 12) + \eta C_2(\overline{U}^{BF} 6) \\
&\quad + \eta' C_2(O^{BF} 6) + \beta C_2(O^{BF} 5) + \gamma C_2(O^{BF} 3) \\
&\quad + \gamma' C_2(SU_s^F 2) + \gamma'' C_2(\mathrm{Spin}^{BF} 3), \tag{3.147}
\end{aligned}
$$

where the quadratic invariant of $\overline{U}^{BF}(6)$ is

$$
\begin{aligned}
C_2(\overline{U}^{BF} 6) &= \bar{G}^{(0)}(0,0) \cdot \bar{G}^{(0)}(0,0) + \sum_{\lambda=0}^{4} \bar{G}^{(\lambda)}(2,2) \cdot \bar{G}^{(\lambda)}(2,2) \\
&\quad + \bar{G}^{(2)}(0,2) \cdot \bar{G}^{(2)}(2,0) + \bar{G}^{(2)}(2,0) \cdot \bar{G}^{(2)}(0,2). \tag{3.148}
\end{aligned}
$$

The eigenvalues are given by

$$
\begin{aligned}
E^{(\mathrm{III}_{4a})}&(N_B{=}N, \bar{N}_F{=}1, [N{+}1{-}i, 1, 1, 1, 1, i], \\
&\quad (\sigma_1, \sigma_2, \sigma_3), (\tau_1, \tau_2), \nu_\Delta, J, M_J) \\
&= E_{04} + \eta[(N{+}1{-}i)(N{+}6{-}i) + (i{-}4)(i{-}1)] \\
&\quad + 2\eta'[\sigma_1(\sigma_1{+}4) + \sigma_2(\sigma_2{+}2) + \sigma_3^2] \\
&\quad + 2\beta[\tau_1(\tau_1{+}3) + \tau_2(\tau_2{+}1)] + 2\gamma L(L{+}1) + 2\gamma'' J(J{+}1), \tag{3.149}
\end{aligned}
$$

with

$$E_{04} = e_0 + e_1 N + e_2 N(N{+}5) + 11 e_3 + 22 e_4 + 11 e_5 N + \tfrac{3}{2}\gamma'. \tag{3.150}$$

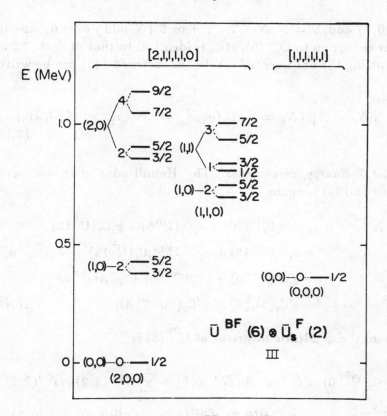

Fig. 3-11   A typical spectrum with $\overline{U}^{BF}(6) \otimes \overline{U}_s^F(2)$ (III) symmetry for $N_B = 1, N_F = 1$. The $\overline{U}^{BF}(6)$ quantum numbers are shown on top (square brackets), the $O^{BF}(6)$ quantum numbers at the bottom, the $O^{BF}(5)$ and $O^{BF}(3)$ quantum numbers to the left and the angular momentum J to the right of each level.

The excitation spectrum associated with (3.149) is shown in Fig. 3.11.

One can repeat the same procedure for route b and find a similar result. Since no experimental evidence for chains of this type has been reported yet, we shall not continue the discussion of their properties.

3.2.5 Spin$'^{\mathrm{BF}}$(6) $\otimes$ SU$^{\mathrm{F}}_{k'}$(3) *(III$_5$)*

For completeness, we now discuss an alternative splitting of the fermion algebra U^{F}(12) for $j = 1/2, 3/2, 5/2$ into a pseudo-orbital part $k' = 1$ and a pseudo-spin part $s' = 3/2$. We limit our discussion to the lattice of algebras, basis states and energy eigenvalues since at present there does not appear to be any experimental evidence for the occurrence of this symmetry in nuclei.

3.2.5.1 Lattice of algebras   The lattice of algebras is here:

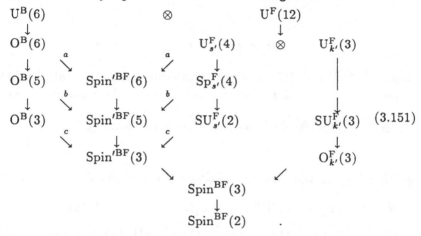

$$(3.151)$$

The generators of the algebras in (3.151) are given by Bijker (1984).

3.2.5.2 Basis states   The basis states for the chain (3.151), route *a*, can be obtained in the usual way. They are given by

$$
\begin{aligned}
&\left| \begin{array}{ccccc}
\mathrm{U^B(6)} \otimes \mathrm{U^F(12)} \supset & \mathrm{O^B(6)} \otimes & \mathrm{SU^F_{s'}(4)} & \otimes \mathrm{SU^F_{k'}(3)} \\
\downarrow \quad\quad\quad \downarrow & \downarrow & \downarrow & \downarrow \\
[N_\mathrm{B}] \quad \{N_\mathrm{F}\} & (\Sigma, 0, 0) & (n_1, n_2, n_3) & (\lambda, \mu)
\end{array} \right. \\[2ex]
&\quad \supset \mathrm{Spin}'^{\mathrm{BF}}(6) \otimes \mathrm{SU^F_{k'}(3)} \supset \mathrm{Spin}'^{\mathrm{BF}}(5) \otimes \mathrm{SU^F_{k'}(3)} \\
&\qquad\qquad \downarrow \qquad\qquad\qquad\qquad\qquad\qquad \downarrow \\
&\qquad\quad (\sigma_1, \sigma_2, \sigma_3) \qquad\qquad\qquad\qquad (\tau_1, \tau_2) \\[2ex]
&\quad \supset \mathrm{Spin}'^{\mathrm{BF}}(3) \otimes \mathrm{O^F_{k'}(3)} \supset \mathrm{Spin}^{\mathrm{BF}}(3) \supset \mathrm{Spin}^{\mathrm{BF}}(2) \\
&\qquad\quad \downarrow \qquad\qquad \downarrow \qquad\qquad \downarrow \qquad\qquad \downarrow \\
&\qquad\quad \nu_\Delta, J' \qquad\quad I' \qquad\qquad J \qquad\qquad M_J
\end{aligned} \right\rangle .
$$

$$(3.152)$$

In the case $N_F = 1$, $(n_1, n_2, n_3) = (1, 0, 0)$, $(\lambda, \mu) = (1, 0)$ and $I' = 1$ for all states. For a given value of N, one has $\Sigma = N, N-2, \ldots, 1$ or 0 (N odd or even), as before. The values of $(\sigma_1, \sigma_2, \sigma_3)$ for $N_F = 1$ are then obtained from

$$(\Sigma, 0, 0) \otimes (\tfrac{1}{2}, \tfrac{1}{2}, \tfrac{1}{2}) = (\Sigma + \tfrac{1}{2}, \tfrac{1}{2}, \tfrac{1}{2}) \oplus (\Sigma - \tfrac{1}{2}, \tfrac{1}{2}, -\tfrac{1}{2}). \quad (3.153)$$

The branching from $\mathrm{Spin}'^{BF}(6)$ to $\mathrm{Spin}'^{BF}(5)$ and $\mathrm{Spin}'^{BF}(3)$ is discussed in Sect. 3.2.1.2. This branching gives the allowed values of J'. Since $I' = 1$, one further obtains

$$\begin{aligned} J &= J'-1, J', J'+1; & J' \neq 1/2, \\ J &= J', J'+1; & J' = 1/2, \end{aligned} \quad (3.154)$$

and $-J \leq M_J \leq +J$ as usual. Omitting the nonessential labels n_1, n_2, n_3, (λ, μ) and I', one can characterize the basis states as

$$|[N_B = N], \{N_F = 1\}, \Sigma, (\sigma_1, \sigma_2, \sigma_3), (\tau_1, \tau_2), \nu_\Delta, J', J, M_J\rangle. \quad (3.155)$$

3.2.5.3 Energy eigenvalues Writing the Hamiltonian as

$$\begin{aligned} H^{(\mathrm{III}_{5a})} = {}& e_0 + e_1 \mathcal{C}_1(\mathrm{U}^B 6) + e_2 \mathcal{C}_2(\mathrm{U}^B 6) + e_3 \mathcal{C}_1(\mathrm{U}^F 12) \\ & + e_4 \mathcal{C}_2(\mathrm{U}^F 12) + e_5 \mathcal{C}_1(\mathrm{U}^B 6)\mathcal{C}_1(\mathrm{U}^F 12) + \eta \mathcal{C}_2(\mathrm{O}^B 6) \\ & + \eta' \mathcal{C}_2(\mathrm{Spin}'^{BF} 6) + \beta \mathcal{C}_2(\mathrm{Spin}'^{BF} 5) + \gamma' \mathcal{C}_2(\mathrm{Spin}'^{BF} 3) \\ & + \gamma \mathcal{C}_2(\mathrm{Spin}^{BF} 3), \end{aligned} \quad (3.156)$$

where we have included in e_0 some terms equal for all states, we can obtain the energy eigenvalues

$$\begin{aligned} & E^{(\mathrm{III}_{5a})}(N_B = N, N_F = 1, \Sigma, (\sigma_1, \sigma_2, \sigma_3), (\tau_1, \tau_2), \nu_\Delta, J', J, M_J) \\ & = E_{05} + 2\eta \Sigma(\Sigma+4) + 2\eta'[\sigma_1(\sigma_1+4) + \sigma_2(\sigma_2+2) + \sigma_3^2] \\ & + 2\beta[\tau_1(\tau_1+3) + \tau_2(\tau_2+1)] + 2\gamma' J'(J'+1) + 2\gamma J(J+1), \end{aligned} \quad (3.157)$$

with

$$E_{05} = e_0 + e_1 N + e_2 N(N+5) + e_3 + 12 e_4 + e_5 N. \quad (3.158)$$

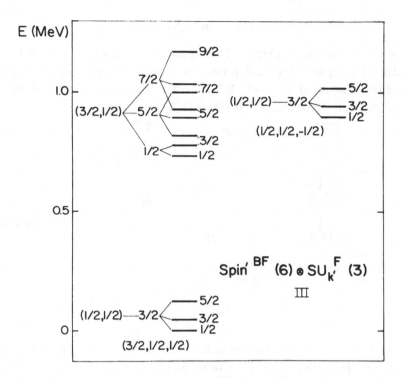

Fig. 3-12 A typical spectrum with Spin$'^{\mathrm{BF}}(6) \otimes \mathrm{SU}^{\mathrm{F}}_{k'}(3)$ (III) symmetry for $N_{\mathrm{B}} = 1$, $N_{\mathrm{F}} = 1$. The Spin$'^{\mathrm{BF}}(6)$ quantum numbers are shown at the bottom, the Spin$'^{\mathrm{BF}}(5)$ and Spin$'^{\mathrm{BF}}(3)$ quantum numbers to the left and the angular momentum J to the right of each level.

The excitation spectrum of (3.157) is shown in Fig. 3.12. No further properties of this chain will be discussed here.

3.3 Symmetries associated with U(5)

The study of Bose–Fermi symmetries presented in Sect. 3.2 can be repeated for symmetries associated with $\mathrm{U}^{\mathrm{B}}(5)$. Since both $\mathrm{O}^{\mathrm{B}}(6)$ and $\mathrm{U}^{\mathrm{B}}(5)$ contain $\mathrm{O}^{\mathrm{B}}(5) \supset \mathrm{O}^{\mathrm{B}}(3) \supset \mathrm{O}^{\mathrm{B}}(2)$, the material presented in this section parallels closely that of the previous section.

$$\textbf{3.3.1 } \text{Spin}^{\text{BF}}(5) \; (I_1)$$

3.3.1.1 Lattice of algebras This symmetry corresponds to bosons
with U(5) symmetry and fermions with $j = 3/2$. It is similar
to that discussed in Sect. 3.2.1 except that boson and fermion
algebras are joined at the level of O(5),

$$
\begin{array}{ccccc}
\text{U}^{\text{B}}(6) & & \otimes & & \text{U}^{\text{F}}(4) \\
\downarrow & & & & \downarrow \\
\text{U}^{\text{B}}(5) & & & & \text{SU}^{\text{F}}(4) \\
\downarrow & & & & \downarrow \\
\text{O}^{\text{B}}(5) & & & & \text{Sp}^{\text{F}}(4) \\
\downarrow \quad \searrow^{a} & & & \nearrow^{a} \quad \downarrow & \\
\text{O}^{\text{B}}(3) & \text{Spin}^{\text{BF}}(5) & & \text{SU}^{\text{F}}(2) & \\
\searrow^{b} & \downarrow & \nearrow^{b} & & \\
& \text{Spin}^{\text{BF}}(3) & & & \\
& \downarrow & & & \\
& \text{Spin}^{\text{BF}}(2) & & & .
\end{array}
\tag{3.159}
$$

3.3.1.2 Basis states We consider here only route a and the case
with one fermion, $N_{\text{F}} = 1$. The basis states are labelled as

$$
\left|
\begin{array}{ccccc}
\text{U}^{\text{B}}(6) & \otimes & \text{U}^{\text{F}}(4) & \supset \text{U}^{\text{B}}(5) \otimes \text{SU}^{\text{F}}(4) \\
\downarrow & & \downarrow & \downarrow \\
[N_{\text{B}} = N] & \{N_{\text{F}} = 1\} & & n_d \\
\end{array}
\right.
$$
$$
\left.
\begin{array}{ccccc}
\supset \text{O}^{\text{B}}(5) \otimes \text{Sp}^{\text{F}}(4) & \supset \text{Spin}^{\text{BF}}(5) & \supset \text{Spin}^{\text{BF}}(3) & \supset \text{Spin}^{\text{BF}}(2) \\
\downarrow & \downarrow & \downarrow & \downarrow & \downarrow \\
v & (n_1, n_2) & (v_1, v_2) & n_\Delta, J & M_J
\end{array}
\right\rangle .
\tag{3.160}
$$

For $N_{\text{F}} = 1$ the Sp(4) labels in Spin(5) notation are $(v_1 = \frac{1}{2}, v_2 = \frac{1}{2})$. The values of n_d and v contained in a representation $[N]$ of $\text{U}^{\text{B}}(6)$ are given in Eqs. (2.29) and (2.30) of Volume 1,

$$[N] : n_d = N, N-1, \ldots, 0;$$

$$n_d : v = n_d, n_d - 2, \ldots, 1 \text{ or } 0; \; (n_d = \text{odd or even}). \tag{3.161}$$

Those of v_1 and v_2 can be obtained by taking the outer product
of $(v, 0)_{\text{B}}$ and $(\frac{1}{2}, \frac{1}{2})_{\text{F}}$,

$$(v, 0) \otimes (\tfrac{1}{2}, \tfrac{1}{2}) = (v - \tfrac{1}{2}, \tfrac{1}{2}) \oplus (v + \tfrac{1}{2}, \tfrac{1}{2}). \tag{3.162}$$

Finally, the angular momenta contained in $(v_1, v_2 = \frac{1}{2})$ are given by (3.12),

$$(v_1, \tfrac{1}{2}) : J = 2v_1 - 6n_\Delta + \tfrac{1}{2}, 2v_1 - 6n_\Delta - \tfrac{1}{2}, \ldots,$$
$$v_1 - 3n_\Delta + 1 - \tfrac{1}{4}[1 - (-)^{2n_\Delta}], \quad (3.163)$$

with $n_\Delta = 0, \frac{1}{2}, 1, \ldots$; $3n_\Delta \leq v_1$, introduced to classify the states uniquely. Summarizing, the basis states can be characterized by

$$|[N_B = N], \{N_F = 1\}, n_d, v, (v_1, v_2), n_\Delta, J, M_J\rangle. \quad (3.164)$$

3.3.1.3 Energy eigenvalues These are obtained by writing the Hamiltonian as

$$
\begin{aligned}
H^{(\mathrm{I}_{1a})} &= e_0 + e_1 \mathcal{C}_1(U^B 6) + e_2 \mathcal{C}_2(U^B 6) + e_3 \mathcal{C}_1(U^F 4) + e_4 \mathcal{C}_2(U^F 4) \\
&\quad + e_5 \mathcal{C}_1(U^B 6)\mathcal{C}_1(U^F 4) + \epsilon \mathcal{C}_1(U^B 5) + \epsilon' \mathcal{C}_1(U^B 5)\mathcal{C}_1(U^F 4) \\
&\quad + \alpha \mathcal{C}_2(U^B 5) + \beta \mathcal{C}_2(O^B 5) + \beta' \mathcal{C}_2(\mathrm{Spin}^{BF} 5) \\
&\quad + \gamma \mathcal{C}_2(\mathrm{Spin}^{BF} 3). \quad (3.165)
\end{aligned}
$$

Eigenvalues are given by

$$
\begin{aligned}
E^{(\mathrm{I}_{1a})}&(N_B = N, N_F = 1, n_d, v, (v_1, v_2), n_\Delta, J, M_J) \\
&= E'_{01} + (\epsilon + \epsilon')n_d + \alpha n_d(n_d + 4) + 2\beta v(v + 3) \\
&\quad + 2\beta'[v_1(v_1 + 3) + v_2(v_2 + 1)] + 2\gamma J(J + 1), \quad (3.166)
\end{aligned}
$$

with

$$E'_{01} = e_0 + e_1 N + e_2 N(N + 5) + e_3 + 4e_4 + e_5 N. \quad (3.167)$$

The excitation spectrum associated with (3.166) is shown in Fig. 3.13.

3.3.1.4 Wave functions. Isoscalar factors These are obtained as in Sect. 3.2.1.4 by expanding the coupled basis states

$$|[N_B = N], \{N_F = 1\}, n_d, v, (v_1, v_2 = \tfrac{1}{2}), n_\Delta, J, M_J\rangle,$$

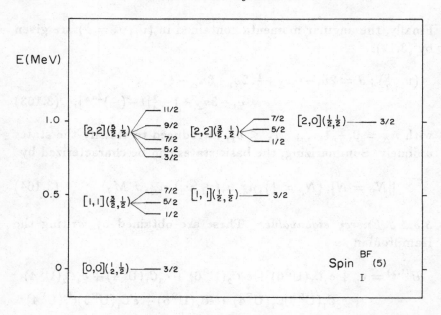

Fig. 3-13   A typical spectrum with $\text{Spin}^{\text{BF}}(5)$ (I) symmetry for $N_B = 2, N_F = 1$. The numbers in square brackets are the $U^B(5)$ and $O^B(5)$ quantum numbers $[n_d, v]$. The numbers in parentheses are the $\text{Spin}^{\text{BF}}(5)$ quantum numbers (v_1, v_2). The angular momentum J is shown to the right of each level.

denoted in short as $|v; v_1; J\rangle$, into products of wave functions for $U^B(5)$,

$$|[N_B = N], \{N_F = 0\}, n_d, v, (v_1 = v, v_2 = 0), n_\Delta, L, M_L\rangle,$$

denoted as $|v; L\rangle$, and fermion wave functions

$$|[N_B = 0], \{N_F = 1\}, n_d = 0, v = 0, (v_1 = \tfrac{1}{2}, v_2 = \tfrac{1}{2}), n_\Delta, J = 3/2, M_J\rangle,$$

denoted as $|\tfrac{1}{2}; 3/2\rangle$. We only consider the angular momenta $J = 2v_1 + \tfrac{1}{2}$, $J = 2v_1 - \tfrac{1}{2}$ and $J = 2v_1 - \tfrac{3}{2}$. For these states the label n_Δ is not needed and will be dropped. In this notation, the expansion can be written as

$$|v; v_1; J\rangle = \sum_L \eta_{v_1, J}^{v, L} |v, L; \tfrac{1}{2}, 3/2; J\rangle. \qquad (3.168)$$

The η-coefficients are $\mathrm{Sp}(4) \supset \mathrm{SU}(2)$ isoscalar factors,

$$\eta_{v_1,J}^{v,L} = \left\langle \begin{array}{cc} (v,0) & (\tfrac{1}{2},\tfrac{1}{2}) \\ L & 3/2 \end{array} \middle| \begin{array}{c} (v_1,\tfrac{1}{2}) \\ J \end{array} \right\rangle , \qquad (3.169)$$

which have been discussed in Sect. 3.2.1.4. The cases of interest here are given in Table 3.3.

3.3.1.5 Electromagnetic transitions and moments; E2 The E2 transition operator is given by (3.31). Using the isoscalar factors determined previously and the results of Volume 1 for the matrix elements of the boson operator $T_{\mathrm{B}}^{(\mathrm{E2})}$, one can derive $B(\mathrm{E2})$ values and quadrupole moments. In Table 3.23 we present a summary of the results.

3.3.1.6 Electromagnetic transitions and moments; M1 The M1 transition operator is given by (3.39). For convenience, we rewrite it here as

$$T_\mu^{(\mathrm{M1})} = \sqrt{\frac{3}{4\pi}} \left(g_{\mathrm{B}} \sqrt{10} [d^\dagger \times \tilde{d}]_\mu^{(1)} - g_{\mathrm{F}} \sqrt{5} [a_{3/2}^\dagger \times \tilde{a}_{3/2}]_\mu^{(1)} \right), \quad (3.170)$$

where g_{B} and g_{F} are the boson and fermion g-factors. The operator (3.170) has the selection rules $\Delta n_{\mathrm{d}} = \Delta v = 0$, $\Delta v_1 = \Delta v_2 = 0$ or $\Delta v_1 = \pm 1, \Delta v_2 = 0$. The reduced matrix elements, $B(\mathrm{M1})$ values and magnetic moments can be calculated in the usual way. A summary is given in Table 3.24.

3.3.1.7 One-nucleon transfer intensities For transfer reactions of the type $N_{\mathrm{B}} = N, N_{\mathrm{F}} = 0 \rightleftharpoons N_{\mathrm{B}} = N, N_{\mathrm{F}} = 1$, the appropriate transfer operator is given by (3.57). The selection rules are $\Delta n_{\mathrm{d}} = \Delta v = 0$ and $\Delta v_1 = \Delta v_2 = \pm\tfrac{1}{2}$. By calculating the reduced matrix elements of $P_+^{(3/2)}$, one obtains the following intensities of allowed transitions:

(i) even–even to odd–even,

$$I([N]; \{0\}; 0; 0; 0; 0 \to [N]; \{1\}; 0; 0; \tfrac{1}{2}; 3/2) = 4\zeta^2; \qquad (3.171)$$

(ii) odd–even to even–even,

$$I([N]; \{1\}; 0; 0; \tfrac{1}{2}; 3/2 \to [N]; \{0\}; 0; 0; 0; 0) = 4\zeta^2, \qquad (3.172)$$

Table 3-23 Some $B(E2)$ values and electric quadrupole moments in the Spin$^{\text{BF}}(5)$ limit

$n_{\text{d}}', v', v_1', J'$	$\rightarrow$	n_{d}, v, v_1, J	$B(E2; J' \rightarrow J)$
$1, 1, \frac{3}{2}, 1/2$		$0, 0, \frac{1}{2}, 3/2$	$\alpha_2^2 N$
$1, 1, \frac{3}{2}, 5/2$		$0, 0, \frac{1}{2}, 3/2$	$\alpha_2^2 N$
$1, 1, \frac{3}{2}, 7/2$		$0, 0, \frac{1}{2}, 3/2$	$\alpha_2^2 N$
$1, 1, \frac{1}{2}, 3/2$		$0, 0, \frac{1}{2}, 3/2$	$\alpha_2^2 N$
$1, 1, \frac{3}{2}, 5/2$		$1, 1, \frac{3}{2}, 1/2$	$\frac{1}{6}\left(\sqrt{\frac{3}{5}}\beta_2 - \sqrt{\frac{21}{10}}f_2\right)^2$
$1, 1, \frac{1}{2}, 3/2$		$1, 1, \frac{3}{2}, 1/2$	$\frac{1}{4}\left(\sqrt{\frac{7}{5}}\beta_2 - \sqrt{\frac{2}{5}}f_2\right)^2$
$1, 1, \frac{3}{2}, 7/2$		$1, 1, \frac{3}{2}, 5/2$	$\frac{1}{8}\left(\sqrt{\frac{648}{245}}\beta_2 + \sqrt{\frac{144}{35}}f_2\right)^2$
$1, 1, \frac{1}{2}, 3/2$		$1, 1, \frac{3}{2}, 5/2$	$\frac{1}{4}\left(\sqrt{\frac{15}{7}}\beta_2 + \sqrt{\frac{6}{5}}f_2\right)^2$
$1, 1, \frac{1}{2}, 3/2$		$1, 1, \frac{3}{2}, 7/2$	$\frac{1}{4}\left(\sqrt{\frac{16}{35}}\beta_2 - \sqrt{\frac{8}{5}}f_2\right)^2$

n_{d}, v, v_1, J	Q_J
$0, 0, \frac{1}{2}, 3/2$	$-\sqrt{\frac{16\pi}{5}}\frac{1}{2}f_2$
$1, 1, \frac{3}{2}, 5/2$	$\sqrt{\frac{16\pi}{5}}\left(\frac{5}{7\sqrt{14}}\beta_2 + \frac{1}{14}f_2\right)$
$1, 1, \frac{3}{2}, 7/2$	$\sqrt{\frac{16\pi}{5}}\left(\sqrt{\frac{2}{7}}\beta_2 - \frac{1}{2}f_2\right)$
$1, 1, \frac{1}{2}, 3/2$	$\sqrt{\frac{16\pi}{5}}\frac{3}{10}f_2$

where $\zeta = p_{3/2}$. In these expressions, states are labelled by $[N_{\text{B}}]; \{N_{\text{F}}\}; n_{\text{d}}; v; v_1; J$.

For reactions of the type $N_{\text{B}} = N + 1, N_{\text{F}} = 0 \rightleftharpoons N_{\text{B}} = N, N_{\text{F}} = 1$, the appropriate transfer operator is given by (3.62). The selection rules for the first term are $\Delta n_{\text{d}} = \Delta v = 0$ and $\Delta v_1 = \Delta v_2 = \pm\frac{1}{2}$, while the second term has $\Delta n_{\text{d}} = \pm 1, \Delta v = \pm 1$ and $\Delta v_1 = \Delta v_2 = \pm\frac{1}{2}$. The transfer intensities for the allowed transitions can be calculated as:

Table 3-24 Some B(M1) values and magnetic dipole moments in the Spin^{BF}(5) limit

n'_d, v', v'_1, J'	$\rightarrow$	n_d, v, v_1, J	$B(\text{M1}; J' \rightarrow J)$
$1, 1, \frac{1}{2}, 3/2$		$1, 1, \frac{3}{2}, 1/2$	$\frac{3}{2}(g_B - g_F)^2 \frac{3}{4\pi}$
$1, 1, \frac{3}{2}, 7/2$		$1, 1, \frac{3}{2}, 5/2$	$\frac{6}{7}(g_B - g_F)^2 \frac{3}{4\pi}$
$1, 1, \frac{3}{2}, 5/2$		$1, 1, \frac{1}{2}, 3/2$	$\frac{7}{5}(g_B - g_F)^2 \frac{3}{4\pi}$

n_d, v, v_1, J	μ_J
$0, 0, \frac{1}{2}, 3/2$	$\frac{3}{2}g_F$
$1, 1, \frac{3}{2}, 1/2$	$g_B - \frac{1}{2}g_F$
$1, 1, \frac{3}{2}, 5/2$	$\frac{11}{7}g_B + \frac{13}{14}g_F$
$1, 1, \frac{3}{2}, 7/2$	$2g_B + \frac{3}{2}g_F$
$1, 1, \frac{1}{2}, 3/2$	$\frac{6}{5}g_B + \frac{3}{10}g_F$

(i) even–even to odd–even,

$$I([N+1]; \{0\}; 0; 0; 0; 0 \rightarrow [N]; \{1\}; 0; 0; \tfrac{1}{2}; 3/2) = 4\theta^2(N+1);$$

(3.173)

(ii) odd–even to even–even,

$$I([N]; \{1\}; 0; 0; \tfrac{1}{2}; 3/2 \rightarrow [N+1]; \{0\}; 0; 0; 0; 0) = 4\theta^2(N+1),$$

$$I([N]; \{1\}; 0; 0; \tfrac{1}{2}; 3/2 \rightarrow [N+1]; \{0\}; 1; 1; 1; 2) = 4\theta'^2, \quad (3.174)$$

where $\theta = p'^{(3/2)}_{0,3/2}$ and $\theta' = p'^{(3/2)}_{2,3/2}$.

3.3.1.8 Two-nucleon transfer intensities The transfer operators are given by (3.65). They have selection rules $\Delta n_d = \Delta v = \Delta v_1 = \Delta v_2 = 0$. Therefore, only one transition from ground state to ground state is allowed with spectroscopic strength

$$S([N]; \{1\}; 0; 0; \tfrac{1}{2}; 3/2 \rightarrow [N+1]; \{1\}; 0; 0; \tfrac{1}{2}; 3/2)$$
$$= \alpha_\rho^2(N_\rho + 1)\left(\Omega_\rho - N_\rho\right), \qquad \rho = \nu, \pi. \quad (3.175)$$

3.3.1.9 Examples of spectra with Spin^{BF}(5) symmetry In order to have spectra with this symmetry, two conditions must be met: (a) the adjacent even–even nucleus must have U^B(5) symmetry; (b)

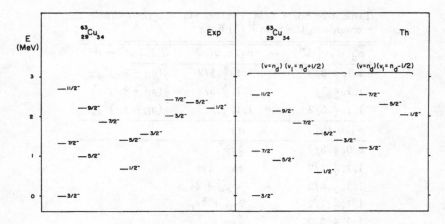

Fig. 3-14   An example of a spectrum with Spin^{BF}(5) (I) symmetry: $^{63}_{29}$Cu$_{34}$ ($N_B = 3$, $N_F = 1$). The theoretical spectrum is calculated using (3.166) with $\epsilon + \epsilon' = 690$ KeV, $\alpha = 100$ KeV, $2\beta = 0$, $2\beta' = -100$ KeV and $2\gamma = 35$ KeV.

the odd nucleon occupies a single-particle level with $j = 3/2$. The regions of the periodic table where examples of U^B(5) symmetry have been found are shown in Fig. 3.5. These regions have been extended down to mass number $A \approx 60$, although most of the studies using the interacting boson model so far have been limited to $A \geq 80$. Inspection of the single-particle levels indicates that a Spin^{BF}(5) symmetry could occur for odd-proton nuclei in the Zn region where the odd proton occupies the $2p_{3/2}$ level.

An example of a spectrum with Spin^{BF}(5) symmetry, $^{63}_{29}$Cu$_{34}$, is shown in Fig. 3.14 (Bijker and Kota, 1984).

Another example is provided by $^{65}_{29}$Cu$_{36}$. One can also study electromagnetic transition properties of these nuclei. The $\Delta n_d = \pm 1$ transitions only depend on the parameter α_2 of (3.32). Those with $\Delta n_d = 0$ depend on β_2 and f_2. With the experimental information presently known, it is not possible to determine β_2 and f_2. Thus, only $\Delta n_d = \pm 1$ transitions are shown in Table 3.25.

Good agreement between theory and experiment is seen. It must be remarked, however, that the results of (^{3}He, d) and (d, ^{3}He) reactions in Ni, Cu and Zn isotopes show that the low-lying $1/2^-$ and $5/2^-$ levels in $^{63}_{29}$Cu$_{34}$ have sizeable admixtures

Table 3-25 Comparison between experimental and calculated $B(E2)$ values in ${}^{63}_{29}\text{Cu}_{34}$

E_i (KeV)	$(n_d, v, v_1, J)_i$	$\rightarrow$	E_f (KeV)	$(n_d, v, v_1, J)_f$	$B(E2)(10^{-2}\ e^2 b^2)$ Exp[a]	Th[b]
670	$1,1,\frac{3}{2},1/2$		0	$0,0,\frac{1}{2},3/2$	2.28(12)	2.38
962	$1,1,\frac{3}{2},5/2$		0	$0,0,\frac{1}{2},3/2$	2.40(20)	2.38
1327	$1,1,\frac{3}{2},7/2$		0	$0,0,\frac{1}{2},3/2$	2.55(20)	2.38
2011	$1,1,\frac{1}{2},3/2$		0	$0,0,\frac{1}{2},3/2$	≤ 2.39	2.38
1547	$2,2,\frac{5}{2},3/2$		0	$0,0,\frac{1}{2},3/2$	≤ 0.14	0
1412	$2,2,\frac{5}{2},5/2$		0	$0,0,\frac{1}{2},3/2$	0.16(5)	0
1861	$2,2,\frac{5}{2},7/2$		0	$0,0,\frac{1}{2},3/2$	0.21(5)	0
2062	$2,2,\frac{3}{2},1/2$		0	$0,0,\frac{1}{2},3/2$	≤ 0.09	0
2081	$2,2,\frac{3}{2},5/2$		0	$0,0,\frac{1}{2},3/2$	≤ 0.47	0
2093	$2,2,\frac{3}{2},7/2$		0	$0,0,\frac{1}{2},3/2$	0.03(1)	0
1547	$2,2,\frac{5}{2},3/2$		670	$1,1,\frac{3}{2},1/2$	≤ 2.08	1.11
1412	$2,2,\frac{5}{2},5/2$		670	$1,1,\frac{3}{2},1/2$	1.35(36)	1.90
1547	$2,2,\frac{5}{2},3/2$		962	$1,1,\frac{3}{2},5/2$	≤ 5.20	1.70
1412	$2,2,\frac{5}{2},5/2$		962	$1,1,\frac{3}{2},5/2$	≤ 9.10	0.48
1861	$2,2,\frac{5}{2},7/2$		962	$1,1,\frac{3}{2},5/2$	≤ 8.10	1.62
2208	$2,2,\frac{5}{2},9/2$		962	$1,1,\frac{3}{2},5/2$	2.73(60)	2.49
2208	$2,2,\frac{5}{2},9/2$		1327	$1,1,\frac{3}{2},7/2$	1.48(81)	0.51
2677	$2,2,\frac{5}{2},11/2$		1327	$1,1,\frac{3}{2},7/2$	3.22(83)	3.17

[a] From Auble (1979).
[b] With $\alpha_2 = 0.089\ eb$.

coming from the $2p_{1/2}$ and $1f_{5/2}$ orbits, indicating a breakdown of the $\text{Spin}^{\text{BF}}(5)$ symmetry.

3.3.2 $\text{Spin}^{\text{BF}}(3)$ (I_2)

3.3.2.1 Lattice of algebras This symmetry corresponds to bosons with U(5) symmetry and fermions with $j = 1/2$. It is similar to that discussed in Sect. 3.2.2. The lattice of algebras is:

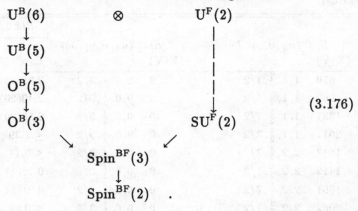

$$(3.176)$$

3.3.2.2 Basis states   The basis states for $N_F = 1$ are of the form:

$$\left| \begin{array}{c} U^B(6) \otimes U^F(2) \supset U^B(5) \otimes SU^F(2) \supset O^B(5) \otimes SU^F(2) \\ \downarrow \qquad \downarrow \qquad \downarrow \qquad \qquad \downarrow \\ [N_B = N] \quad \{N_F = 1\} \quad n_d \qquad \qquad v \\ \\ \supset O^B(3) \otimes SU^F(2) \supset \mathrm{Spin}^{BF}(3) \supset \mathrm{Spin}^{BF}(2) \\ \downarrow \qquad \qquad \downarrow \qquad \qquad \downarrow \\ n_\Delta, L \qquad \qquad J \qquad \qquad M_J \end{array} \right\rangle .$$

$$(3.177)$$

The values of the quantum numbers n_d, v, n_Δ, L are the same as those discussed in Volume 1 for chain I. The quantum number J is given by angular momentum coupling, as in (3.70).

3.3.2.3 Energy eigenvalues   Expanding the Hamiltonian in Casimir invariants

$$
\begin{aligned}
H^{(I_2)} &= e_0 + e_1 \mathcal{C}_1(U^B 6) + e_2 \mathcal{C}_2(U^B 6) + e_3 \mathcal{C}_1(U^F 2) + e_4 \mathcal{C}_2(U^F 2) \\
&\quad + e_5 \mathcal{C}_1(U^B 6)\mathcal{C}_1(U^F 2) + \epsilon \mathcal{C}_1(U^B 5) + \epsilon' \mathcal{C}_1(U^B 5)\mathcal{C}_1(U^F 2) \\
&\quad + \alpha \mathcal{C}_2(U^B 5) + \beta \mathcal{C}_2(O^B 5) + \gamma \mathcal{C}_2(O^B 3) + \gamma' \mathcal{C}_2(\mathrm{Spin}^{BF} 3),
\end{aligned}
$$

$$(3.178)$$

one obtains

$$
\begin{aligned}
E^{(I_2)}(N_B &= N, N_F = 1, n_d, v, n_\Delta, L, J, M_J) \\
&= E_{02}' + (\epsilon + \epsilon')n_d + \alpha n_d(n_d + 4) + 2\beta v(v + 3) \\
&\quad + 2\gamma L(L + 1) + 2\gamma' J(J + 1),
\end{aligned}
$$

$$(3.179)$$

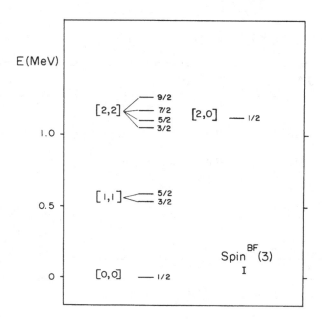

Fig. 3-15 A typical spectrum with Spin^{BF}(3) (I) symmetry for $N_B = 2, N_F = 1$. The numbers in square brackets are the $U^B(5)$ and $O^B(5)$ quantum numbers $[n_d, v]$. The angular momentum J is shown to the right of each level.

with

$$E'_{02} = e_0 + e_1 N + e_2 N(N + 5) + e_3 + 2e_4 + e_5 N. \qquad (3.180)$$

The energy spectrum associated with (3.179) is shown in Fig. 3.15.

3.3.2.4 Electromagnetic transitions and moments; E2 In the present case, the E2 transitions are completely determined by the collective part of the E2 operator,

$$T_\mu^{(E2)} = \alpha_2 [s^\dagger \times \tilde{d} + d^\dagger \times \tilde{s}]_\mu^{(2)} + \beta_2 [d^\dagger \times \tilde{d}]_\mu^{(2)}. \qquad (3.181)$$

The second term obeys the selection rule $\Delta n_d = 0$, while the first term has $\Delta n_d = \pm 1$. The matrix elements of the E2 transition operator between states in the odd–even nucleus are related to

Table 3-26 Some $B(E2)$ values and electric quadrupole moments in the $\mathrm{Spin}^{BF}(3)$ limit

L'	J'^a	$\to$	L	J^b	$\dfrac{B(E2; J' \to J)}{N - n_\mathrm{d}}$
$2n_\mathrm{d} + 2$	$2n_\mathrm{d} + \frac{5}{2}$		$2n_\mathrm{d}$	$2n_\mathrm{d} + \frac{1}{2}$	$\alpha_2^2(n_\mathrm{d} + 1)$
$2n_\mathrm{d} + 2$	$2n_\mathrm{d} + \frac{3}{2}$		$2n_\mathrm{d}$	$2n_\mathrm{d} + \frac{1}{2}$	$\alpha_2^2 \dfrac{1}{4n_\mathrm{d} + 1}$
$2n_\mathrm{d} + 2$	$2n_\mathrm{d} + \frac{3}{2}$		$2n_\mathrm{d}$	$2n_\mathrm{d} - \frac{1}{2}$	$\alpha_2^2 \dfrac{n_\mathrm{d}(4n_\mathrm{d} + 5)}{4n_\mathrm{d} + 1}$
$2n_\mathrm{d}$	$2n_\mathrm{d} + \frac{1}{2}$		$2n_\mathrm{d}$	$2n_\mathrm{d} + \frac{1}{2}$	$\alpha_2^2 \dfrac{4(n_\mathrm{d} + 1)}{4n_\mathrm{d} + 1}$
$2n_\mathrm{d}$	$2n_\mathrm{d} + \frac{1}{2}$		$2n_\mathrm{d}$	$2n_\mathrm{d} - \frac{1}{2}$	$\alpha_2^2 \dfrac{6}{(4n_\mathrm{d} - 1)(4n_\mathrm{d} + 1)}$
$2n_\mathrm{d}$	$2n_\mathrm{d} - \frac{1}{2}$		$2n_\mathrm{d}$	$2n_\mathrm{d} - \frac{1}{2}$	$\alpha_2^2 \dfrac{(2n_\mathrm{d} - 1)(2n_\mathrm{d} + 1)(4n_\mathrm{d} + 3)}{n_\mathrm{d}(4n_\mathrm{d} - 1)(4n_\mathrm{d} + 1)}$

L	J	Q_J
2	3/2	$\sqrt{\frac{14\pi}{5}} \frac{2}{5} \beta_2$
2	5/2	$\sqrt{\frac{14\pi}{5}} \frac{4}{7} \beta_2$

[a] Complete labelling is $|[N]; \{1\}; n_\mathrm{d}' = n_\mathrm{d} + 1, v' = n_\mathrm{d} + 1; n_\Delta' = 0; L'; J'\rangle$.
[b] Complete labelling is $|[N]; \{1\}; n_\mathrm{d}, v = n_\mathrm{d}; n_\Delta = 0; L; J\rangle$.

the corresponding matrix elements in the even–even nucleus by a simple geometric factor,

$$\langle [N]; \{1\}; n_\mathrm{d}; v; n_\Delta; L; J \parallel T^{(E2)} \parallel [N]; \{1\}; n_\mathrm{d}'; v'; n_\Delta'; L'; J'\rangle$$
$$= (-)^{L+1/2+J'} \sqrt{(2J+1)(2J'+1)} \begin{Bmatrix} L & J & 1/2 \\ J' & L' & 2 \end{Bmatrix}$$
$$\times \langle [N], n_\mathrm{d}, v, n_\Delta, L \parallel T^{(E2)} \parallel [N], n_\mathrm{d}', v', n_\Delta', L'\rangle, \quad (3.182)$$

Using the results of Volume 1, one can derive the $B(E2)$ values and quadrupole moments shown in Table 3.26.

3.3.2.5 Electromagnetic transitions and moments; M1 The M1 transition operator is given by

$$T_\mu^{(M1)} = \sqrt{\frac{3}{4\pi}} \left(g_\mathrm{B} \sqrt{10}[d^\dagger \times \tilde{d}]_\mu^{(1)} - \frac{g_\mathrm{F}}{\sqrt{2}}[a_{1/2}^\dagger \times \tilde{a}_{1/2}]_\mu^{(1)} \right), \quad (3.183)$$

and has the selection rules $\Delta n_d = \Delta v = \Delta n_\Delta = \Delta L = 0$. The only non-vanishing M1 transitions are those between states with $J = L + 1/2$ and $J = L - 1/2$,

$$B(M1; [N]; \{1\}; n'_d; v'; n'_\Delta; L'; J' = L + 1/2 \rightarrow$$
$$[N]; \{1\}; n_d; v; n_\Delta; L; J = L - 1/2)$$
$$= \frac{L}{2L + 1} (g_B - g_F)^2 \frac{3}{4\pi} \delta_{n'_d, n_d} \delta_{v', v} \delta_{n'_\Delta, n_\Delta} \delta_{L', L}. \qquad (3.184)$$

Magnetic moments are given by

$$\mu_{J=L\pm1/2} = g_B L \pm \tfrac{1}{2} g_F. \qquad (3.185)$$

3.3.2.6 One-nucleon transfer intensities   For transfer reactions of the type $N_B = N, N_F = 0 \rightleftharpoons N_B = N, N_F = 1$, the transfer operator is:

$$P_{+,m}^{(1/2)} = p_{1/2} a^\dagger_{1/2,m}. \qquad (3.186)$$

This operator has the selection rules $\Delta n_d = \Delta v = \Delta n_\Delta = \Delta L = 0$ and therefore only the ground state with quantum numbers $n_d = v = n_\Delta = L = 0$ can be populated with intensity

$$I([N]; \{0\}; 0; 0; 0; 0 \rightleftharpoons [N]; \{1\}; 0; 0; 0; 1/2) = 2\zeta^2, \qquad (3.187)$$

where $\zeta = p_{1/2}$.

For reactions of the type $N_B = N + 1, N_F = 0 \rightleftharpoons N_B = N, N_F = 1$, the transfer operator is

$$P'^{(1/2)}_{+,m} = p'_{1/2} [s^\dagger \times \tilde{a}_{1/2}]^{(1/2)}_m. \qquad (3.188)$$

This operator has the same selection rules as (3.186). Thus, only the ground state can be excited with intensity

$$I([N + 1]; \{0\}; 0; 0; 0; 0 \rightleftharpoons [N]; \{1\}; 0; 0; 0; 1/2) = 2\theta^2 (N + 1), \qquad (3.189)$$

where $\theta = p'_{1/2}$.

3.3.2.7 Examples of spectra with $\text{Spin}^{BF}(3)$ symmetry   In order to have spectra with this symmetry, the bosons must have $U^B(5)$

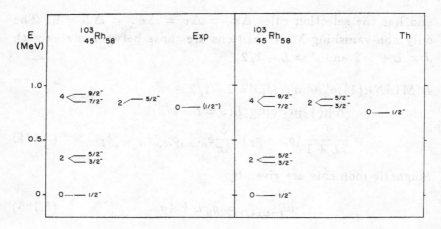

Fig. 3-16   An example of a spectrum with Spin^{BF}(3) (I) symmetry: $^{103}_{45}$Rh$_{58}$ ($N_{\text{B}} = 6, N_{\text{F}} = 1$). The theoretical spectrum is calculated using (3.179) with $\epsilon + \epsilon' = 340$ KeV, $\alpha = 0$, $2\beta = 13$ KeV, $2\gamma = -8$ KeV and $2\gamma' = 11$ KeV.

symmetry and the fermions must occupy a state with $j = 1/2$. The regions of the periodic table where examples of $U^{\text{B}}(5)$ symmetry have been found are shown in Fig. 3.5. Inspection of the single-particle levels indicates that a Spin^{BF}(3) symmetry could occur for odd-proton nuclei in the Ru–Pd region.

An example of Spin^{BF}(3) symmetry, $^{103}_{45}$Rh$_{58}$, is shown in Fig. 3.16 (Vervier and Janssens, 1982). In Table 3.27 the experimental data on E2 transition rates in this nucleus are compared with the predictions of the Spin^{BF}(3) symmetry for $\Delta n_{\text{d}} = \pm 1$ transitions which depend on the parameter α_2. Other examples, $^{107}_{47}$Ag$_{60}$ and $^{109}_{47}$Ag$_{62}$, have been discussed by Vervier and Janssens (1982). Again, however, it must be remarked that the presence of single-particle orbits other than the one with $j = 1/2$ will limit the applicability of the Spin^{BF}(3) symmetry.

3.3.3 $U^{\text{BF}}(6) \otimes U_s^{\text{F}}(2)$ (I_3)

3.3.3.1 Lattice of algebras   This symmetry is similar to that discussed in Sect. 3.2.3 and we only give here its essential properties. It corresponds to bosons with $U(5)$ symmetry and fermions with $j = 1/2, 3/2, 5/2$. The lattice of algebras is:

Table 3-27 Comparison between experimental and calculated $B(E2)$ values in $^{103}_{45}\text{Rh}_{58}$

E_i (KeV)	$(n_d, v, L, J)_i$	$\rightarrow$	E_f (KeV)	$(n_d, v, L, J)_f$	$B(E2)(e^2b^2)$ Exp[a]	Th[b]
295	1,1,2,3/2		0	0,0,0,1/2	0.109(8)	0.111
357	1,1,2,5/2		0	0,0,0,1/2	0.111(3)	0.111
881	2,2,2,5/2		0	0,0,0,1/2	0.0044(3)	0
881	2,2,2,5/2		295	1,1,2,3/2	0.077(10)	0.037
					0.0073(13)	
848	2,2,4,7/2		295	1,1,2,3/2	0.130(20)	0.166
881	2,2,2,5/2		357	1,1,2,5/2	0.225(37)	0.148
					0.015(4)	
920	2,2,4,9/2		357	1,1,2,5/2	0.181(15)	0.185

[a] From Vervier and Janssens (1982) and Vervier (1987).
[b] With $\alpha_2 = 0.136\ eb$.

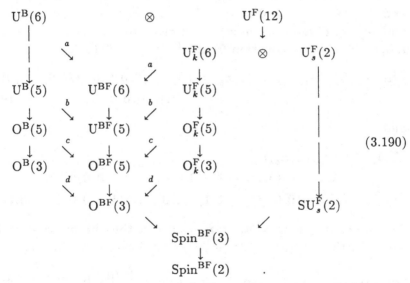

$$(3.190)$$

The generators of the fermion algebras in (3.190) are given by (3.75) and

$$\{G\}_{U^F_k(5)} = \{K^{(\lambda)}_\mu(2,2); \lambda = 0,\ldots,4\}. \qquad (3.191)$$

The generators of the Bose–Fermi algebras are specified in (3.77) and

$$\{G\}_{U^{BF}(5)} = \{G^{(\lambda)}_\mu(2,2); \lambda = 0,\ldots,4\}, \qquad (3.192)$$

with the G-operators as in (3.76). In the following subsections we shall discuss only route a.

3.3.3.2 Basis states The basis states in this chain are labelled by

$$
\left|
\begin{array}{l}
U^B(6) \otimes U^F(12) \supset U^B(6) \otimes \quad U^F_k(6) \quad \otimes U^F_s(2) \\
\quad \downarrow \qquad\quad \downarrow \qquad\qquad\qquad \downarrow \\
\quad [N_B] \qquad \{N_F\} \qquad\qquad\quad [N'_1,\ldots,N'_6] \\[4pt]
\supset \quad U^{BF}(6) \quad \otimes U^F_s(2) \supset \quad U^{BF}(5) \quad \otimes U^F_s(2) \supset O^{BF}(5) \otimes U^F_s(2) \\
\qquad\quad \downarrow \qquad\qquad\qquad\qquad\quad \downarrow \qquad\qquad\qquad\quad \downarrow \\
\quad [N_1,\ldots,N_6] \qquad\qquad\quad (n_1,\ldots,n_5) \qquad\qquad (v_1,v_2) \\[4pt]
\supset O^{BF}(3) \otimes SU^F_s(2) \supset \mathrm{Spin}^{BF}(3) \supset \mathrm{Spin}^{BF}(2) \\
\qquad \downarrow \qquad\quad \downarrow \qquad\qquad\quad \downarrow \qquad\qquad\quad \downarrow \\
\qquad n_\Delta, L \qquad\quad S \qquad\qquad\quad J \qquad\qquad\quad M_J
\end{array}
\right\rangle .
$$
$$(3.193)$$

We consider again the case $N_B = N$ and $N_F = 1$, for which $N'_1 = 1$ and $N'_2 = \cdots = N'_6 = 0$ and, as discussed in Sect. 3.2.3.2, only N_1 and N_2 are different from zero and given by $[N_1, N_2] = [N + 1 - i, i], i = 0, 1$. The reduction from $U^{BF}(6)$ to $U^{BF}(5)$ gives

$$
[N+1,0] : (n_1,n_2,n_3,n_4,n_5) = (N+1,0,0,0,0),(N,0,0,0,0),
$$
$$
\ldots,(0,0,0,0,0), \qquad (3.194)
$$

and

$$
[N,1] : (n_1,n_2,n_3,n_4,n_5)
$$
$$
= (N,0,0,0,0),(N-1,0,0,0,0),\ldots,(1,0,0,0,0),
$$
$$
(N,1,0,0,0),(N-1,1,0,0,0),\ldots,(1,1,0,0,0). \quad (3.195)
$$

The three labels $n_3 = n_4 = n_5 = 0$ can thus be omitted. The values of the $O^{BF}(5)$ labels (v_1, v_2) contained in (n_1, n_2) are

$$
(n_1,0) : (v_1,v_2) = (n_1,0),(n_1-2,0),\ldots,\begin{cases} (0,0), & n_1 = \text{even}, \\ (1,0), & n_1 = \text{odd}, \end{cases}
$$
$$(3.196)$$

and

$$
(n_1,1) : (v_1,v_2) = (n_1,1),(n_1-1,0),(n_1-2,1),
$$
$$
\ldots,\begin{cases} (1,0), & n_1 = \text{even}, \\ (1,1), & n_1 = \text{odd}. \end{cases} \qquad (3.197)
$$

The allowed values of L and J are then given by (3.85)–(3.86). Basis states can be labelled by

$$|[N_{\rm B}=N],\{N_{\rm F}=1\},[N+1-i,i],(n_1,n_2),(v_1,v_2),n_\Delta,L,J,M_J\rangle.$$
(3.198)

3.3.3.3 Energy eigenvalues Energy eigenvalues can be obtained by expanding the Hamiltonian as

$$
\begin{aligned}
H^{(\rm I_{3a})} &= e_0+e_1C_1({\rm U^B}6)+e_2C_2({\rm U^B}6)+e_3C_1({\rm U^F}12)+e_4C_2({\rm U^F}12)\\
&\quad + e_5C_1({\rm U^B}6)C_1({\rm U^F}12)+\eta C_2({\rm U^{BF}}6)+\epsilon C_1({\rm U^{BF}}5)\\
&\quad + \alpha C_2({\rm U^{BF}}5)+\beta C_2({\rm O^{BF}}5)+\gamma C_2({\rm O^{BF}}3)+\gamma' C_2({\rm SU}_s^{\rm F}2)\\
&\quad + \gamma'' C_2({\rm Spin^{BF}}3).
\end{aligned}
$$
(3.199)

The form of the Casimir invariants of the coupled algebras ${\rm O^{BF}}(5)$, ${\rm O^{BF}}(3)$ and ${\rm Spin^{BF}}(3)$ is specified in (3.89) and furthermore we have:

$$
\begin{aligned}
C_1({\rm U^{BF}}5) &= \sqrt{5}G_0^{(0)}(2,2),\\
C_2({\rm U^{BF}}5) &= \sum_{\lambda=0}^4 G^{(\lambda)}(2,2)\cdot G^{(\lambda)}(2,2).
\end{aligned}
$$
(3.200)

From (3.199) one can obtain the energy eigenvalues

$$
\begin{aligned}
&E^{(\rm I_{3a})}(N_{\rm B}=N,N_{\rm F}=1,[N+1-i,i],(n_1,n_2),(v_1,v_2),n_\Delta,J,M_J)\\
&= E_{03}' + \eta[(N+1-i)(N+6-i)+i(i+3)]\\
&\quad + \epsilon(n_1+n_2)+\alpha[n_1(n_1+4)+n_2(n_2+2)]\\
&\quad + 2\beta[v_1(v_1+3)+v_2(v_2+1)]+2\gamma L(L+1)+2\gamma'' J(J+1),
\end{aligned}
$$
(3.201)

with

$$E_{03}' = e_0 + e_1 N + e_2 N(N+5) + e_3 + 12e_4 + e_5 N + \tfrac{3}{2}\gamma'. \quad (3.202)$$

The excitation spectrum associated with (3.201) is shown in Fig. 3.17.

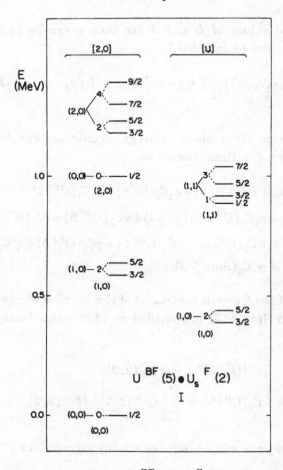

Fig. 3-17 A typical spectrum with $U^{BF}(6) \otimes U^F_s(2)$ (I) symmetry for $N_B = 1, N_F = 1$. The $U^{BF}(6)$ quantum numbers are shown on top (square brackets), the $U^{BF}(5)$ quantum numbers at the bottom, the $O^{BF}(5)$ and $O^{BF}(3)$ quantum numbers to the left and the angular momentum J to the right of each level.

3.3.3.4 Wave functions. Isoscalar factors   For completeness, we give a brief discussion of the wave functions for route a, from which those for the other routes can be derived. Again, the wave functions can be obtained by expanding $U^{BF}(6)$ states into states of the product $U^B(6) \otimes U^F_k(6)$,

$$|[N_1, N_2], (n_1, n_2), (v_1, v_2), L\rangle$$

Table 3-28 General expressions for the $U(5) \supset O(5)$ isoscalar factors

$$\left\langle \begin{array}{cc|c} (n_{\rm d},0) & (1,0) & (n_1,n_2) \\ (v,0) & (1,0) & (v_1,v_2) \end{array} \right\rangle$$

	$(n_1,n_2) = (n_{\rm d}+1,0)$	$(n_1,n_2) = (n_{\rm d},1)$
$(v_1,v_2) = (v-1,0)$	$\left[\dfrac{(v_1+3)(n_{\rm d}-v_1+1)}{(2v_1+3)(n_{\rm d}+1)}\right]^{\frac{1}{2}}$	$-\left[\dfrac{v_1(n_{\rm d}+v_1+4)}{(2v_1+3)(n_{\rm d}+1)}\right]^{\frac{1}{2}}$
$(v_1,v_2) = (v+1,0)$	$\left[\dfrac{v_1(n_{\rm d}+v_1+4)}{(2v_1+3)(n_{\rm d}+1)}\right]^{\frac{1}{2}}$	$\left[\dfrac{(v_1+3)(n_{\rm d}-v_1+1)}{(2v_1+3)(n_{\rm d}+1)}\right]^{\frac{1}{2}}$
$(v_1,v_2) = (v,1)$		1

$$= \sum_{\substack{n'_{\rm d},v',L' \\ n''_{\rm d},v'',L''}} \eta^{[N],(n'_{\rm d},0),(v',0),L'}_{[N_1,N_2],(n_1,n_2),(v_1,v_2),L}$$

$$\times \, |[N],(n'_{\rm d},0),(v',0),L';[1],(n''_{\rm d},0),(v'',0),L'';L\rangle,$$

$$(3.203)$$

where we have omitted the quantum number n_Δ. In this summation, the possible values of $(n''_{\rm d},v'',L'')$ are $(0,0,0)$ and $(1,1,2)$ and furthermore we have $n'_{\rm d} + n''_{\rm d} = n_1 + n_2$. The η-coefficients can be written as a product of three isoscalar factors,

$$\eta^{[N],(n'_{\rm d},0),(v',0),L'}_{[N_1,N_2],(n_1,n_2),(v_1,v_2),L} = \left\langle \begin{array}{cc|c} [N] & [1] & [N_1,N_2] \\ (n'_{\rm d},0) & (n''_{\rm d},0) & (n_1,n_2) \end{array} \right\rangle$$

$$\times \left\langle \begin{array}{cc|c} (n'_{\rm d},0) & (n''_{\rm d},0) & (n_1,n_2) \\ (v',0) & (v'',0) & (v_1,v_2) \end{array} \right\rangle \left\langle \begin{array}{cc|c} (v',0) & (v'',0) & (v_1,v_2) \\ L' & L'' & L \end{array} \right\rangle.$$

$$(3.204)$$

The last factor, associated with $O(5) \supset O(3)$, is discussed in Sect. 3.2.3.6 and some cases are listed in Table 3.13. For the two other isoscalar factors in (3.204), associated with $U(6) \supset U(5)$ and $U(5) \supset O(5)$, general expressions can be obtained with techniques similar to those described in Sect. 3.2.3.6. The results are summarized in Tables 3.28 and 3.29.

The total wave function (3.198) can now be obtained from the expansion (3.203) and a recoupling of angular momenta from

Table 3-29 General expressions for the $U(6) \supset U(5)$ isoscalar factors

$$\left\langle \begin{array}{cc|c} [N] & [1] & [N_1, N_2] \\ (n'_d, 0) & (n''_d, 0) & (n_1, n_2) \end{array} \right\rangle$$

	$[N_1, N_2] = [N+1, 0]$	$[N_1, N_2] = [N, 1]$
$n''_d = 0, (n_1, n_2) = (n'_d, 0)$	$\left[\dfrac{N - n_1 + 1}{N + 1}\right]^{\frac{1}{2}}$	$-\left[\dfrac{n_1}{N + 1}\right]^{\frac{1}{2}}$
$n''_d = 1, (n_1, n_2) = (n'_d + 1, 0)$	$\left[\dfrac{n_1}{N + 1}\right]^{\frac{1}{2}}$	$\left[\dfrac{N - n_1 + 1}{N + 1}\right]^{\frac{1}{2}}$
$n''_d = 1, (n_1, n_2) = (n'_d, 1)$		1

$(L', L'')L, 1/2, J$ to $L', (L'', 1/2)j, J$,

$$
\begin{aligned}
&|[N_1, N_2], (n_1, n_2), (v_1, v_2), L, J\rangle \\
&= \sum_{\substack{n'_d, v', L' \\ n''_d, v'', L''}} \eta^{[N],(n'_d,0),(v',0),L'}_{[N_1,N_2],(n_1,n_2),(v_1,v_2),L} \sum_j (-)^{L'+L''+1/2+J} \\
&\quad \times \sqrt{(2L+1)(2j+1)} \left\{ \begin{array}{ccc} L' & L'' & L \\ 1/2 & J & j \end{array} \right\} |[N], n'_d, v', L'; j; J\rangle.
\end{aligned}
$$

$$(3.205)$$

3.3.3.5 Examples of spectra with $U^{BF}(6) \otimes U_s^F(2)$ *symmetry* In this case one needs bosons with $U^B(5)$ symmetry and fermions occupying single-particle levels with $j = 1/2, 3/2, 5/2$. Examples could occur for odd-proton or odd-neutron nuclei in the Zn–Ge–Se region, where the odd particle occupies the $2p_{1/2}$, $2p_{3/2}$ and $1f_{5/2}$ orbits.

An example of this symmetry, $^{63}_{30}\text{Zn}_{33}$, is shown in Fig. 3.18 (Bijker and Kota, 1984).

Another example, $^{75}_{42}\text{As}_{33}$, has been discussed by Vervier *et al.* (1985).

$$3.3.4 \ \overline{U}^{BF}(6) \otimes \overline{U}_s^F(2) \ (I_4)$$

3.3.4.1 Lattice of algebras This symmetry is similar to that discussed in Sect. 3.2.4. It is the conjugate symmetry of that in Sect. 3.3.3. The lattice of algebras is:

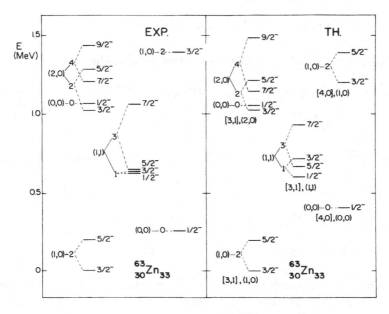

Fig. 3-18   An example of a spectrum with $U^{BF}(6) \otimes U_s^F(2)$ (I) symmetry: $_{30}^{63}Zn_{33}$ ($N_B = 3, N_F = 1$). The theoretical spectrum is calculated using (3.201) with $\eta = 150$ KeV, $\epsilon = 325$ KeV, $\alpha = 100$ KeV, $2\beta = 0$, $2\gamma = -24$ KeV and $2\gamma'' = 38$ KeV.

$$
\begin{array}{ccc}
U^B(6) & \otimes & \overline{U}^F(12) \\
\end{array}
$$

$$(3.206)$$

In the following, we shall discuss only route a.

3.3.4.2 Basis states Basis states can be classified as

$$
\left|
\begin{aligned}
& U^B(6) \otimes \overline{U}^F(12) \supset U^B(6) \otimes \quad \overline{U}^F_k(6) \quad \otimes \overline{U}^F_s(2) \\
& \quad \downarrow \qquad\quad \downarrow \qquad\qquad\qquad\qquad \downarrow \\
& \quad [N_B] \qquad \{\bar{N}_F\} \qquad\qquad [N'_1,\dots,N'_6] \\[4pt]
& \supset \quad \overline{U}^{BF}(6) \ \otimes \overline{U}^F_s(2) \supset \ \overline{U}^{BF}(5) \ \otimes \overline{U}^F_s(2) \supset O^{BF}(5) \otimes \overline{U}^F_s(2) \\
& \qquad \downarrow \qquad\qquad\qquad\qquad \downarrow \qquad\qquad\qquad\qquad \downarrow \\
& \quad [N_1,\dots,N_6] \qquad\qquad (n_1,\dots,n_5) \qquad\qquad (v_1,v_2) \\[4pt]
& \supset O^{BF}(3) \otimes SU^F_s(2) \supset \mathrm{Spin}^{BF}(3) \supset \mathrm{Spin}^{BF}(2) \\
& \qquad \downarrow \qquad\quad \downarrow \qquad\qquad \downarrow \qquad\qquad \downarrow \\
& \qquad n_\Delta, L \qquad\ S \qquad\qquad J \qquad\qquad M_J
\end{aligned}
\right\rangle .
$$
(3.207)

When $\bar{N}_F = 1$, which is the case we are interested here, the representations of $\overline{U}^F_k(6)$ are of the type $[N+1-i,1,1,1,1,i]$, $i = 0,1$. The decomposition to $\overline{U}^{BF}(5)$ is then given by

$$
[N,1,1,1,1,1] : (n_1,n_2,n_3,n_4,n_5)
$$
$$
= (N,1,1,1,1),(N-1,1,1,1,1),\dots,(1,1,1,1,1), \quad (3.208)
$$

and

$$
[N+1,1,1,1,1,0] : (n_1,n_2,n_3,n_4,n_5)
$$
$$
= (N+1,1,1,1,1),(N,1,1,1,1),\dots,(1,1,1,1,1),
$$
$$
(N+1,1,1,1,0),(N,1,1,1,0),\dots,(1,1,1,1,0). \quad (3.209)
$$

The allowed values of (v_1,v_2) contained in (n_1,n_2,n_3,n_4,n_5) are:

$$
(n_1,1,1,1,1) : (v_1,v_2) = (n_1-1,0),(n_1-3,0),\dots,
\begin{cases}
(0,0), & n_1 = \text{odd}, \\
(1,0), & n_1 = \text{even},
\end{cases}
$$
(3.210)

and

$$
(n_1,1,1,1,0) : (v_1,v_2) = (n_1,0),(n_1-1,1),(n_1-2,0),
$$
$$
\dots,
\begin{cases}
(1,0), & n_1 = \text{odd}, \\
(1,1), & n_1 = \text{even}.
\end{cases}
$$
(3.211)

The allowed values of L and J are then given by (3.85)–(3.86). The basis states can be labelled by

$$|[N_B = N], \{\bar{N}_F = 1\}, [N+1-i,1,1,1,1,i],$$
$$(n_1,\ldots,n_5),(v_1,v_2),n_\triangle,L,J,M_J\rangle. \quad (3.212)$$

3.3.4.3 Energy eigenvalues These can be obtained by considering the Hamiltonian

$$\begin{aligned}
H^{(I_{4a})} &= e_0+e_1 C_1(U^B 6)+e_2 C_2(U^B 6)+e_3 C_1(\bar{U}^F 12)+e_4 C_2(\bar{U}^F 12) \\
&\quad + e_5 C_1(U^B 6)C_1(\bar{U}^F 12)+\eta C_2(\bar{U}^{BF} 6)+\epsilon C_1(\bar{U}^{BF} 5) \\
&\quad + \alpha C_2(\bar{U}^{BF} 5)+\beta C_2(O^{BF} 5)+\gamma C_2(O^{BF} 3)+\gamma' C_2(SU_s^F 2) \\
&\quad + \gamma'' C_2(Spin^{BF} 3).
\end{aligned} \quad (3.213)$$

and are given by

$$\begin{aligned}
E^{(I_{4a})}&(N_B{=}N, \bar{N}_F{=}1, [N+1-i,1,1,1,1,i], \\
&(n_1,\ldots,n_5),(v_1,v_2),n_\triangle,J,M_J) \\
&= E'_{04} + \eta[(N+1-i)(N+6-i) + (i-4)(i-1)] \\
&\quad + \epsilon(n_1+n_2+n_3+n_4+n_5) \\
&\quad + \alpha[n_1(n_1+4) + n_2(n_2+2) + n_3^2 + n_4(n_4-2) + n_5(n_5-4)] \\
&\quad + 2\beta[v_1(v_1+3) + v_2(v_2+1)] + 2\gamma L(L+1) + 2\gamma'' J(J+1),
\end{aligned} \quad (3.214)$$

where

$$E'_{04} = e_0+e_1 N+e_2 N(N+5)+11e_3+22e_4+11e_5 N+\tfrac{3}{2}\gamma'. \quad (3.215)$$

The excitation spectrum associated with (3.215) is shown in Fig. 3.19.

3.4 Symmetries associated with SU(3)

In this final section we discuss Bose–Fermi symmetries associated with $SU^B(3)$.

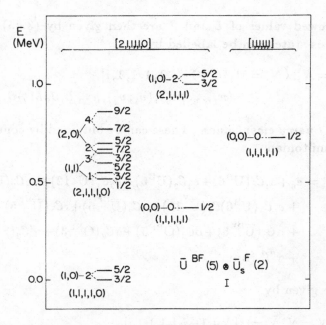

Fig. 3-19   A typical spectrum with $\overline{\mathrm{U}}^{\mathrm{BF}}(6) \otimes \overline{\mathrm{U}}^{\mathrm{F}}_s(2)$ (I) symmetry for $N_{\mathrm{B}} = 1, N_{\mathrm{F}} = 1$. The $\overline{\mathrm{U}}^{\mathrm{BF}}(6)$ quantum numbers are shown on top (square brackets), the $\mathrm{U}^{\mathrm{BF}}(5)$ quantum numbers at the bottom, the $\mathrm{O}^{\mathrm{BF}}(5)$ and $\mathrm{O}^{\mathrm{BF}}(3)$ quantum numbers to the left and the angular momentum J to the right of each level.

3.4.1 $\mathrm{SU}^{\mathrm{BF}}(3) \otimes \mathrm{U}^{\mathrm{F}}_s(2)$ (II_1)

3.4.1.1 Lattice of algebras   A symmetry of considerable practical importance arises every time the fermion angular momenta can be split into a pseudo-orbital part k and a pseudo-spin part $s = 1/2$, with the k-values corresponding to an irreducible representation $(\lambda_{\mathrm{F}}, \mu_{\mathrm{F}})$ of SU(3). We list in Table 3.30 some cases of interest.

We begin with a discussion of this problem for general $(\lambda_{\mathrm{F}}, \mu_{\mathrm{F}})$. We consider a situation in which the bosons have SU(3) symmetry and the fermions occupy orbits with $j = 1/2, 3/2, \ldots, n+1/2$. The corresponding fermion algebra is $\mathrm{U}^{\mathrm{F}}(\sum_j(2j+1)) = \mathrm{U}^{\mathrm{F}}((n+1)(n+$

Table 3-30 Splitting of j into k and s for some
cases of interest in nuclei

(λ_F, μ_F)	k	s	j
(0,0)	0	1/2	1/2
(1,0)	1	1/2	1/2,3/2
(2,0)	0,2	1/2	1/2,3/2,5/2
(3,0)	1,3	1/2	1/2,3/2,5/2,7/2
(4,0)	0,2,4	1/2	1/2,3/2,5/2,7/2,9/2

2)). The lattice of algebras is:

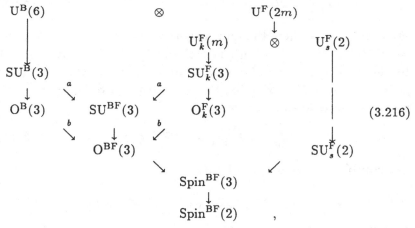

(3.216)

where the notation $m = (n+1)(n+2)/2$ is introduced. The general
expression for the generators of the fermion algebra $SU^F(3)$ was
given by Elliott (1958),

$$Q^{(2)}_{F,\mu} = \sum_{l,l'} q_{l,l'}(n) K^{(2)}_{\mu}(l,l'),$$

$$L^{(1)}_{F,\mu} = \sum_{l} \sqrt{\frac{l(l+1)(2l+1)}{3}} K^{(1)}_{\mu}(l,l), \qquad (3.217)$$

with $l, l' = n, n-2, \ldots, 1$ or 0 and

$$q_{l,l}(n) = -(2n+3) \left[\frac{l(l+1)(2l+1)}{40(2l-1)(2l+3)} \right]^{\frac{1}{2}},$$

$$q_{l,l+2}(n) = q_{l+2,l}(n) = \left[\frac{3(l+1)(l+2)(n-l)(n+l+3)}{20(2l+3)} \right]^{\frac{1}{2}}. \qquad (3.218)$$

With the help of (2.25), the operators $K_\mu^{(\lambda)}(l, l')$ can be expressed in terms of the bilinear products $A_\mu^{(\lambda)}(j, j')$. The generators of the coupled algebra $\mathrm{SU}^{\mathrm{BF}}(3)$ are:

$$Q_{\mathrm{BF},\mu}^{(2)} = Q_{\mathrm{B},\mu}^{(2)} + Q_{\mathrm{F},\mu}^{(2)},$$
$$L_{\mathrm{BF},\mu}^{(1)} = L_{\mathrm{B},\mu}^{(1)} + L_{\mathrm{F},\mu}^{(1)}, \tag{3.219}$$

with

$$Q_{\mathrm{B},\mu}^{(2)} = [s^\dagger \times \tilde{d} + d^\dagger \times \tilde{s}]_\mu^{(2)} - \frac{\sqrt{7}}{2}[d^\dagger \times \tilde{d}]_\mu^{(2)},$$
$$L_{\mathrm{B},\mu}^{(1)} = \sqrt{10}[d^\dagger \times \tilde{d}]_\mu^{(1)}. \tag{3.220}$$

We only consider here the minus sign in front of $\sqrt{7}/2$ corresponding to prolate deformation (see Volume 1). The positive sign can be dealt with in a similar way. The coupled algebra $\mathrm{O}^{\mathrm{BF}}(3)$ is generated by $L_{\mathrm{BF},\mu}^{(1)}$. Finally, the generators of $\mathrm{Spin}^{\mathrm{BF}}(3)$ are the angular momentum operators

$$J_\mu^{(1)} = L_{\mathrm{BF},\mu}^{(1)} - \frac{1}{\sqrt{2}}S_\mu^{(1)}(\tfrac{1}{2}, \tfrac{1}{2}), \tag{3.221}$$

and $\mathrm{Spin}^{\mathrm{BF}}(2)$ is generated by $J_0^{(1)}$.

3.4.1.2 Basis states   The basis states are given by

$$
\left|
\begin{array}{cccc}
\mathrm{U}^{\mathrm{B}}(6) \otimes \mathrm{U}^{\mathrm{F}}(2m) \supset \mathrm{U}^{\mathrm{B}}(6) \otimes & \mathrm{U}_k^{\mathrm{F}}(m) & \otimes \mathrm{U}_s^{\mathrm{F}}(2) \\
\downarrow \qquad\quad \downarrow & \downarrow & \\
[N_{\mathrm{B}}] \qquad \{N_{\mathrm{F}}\} & [N_1', \ldots, N_m'] & \\[2mm]
\supset \mathrm{SU}^{\mathrm{B}}(3) \otimes \mathrm{SU}^{\mathrm{F}}(3) \otimes \mathrm{U}_s^{\mathrm{F}}(2) \supset \mathrm{SU}^{\mathrm{BF}}(3) \otimes \mathrm{U}_s^{\mathrm{F}}(2) \\
\downarrow \qquad\qquad \downarrow \qquad\qquad\quad \downarrow \\
(\lambda_{\mathrm{B}}, \mu_{\mathrm{B}}) \quad (\lambda_{\mathrm{F}}, \mu_{\mathrm{F}}) \qquad\qquad (\lambda, \mu) \\[2mm]
\supset \mathrm{O}^{\mathrm{BF}}(3) \otimes \mathrm{SU}_s^{\mathrm{F}}(2) \supset \mathrm{Spin}^{\mathrm{BF}}(3) \supset \mathrm{Spin}^{\mathrm{BF}}(2) \\
\downarrow \qquad\quad \downarrow \qquad\quad \downarrow \qquad\qquad \downarrow \\
\chi, L \qquad\quad S \qquad\quad J \qquad\qquad M_J
\end{array}
\right\rangle.
$$

$$\tag{3.222}$$

When $N_{\mathrm{F}} = 1$, one has $[N_1', \ldots, N_m'] = [1, 0, \ldots, 0]$, $(\lambda_{\mathrm{F}}, \mu_{\mathrm{F}}) = (n, 0)$ and $S = 1/2$. The values of $(\lambda_{\mathrm{B}}, \mu_{\mathrm{B}})$ (Elliott quantum numbers) contained in a representation $[N_B = N]$ are given in Eq. (2.41) of Volume 1,

$$[N] : (\lambda_B, \mu_B) = \{(2N - 6p - 4q, 2q)\}, \tag{3.223}$$

where $\{\cdots\}$ denotes the set of all possible $SU^B(3)$ representations with $p, q = 0, 1, \ldots$, and $\lambda_B \geq 0$. The values of (λ, μ) can be obtained by taking the product $(\lambda_B, \mu_B) \otimes (\lambda_F, \mu_F)$. For $N_F = 1$, this product becomes (Bijker and Kota, 1988):

$$(\lambda_B, \mu_B) \otimes (n, 0) = \oplus \sum_{r,s} (\lambda_B + n - 2r - s, \mu_B + r - s), \tag{3.224}$$

with $r = 0, 1, \ldots, \min\{\lambda_B, n\}$ and $s = 0, 1, \ldots, \min\{n - r, \mu_B\}$. The angular momentum content of the SU(3) representation (λ, μ) was given by Elliott (1958),

$$(\lambda, \mu) : L = K, K + 1, \ldots, K + \max\{\lambda, \mu\}, \tag{3.225}$$

with

$$K = \min\{\lambda, \mu\}, \min\{\lambda, \mu\} - 2, \ldots, \begin{cases} 0, & \min\{\lambda, \mu\} \text{ even}, \\ 1, & \min\{\lambda, \mu\} \text{ odd}, \end{cases} \tag{3.226}$$

with the exception of $K = 0$ for which

$$(\lambda, \mu) : L = \max\{\lambda, \mu\}, \max\{\lambda, \mu\} - 2, \ldots, \begin{cases} 0, & \max\{\lambda, \mu\} \text{ even}, \\ 1, & \max\{\lambda, \mu\} \text{ odd}. \end{cases} \tag{3.227}$$

As discussed in Volume 1, p. 30, instead of Elliott's label K it is convenient to introduce Vergados's label χ (Vergados, 1968). The values taken by χ are the same as those of K but the values of L for each χ are different. If a value of L occurs only once, it belongs to the lowest χ. If it occurs twice, it belongs to the two lowest χs, etc. The only exception is when $\chi = 0$ in which case χ coincides with K. The total angular momentum J is simply given by $J = L \pm 1/2$, except for $L = 0$ when $J = 1/2$ only. In conclusion, the basis states can be labelled by

$$|[N_B = N], \{N_F = 1\}, (\lambda_B, \mu_B), (n, 0), (\lambda, \mu), \chi, L, J, M_J\rangle. \tag{3.228}$$

3.4.1.3 Energy eigenvalues   Energy eigenvalues can be obtained by writing the Hamiltonian as:

$$
\begin{aligned}
H^{(\mathrm{II}_{1a})} &= e_0 + e_1 \mathcal{C}_1(\mathrm{U}^{\mathrm{B}}6) + e_2 \mathcal{C}_2(\mathrm{U}^{\mathrm{B}}6) + e_3 \mathcal{C}_1(\mathrm{U}^{\mathrm{F}}2m) + e_4 \mathcal{C}_2(\mathrm{U}^{\mathrm{F}}2m) \\
&\quad + e_5 \mathcal{C}_1(\mathrm{U}^{\mathrm{B}}6)\mathcal{C}_1(\mathrm{U}^{\mathrm{F}}2m) + \delta \mathcal{C}_2(\mathrm{SU}^{\mathrm{B}}3) + \delta' \mathcal{C}_2(\mathrm{SU}^{\mathrm{F}}3) \\
&\quad + \delta'' \mathcal{C}_2(\mathrm{SU}^{\mathrm{BF}}3) + \gamma \mathcal{C}_2(\mathrm{O}^{\mathrm{BF}}3) + \gamma' \mathcal{C}_2(\mathrm{SU}^{\mathrm{F}}_s 2) \\
&\quad + \gamma'' \mathcal{C}_2(\mathrm{Spin}^{\mathrm{BF}}3).
\end{aligned}
\tag{3.229}
$$

The Casimir operators for the coupled algebras are given by

$$
\begin{aligned}
\mathcal{C}_2(\mathrm{SU}^{\mathrm{BF}}3) &= \tfrac{2}{3}\left[2 Q_{\mathrm{BF}}^{(2)} \cdot Q_{\mathrm{BF}}^{(2)} + \tfrac{3}{4} L_{\mathrm{BF}}^{(1)} \cdot L_{\mathrm{BF}}^{(1)} \right], \\
\mathcal{C}_2(\mathrm{O}^{\mathrm{BF}}3) &= 2 L_{\mathrm{BF}}^{(1)} \cdot L_{\mathrm{BF}}^{(1)}, \\
\mathcal{C}_2(\mathrm{Spin}^{\mathrm{BF}}3) &= 2 J^{(1)} \cdot J^{(1)}.
\end{aligned}
\tag{3.230}
$$

The energy eigenvalues are then

$$
\begin{aligned}
E^{(\mathrm{II}_{1a})}&(N_{\mathrm{B}} = N, N_{\mathrm{F}} = 1, (\lambda_{\mathrm{B}}, \mu_{\mathrm{B}}), (n, 0), (\lambda, \mu), \chi, L, J, M_J) \\
&= E''_{01} + \tfrac{2}{3}\delta C(\lambda_{\mathrm{B}}, \mu_{\mathrm{B}}) + \tfrac{2}{3}\delta'' C(\lambda, \mu) + 2\gamma L(L+1) + 2\gamma'' J(J+1),
\end{aligned}
\tag{3.231}
$$

with

$$
C(\lambda, \mu) = \lambda^2 + \mu^2 + \lambda\mu + 3\lambda + 3\mu, \tag{3.232}
$$

and

$$
E''_{01} = e_0 + e_1 N + e_2 N(N+5) + e_3 + 2m e_4 + e_5 N + n(n+3)\delta' + \tfrac{3}{2}\gamma'. \tag{3.233}
$$

Notice that E''_{01} includes the contribution coming from $\mathcal{C}_2(\mathrm{SU}^{\mathrm{F}}3)$, since n is constant for a given nucleus. The excitation spectrum associated with (3.231) is shown in Fig. 3.20 for $n = 1$.

3.4.1.4 Wave functions. Isoscalar factors   The wave functions for this chain can be obtained by expanding the coupled states into

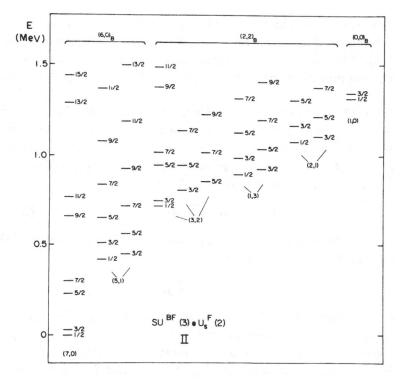

Fig. 3-20   A typical spectrum with $\mathrm{SU}^{\mathrm{BF}}(3)\otimes\mathrm{U}^{\mathrm{F}}_s(2)$ (II) symmetry for $N_{\mathrm{B}} = 2, N_{\mathrm{F}} = 1$. The $\mathrm{SU}^{\mathrm{B}}(3)$ quantum numbers are shown on top, the $\mathrm{SU}^{\mathrm{BF}}(3)$ quantum numbers at the bottom and the angular momentum J to the right of each level.

product states,

$$|(\lambda_{\mathrm{B}},\mu_{\mathrm{B}}),(n,0),(\lambda,\mu),\chi,L,J\rangle$$

$$= \sum_{\substack{\chi',L'\\L''}} \left\langle \begin{array}{cc} (\lambda_{\mathrm{B}},\mu_{\mathrm{B}}) & (n,0) \\ \chi',L' & L'' \end{array} \middle| \begin{array}{c} (\lambda,\mu) \\ \chi,L \end{array} \right\rangle \sum_j (-)^{L'+L''+1/2+J}$$

$$\times \sqrt{(2L+1)(2j+1)} \left\{ \begin{array}{ccc} L' & L'' & L \\ 1/2 & J & j \end{array} \right\} |(\lambda_{\mathrm{B}},\mu_{\mathrm{B}}),\chi',L';j;J\rangle.$$

$$(3.234)$$

The first factor in (3.234) is the Wigner coefficient (or isoscalar factor) for the reduction $\mathrm{SU}(3) \supset \mathrm{O}(3)$. Extensive tables are given by

Vergados (1968), which cover all cases of interest for $n = 1, 2$. Additional isoscalar factors for $n = 3, 4$ are given by Bijker and Kota (1988). The second factor represents a recoupling transformation of the angular momenta from $(L', L'')L, 1/2, J$ to $L', (L'', 1/2)j, J$.

3.4.1.5 Electromagnetic transitions and moments; E2

The E2 transition operator can be written in general as in (3.117). Simplifications in the calculation of its matrix elements occur if it can be written in terms of generators of the boson and fermion algebras appearing in the chain (3.222),

$$T_\mu^{(E2)} = \alpha_2 Q_{B,\mu}^{(2)} + f_2 Q_{F,\mu}^{(2)}. \tag{3.235}$$

When in addition $f_2 = \alpha_2$, the operator $T^{(E2)}$ is a generator of the combined Bose–Fermi algebra $SU^{BF}(3)$,

$$T_\mu^{(E2)} = \alpha_2 Q_{BF,\mu}^{(2)}, \tag{3.236}$$

and its matrix elements can be computed simply. The operator (3.236) satisfies the selection rules $\Delta \lambda_B = \Delta \mu_B = \Delta \lambda = \Delta \mu = 0$. Furthermore, since it is a generator of $SU^{BF}(3)$, its matrix elements will not depend on N, (λ_B, μ_B) and $(n, 0)$, which are labels associated with groups appearing before $SU^{BF}(3)$ in the reduction (3.222). To evaluate the matrix elements of this E2 operator explicitly, one first factors out the pseudo-spin part,

$$\langle (\lambda, \mu), \chi, L, J \parallel Q_{BF}^{(2)} \parallel (\lambda, \mu), \chi', L', J' \rangle$$

$$= (-)^{L+1/2+J'} \sqrt{(2J+1)(2J'+1)} \left\{ \begin{array}{ccc} L & J & 1/2 \\ J' & L' & 2 \end{array} \right\}$$

$$\times \langle (\lambda, \mu), \chi, L \parallel Q_{BF}^{(2)} \parallel (\lambda, \mu), \chi', L' \rangle, \tag{3.237}$$

= where redundant quantum numbers are omitted. The reduced matrix element on the right-hand side of (3.237) can be derived by using the tensorial character $T_{L,M_L}^{(\lambda,\mu)}$ of the generators (3.217) under $SU^{BF}(3) \supset O^{BF}(3) \supset O^{BF}(2)$,

$$Q_{BF,\mu}^{(2)} = \sqrt{\frac{3}{8}} T_{2,\mu}^{(1,1)},$$

$$L_{BF,\mu}^{(1)} = T_{1,\mu}^{(1,1)}, \tag{3.238}$$

from which the reduced matrix element can be calculated as

$$\langle (\lambda,\mu), \chi, L \parallel T_l^{(1,1)} \parallel (\lambda,\mu), \chi', L' \rangle$$
$$= \sum_\alpha \sqrt{2L+1} \left\langle \begin{array}{cc|c} (\lambda,\mu) & (1,1) & (\lambda,\mu) \\ \chi', L' & l & \chi, L \end{array} \right\rangle_\alpha$$
$$\times \langle (\lambda,\mu) \parallel T^{(1,1)} \parallel (\lambda,\mu) \rangle_\alpha. \tag{3.239}$$

The first factor on the right-hand side of (3.239) is an SU(3) $\supset$ O(3) Wigner coefficient for the multiplication $(\lambda,\mu) \otimes (1,1)$, for which general expressions are given by Vergados (1968). Note that an additional index $\alpha = 1,2$ is needed since, in general, the representation (λ,μ) is contained *twice* in the multiplication $(\lambda,\mu) \otimes (1,1)$. The second factor is an SU(3)-reduced matrix element that is non-zero only for $\alpha = 1$ (Biedenharn, 1963). For $l = 1$, (3.239) reduces to the matrix element of the angular momentum operator,

$$\sqrt{L(L+1)} = \left\langle \begin{array}{cc|c} (\lambda,\mu) & (1,1) & (\lambda,\mu) \\ \chi, L & 1 & \chi, L \end{array} \right\rangle_{\alpha=1}$$
$$\times \langle (\lambda,\mu) \parallel T^{(1,1)} \parallel (\lambda,\mu) \rangle_{\alpha=1}, \tag{3.240}$$

and hence provides us with an expression for the SU(3)-reduced matrix element in terms of an SU(3) isoscalar factor. Combining (3.239) and (3.240), one finds (Bijker and Kota, 1988)

$$\langle (\lambda,\mu), \chi, L \parallel Q_{\mathrm{BF}}^{(2)} \parallel (\lambda,\mu), \chi', L' \rangle$$
$$= \sqrt{\frac{3L(L+1)(2L+1)}{8}} \frac{\left\langle \begin{array}{cc|c} (\lambda,\mu) & (1,1) & (\lambda,\mu) \\ \chi', L' & 2 & \chi, L \end{array} \right\rangle_{\alpha=1}}{\left\langle \begin{array}{cc|c} (\lambda,\mu) & (1,1) & (\lambda,\mu) \\ \chi, L & 1 & \chi, L \end{array} \right\rangle_{\alpha=1}}. \tag{3.241}$$

Knowledge of the $(\lambda,\mu) \otimes (1,1) \to (\lambda,\mu)$ isoscalar factors (Vergados, 1968), leads to expressions for reduced matrix elements of $T^{(E2)}$. From these one can calculate the $B(E2)$ values and electric quadrupole moments for states with $\mu = 0$ and 1 shown in Tables 3.31 and 3.32.

Table 3-31 B(E2) values and electric quadrupole moments in the $SU^{BF}(3) \otimes U_s^F(2)$ limit for states with $(\lambda, \mu = 0)$

L'	J'	$\to$	L	J	$B(E2; J' \to J)$
$L+2$	$L+\frac{5}{2}$		L	$L+\frac{1}{2}$	$\alpha_2^2 \dfrac{3(L+1)(L+2)(\lambda-L)(\lambda+L+3)}{4(2L+3)(2L+5)}$
$L+2$	$L+\frac{3}{2}$		L	$L-\frac{1}{2}$	$\alpha_2^2 \dfrac{3L(L+1)(\lambda-L)(\lambda+L+3)}{4(2L+1)(2L+3)}$
$L+2$	$L+\frac{3}{2}$		L	$L+\frac{1}{2}$	$\alpha_2^2 \dfrac{3(L+1)(\lambda-L)(\lambda+L+3)}{2(2L+1)(2L+3)(2L+5)}$
L	$L+\frac{1}{2}$		L	$L-\frac{1}{2}$	$\alpha_2^2 \dfrac{3L(2\lambda+3)^2}{8(2L-1)(2L+1)(2L+3)}$

L	J				Q_J
L	$L\pm\frac{1}{2}$				$-\sqrt{\dfrac{2\pi}{5}}\alpha_2 \dfrac{(2J-1)(2\lambda+3)}{4(J+1)}$

3.4.1.6 Electromagnetic transitions and moments; M1 As in Sect. 3.2.3.8, the magnetic dipole operator cannot, to a good approximation, be written as a linear combination of generators of algebras appearing in (3.222). For a realistic calculation one should use an M1 operator of the form discussed in Chapter 1. An alternative approach (Bijker and Kota, 1988) is to assume that the M1 operator contains also two-body terms, that make it quadratic in the generators of (3.222). In both cases, the resulting expressions for B(M1) values and magnetic dipole moments are rather complicated and will be omitted here.

3.4.1.7 One-nucleon transfer intensities For reactions of the type $N_B = N, N_F = 0 \rightleftharpoons N_B = N, N_F = 1$, the transfer operator is of the form (3.127). It leads to the following expressions for the spectroscopic strengths:
(i) even–even to odd–even,

$$S_j((2N,0),\chi'=0,L'=0 \to (\lambda_B,\mu_B),(\lambda,\mu),\chi,L,J)$$

$$= (2j+1)\left\langle \begin{matrix} (2N,0) & (n,0) \\ 0 & L \end{matrix} \, \middle| \, \begin{matrix} (\lambda,\mu) \\ \chi,L \end{matrix} \right\rangle^2 p_j^2 \delta_{\lambda_B,2N}\delta_{\mu_B,0}\delta_{j,J};$$

$$(3.242)$$

Table 3-32 $B(E2)$ values and electric quadrupole moments in the $SU^{BF}(3) \otimes U_s^F(2)$ limit for states with $(\lambda, \mu = 1)$

L'	J'	$\rightarrow$	L	J	$B(E2; J' \rightarrow J)$
$L+2$	$L+\frac{5}{2}$		L	$L+\frac{1}{2}$	$\alpha_2^2 \dfrac{3L(L+3)a_{L+2,L}}{4(2L+3)(2L+5)}$
$L+2$	$L+\frac{3}{2}$		L	$L-\frac{1}{2}$	$\alpha_2^2 \dfrac{3L^2(L+3)a_{L+2,L}}{4(L+2)(2L+1)(2L+3)}$
$L+2$	$L+\frac{3}{2}$		L	$L+\frac{1}{2}$	$\alpha_2^2 \dfrac{3L(L+3)a_{L+2,L}}{2(L+2)(2L+1)(2L+3)(2L+5)}$
$L+1$	$L+\frac{3}{2}$		L	$L+\frac{1}{2}$	$\alpha_2^2 \dfrac{3L(2L+5)a_{L+1,L}}{2(L+1)(L+2)(2L+1)(2L+3)}$
$L+1$	$L+\frac{3}{2}$		L	$L-\frac{1}{2}$	$\alpha_2^2 \dfrac{3a_{L+1,L}}{(L+1)(L+2)(2L+1)(2L+3)}$
$L+1$	$L+\frac{1}{2}$		L	$L+\frac{1}{2}$	$\alpha_2^2 \dfrac{9a_{L+1,L}}{2(L+1)^2(2L+1)(2L+3)}$
$L+1$	$L+\frac{1}{2}$		L	$L-\frac{1}{2}$	$\alpha_2^2 \dfrac{3(L+2)(2L-1)a_{L+1,L}}{2(L+1)^2(2L+1)(2L+3)}$
L	$L+\frac{1}{2}$		L	$L-\frac{1}{2}$	$\alpha_2^2 \dfrac{3a_{L,L}^2}{8L(L+1)^2(2L-1)(2L+1)(2L+3)}$

L	J				Q_J
L	$L-\frac{1}{2}$				$\sqrt{\dfrac{2\pi}{5}}\alpha_2 \dfrac{(L-1)a_{L,L}}{L(L+1)(2L+1)}$
L	$L+\frac{1}{2}$				$\sqrt{\dfrac{2\pi}{5}}\alpha_2 \dfrac{a_{L,L}}{(L+1)(2L+3)}$

$$a_{L+2,L} = \begin{cases} (\lambda - L)(\lambda + L + 3) & \lambda - L \text{ even} \\ (\lambda - L + 1)(\lambda + L + 4) & \lambda - L \text{ odd} \end{cases}$$

$$a_{L+1,L} = \begin{cases} (\lambda + 2)(\lambda + L + 3) & \lambda - L \text{ even} \\ (\lambda + 2)(\lambda - L + 1) & \lambda - L \text{ odd} \end{cases}$$

$$a_{L,L} = \begin{cases} 6(\lambda + 2) - L(L+1)(2\lambda + 7) & \lambda - L \text{ even} \\ 6(\lambda + 2) - L(L+1)(2\lambda + 1) & \lambda - L \text{ odd} \end{cases}$$

Table 3-33 One-nucleon transfer reaction intensities in the $SU^{BF}(3) \otimes U_s^F(2)$ limit

Final state[a]	Transf j	Intensity
(i) Reaction $N_B = N, N_F = 0 \rightarrow N_B = N, N_F = 1$ for $n = 1$		
$\|(2N,0),(1,0),(2N+1,0),0,1,j\rangle$	1/2,3/2	$\dfrac{(2j+1)(2N+3)}{3(2N+1)}p_j^2$
$\|(2N,0),(1,0),(2N-1,1),1,1,j\rangle$	1/2,3/2	$\dfrac{4(2j+1)N}{3(2N+1)}p_j^2$
(ii) Reaction $N_B = N, N_F = 0 \rightarrow N_B = N, N_F = 1$ for $n = 2$		
$\|(2N,0),(2,0),(2N+2,0),0,0,1/2\rangle$	1/2	$\dfrac{2(2N+3)}{3(2N+1)}p_{1/2}^2$
$\|(2N,0),(2,0),(2N-2,2),0,0,1/2\rangle$	1/2	$\dfrac{8N}{3(2N+1)}p_{1/2}^2$
$\|(2N,0),(2,0),(2N+2,0),0,2,j\rangle$	3/2,5/2	$\dfrac{(2j+1)(2N+3)(2N+5)}{15(N+1)(2N+1)}p_j^2$
$\|(2N,0),(2,0),(2N,1),1,2,j\rangle$	3/2,5/2	$\dfrac{(2j+1)(2N+3)}{5(N+1)}p_j^2$
$\|(2N,0),(2,0),(2N-2,2),0,2,j\rangle$	3/2,5/2	$\dfrac{(2j+1)(2N-3)^2(2N+3)}{15(2N+1)(4N^2-3)}p_j^2$
$\|(2N,0),(2,0),(2N-2,2),2,2,j\rangle$	3/2,5/2	$\dfrac{8(2j+1)(N-1)N}{5(4N^2-3)}p_j^2$

[a] Initial state is $\|[N],(2N,0),\chi = 0, L = 0\rangle$.

(ii) odd–even to even–even,

$$S_j((\lambda_B,\mu_B),(\lambda,\mu),\chi,L,J \rightarrow (\lambda_B',\mu_B'),\chi',L')$$

$$= (2j+1)(2L+1)\left\{ \begin{array}{ccc} L' & L'' & L \\ 1/2 & J & j \end{array} \right\}^2$$

$$\times \left\langle \begin{array}{ccc} (\lambda_B',\mu_B') & (n,0) & | & (\lambda,\mu) \\ \chi',L' & L'' & | & \chi,L \end{array} \right\rangle^2 p_j^2 \delta_{\lambda_B,\lambda_B'} \delta_{\mu_B,\mu_B'} \delta_{\chi,\chi'}.$$

$$(3.243)$$

In Table 3.33 the explicit expressions for even–even to odd–even transfer are summarized for $n = 1$ and $n = 2$.

Using the transfer operator (3.129), expressions can be derived for reactions of the type $N_B = N+1, N_F = 0 \rightleftharpoons N_B = N, N_F = 1$.

Since they have not found applications yet in odd–even nuclei, they will not be discussed here.

3.4.1.8 Examples of spectra with $\mathrm{SU^{BF}}(3) \otimes \mathrm{U}_s^{F}(2)$ *symmetry*   It is somewhat difficult to find experimental examples of this type of symmetry. One must look for regions where the bosons have $\mathrm{SU^{B}}(3)$ symmetry. These are shown in Fig. 3.5. However, this is a necessary but not a sufficient condition for the occurrence of spectra with $\mathrm{SU^{BF}}(3)$ symmetry. One needs, in addition, that the odd nucleon occupies single-particle levels with j values as in one of the combinations shown in Table 3.30. Because of the large mixing between the different single-particle orbits, characteristic of the SU(3) limit, one needs to consider many of them. For example, an analysis of rare-earth nuclei with $50 \leq$ proton number ≤ 82 and $82 \leq$ neutron number ≤ 126 requires the proton orbits $3s_{1/2}$, $2d_{3/2}$, $2d_{5/2}$, $1g_{7/2}$ and the neutron orbits $3p_{1/2}$, $3p_{3/2}$, $2f_{5/2}$, $2f_{7/2}$, $1h_{9/2}$. Nonetheless, a partial analysis can be done using a subset of these levels. Two cases have been discussed: (a) the odd-proton Tm isotopes where the odd proton occupies the $3s_{1/2}$ and $2d_{3/2}$ orbits (Vervier, 1987) and (b) the odd-neutron W isotopes where the odd neutron occupies the $3p_{1/2}$, $3p_{3/2}$ and $2f_{5/2}$ orbits (Warner, 1984). The Tm isotopes have also been described by Bijker and Kota (1988) using *all* proton orbits of the 50–82 shell. Their results for the ground-state band are similar to those obtained by Vervier (1987) and we will confine ourselves to a discussion of the latter. The example of the W isotopes will be discussed in Sect. 3.4.3.7.

In Fig. 3.21 the experimental spectrum of $^{169}_{69}\mathrm{Tm}_{100}$ is compared with the theoretical one, calculated for $n = 1$ ($k = 1$). It should be emphasized, however, that the experimental spectrum of $^{169}_{69}\mathrm{Tm}_{100}$ includes rotational bands other than the ground-state band, which are not shown in Fig. 3.21 and which are outside the scope of a calculation with just two single-particle orbits with $j = 1/2$ and $3/2$. Conversely, if one includes all proton orbits of the 50-82 shell, one finds that more bands are calculated than present in the data (Bijker and Kota, 1988). Thus, it appears that the expression (3.229) can only describe the ground-state band energies

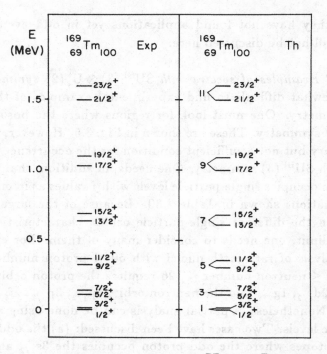

Fig. 3-21 An example of a spectrum with $SU^{BF}(3) \otimes U_s^F(2)$ (II) symmetry:
$^{169}_{69}Tm_{100}$ ($N_B = 15$, $N_F = 1$). The theoretical spectrum is calculated using
(3.231) with $2\gamma = 8$ KeV and $2\gamma'' = 4$ KeV. All states in the figure belong to
the same $SU^B(3)$ and $SU^{BF}(3)$ representations and thus the terms multiplying
δ and δ'' in (3.231) do not contribute to the excitation energies.

of some deformed odd–even nuclei. A comparison between calculated and experimental $B(E2)$ values in $^{169}_{69}Tm_{100}$ is shown in
Fig. 3.22. Again, there is good agreement with experiment but,
as this agreement is confined to the ground-state band, these results are not complete enough to determine the goodness of the
$SU^{BF}(3)$ symmetry in this nucleus.

3.4.2 $\overline{SU}^{BF}(3) \otimes \overline{U}_s^F(2)$ (II₂)

3.4.2.1 Lattice of algebras This symmetry arises from a particle–
hole conjugation applied to the symmetry described in Sect. 3.4.1.

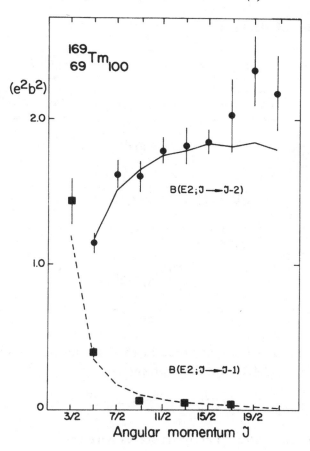

Fig. 3-22   Comparison between calculated and experimental (squares and circles) $B(E2)$ values in $^{169}_{69}\text{Tm}_{100}$ for the $J \to J - 1$ (squares and broken line) and $J \to J - 2$ (circles and solid line). The $B(E2)$ values are calculated using the operator (3.236) with $\alpha_2 = 0.109$ eb. The number of bosons is $N_{\text{B}} = 15$.

The lattice of algebras is:

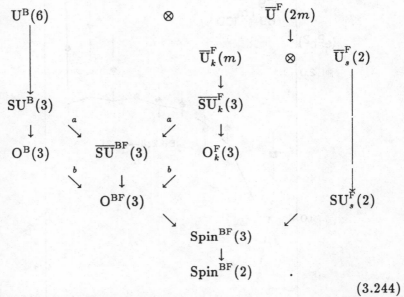

$$(3.244)$$

The generators of the algebras in (3.244) can be obtained from those in Sect. 3.4.1.1 by the replacement

$$Q_{F,\mu}^{(2)} \to \bar{Q}_{F,\mu}^{(2)} = -Q_{F,\mu}^{(2)}. \qquad (3.245)$$

3.4.2.2 Basis states   The basis states here are

$$\left| \begin{array}{l} U^B(6) \otimes \overline{U}^F(2m) \supset U^B(6) \otimes \quad \overline{U}_k^F(m) \quad \otimes \overline{U}_s^F(2) \\ \quad \downarrow \qquad\qquad \downarrow \qquad\qquad\qquad\qquad \downarrow \\ \quad [N_B] \qquad \{\bar{N}_F\} \qquad\qquad\qquad\quad [N_1', \dots, N_m'] \\ \\ \supset SU^B(3) \otimes \overline{SU}^F(3) \otimes \overline{U}_s^F(2) \supset \overline{SU}^{BF}(3) \otimes \overline{U}_s^F(2) \\ \quad \downarrow \qquad\qquad \downarrow \qquad\qquad \downarrow \qquad\qquad\qquad \downarrow \\ \quad (\lambda_B, \mu_B) \quad (\bar{\lambda}_F, \bar{\mu}_F) \qquad\qquad\qquad (\lambda, \mu) \\ \\ \supset O^{BF}(3) \otimes SU_s^F(2) \supset Spin^{BF}(3) \supset Spin^{BF}(2) \\ \quad \downarrow \qquad \downarrow \qquad\qquad \downarrow \qquad\qquad \downarrow \\ \quad \chi, L \qquad S \qquad\qquad J \qquad\qquad M_J \end{array} \right\rangle . $$

$$(3.246)$$

When $\bar{N}_F = 1$, one has $[N_1', \ldots, N_m'] = [1, 1, \ldots, 1, 0]$, $(\bar{\lambda}_F, \bar{\mu}_F) = (0, n)$ and $S = 1/2$. The values of (λ_B, μ_B) are given by (3.223). Those of (λ, μ) can be obtained by considering the product $(\lambda_B, \mu_B) \otimes (\bar{\lambda}_F, \bar{\mu}_F)$. For $\bar{N}_F = 1$, this product becomes (Bijker and Kota, 1988):

$$(\lambda_B, \mu_B) \otimes (0, n) = \oplus \sum_{r,s} (\lambda_B + r - 2s, \mu_B + n - 2r + s), \quad (3.247)$$

with $r = 0, 1, \ldots, \min\{\lambda_B + \mu_B, n\}$ and $s = \max\{r - \mu_B, 0\}, \max\{r - \mu_B, 0\} + 1, \ldots, \min\{\lambda_B, r\}$. The further decomposition to $O^{BF}(3)$ is the same as in Sect. 3.4.1.2. Thus, in conclusion, basis states here are given by

$$|[N_B = N], \{\bar{N}_F = 1\}, (\lambda_B, \mu_B), (0, n), (\lambda, \mu), \chi, L, J, M_J\rangle. \quad (3.248)$$

3.4.2.3 Energy eigenvalues Energy eigenvalues can be obtained by writing the Hamiltonian as

$$
\begin{aligned}
H^{(\text{II}_{2a})} &= e_0 + e_1 \mathcal{C}_1(\text{U}^B 6) + e_2 \mathcal{C}_2(\text{U}^B 6) + e_3 \mathcal{C}_1(\overline{\text{U}}^F 2m) + e_4 \mathcal{C}_2(\overline{\text{U}}^F 2m) \\
&\quad + e_5 \mathcal{C}_1(\text{U}^B 6) \mathcal{C}_1(\overline{\text{U}}^F 2m) + \delta \mathcal{C}_2(\text{SU}^B 3) + \delta' \mathcal{C}_2(\overline{\text{SU}}^F 3) \\
&\quad + \delta'' \mathcal{C}_2(\overline{\text{SU}}^{BF} 3) + \gamma \mathcal{C}_2(\text{O}^{BF} 3) + \gamma' \mathcal{C}_2(\text{SU}_s^F 2) \\
&\quad + \gamma'' \mathcal{C}_2(\text{Spin}^{BF} 3).
\end{aligned}
\quad (3.249)
$$

The Casimir operators of the various algebras are obtained after applying the transformation (3.245) on the expressions given in Sect. 3.4.1.3. The energy eigenvalues of (3.249) are:

$$
\begin{aligned}
&E^{(\text{II}_{2a})}(N_B = N, \bar{N}_F = 1, (\lambda_B, \mu_B), (0, n), (\lambda, \mu), \chi, L, J, M_J) \\
&= E_{02}'' + \tfrac{2}{3}\delta C(\lambda_B, \mu_B) + \tfrac{2}{3}\delta'' C(\lambda, \mu) + 2\gamma L(L + 1) + 2\gamma'' J(J + 1),
\end{aligned}
\quad (3.250)
$$

with

$$
\begin{aligned}
E_{02}'' &= e_0 + e_1 N + e_2 N(N + 5) + (2m - 1)e_3 + 2(2m - 1)e_4 \\
&\quad + (2m - 1)e_5 N + n(n + 3)\delta' + \tfrac{3}{2}\gamma'.
\end{aligned}
\quad (3.251)
$$

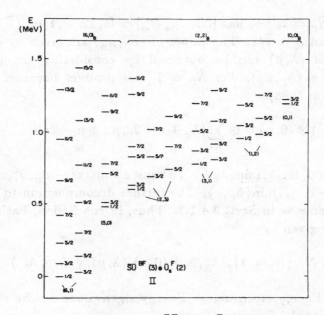

Fig. 3-23   A typical spectrum with $\overline{\mathrm{SU}}^{\mathrm{BF}}(3)\otimes\overline{\mathrm{U}}^{\mathrm{F}}_s(2)$ (II) symmetry for $N_{\mathrm B} = 2$, $N_{\mathrm F} = 1$. The $\mathrm{SU}^{\mathrm B}(3)$ quantum numbers are shown on top, the $\overline{\mathrm{SU}}^{\mathrm{BF}}(3)$ quantum numbers at the bottom and the angular momentum J to the right of each level.

The expression (3.250) for the excitation energies is formally identical to (3.231). However, the classification schemes are rather different and thus (3.250) leads to a different spectrum shown in Fig. 3.23 for $n = 1$.

3.4.2.4 Wave functions. Isoscalar factors   The wave functions for this chain can be obtained from Sect. 3.4.1.4 by the replacement $(n, 0) \to (0, n)$ which corresponds to a particle–hole conjugation,

$$|(\lambda_{\mathrm B}, \mu_{\mathrm B}), (0, n), (\lambda, \mu), \chi, L, J\rangle$$

$$= \sum_{\substack{\chi', L' \\ L''}} \left\langle \begin{array}{cc} (\lambda_{\mathrm B}, \mu_{\mathrm B}) & (0, n) \\ \chi', L' & L'' \end{array} \middle| \begin{array}{c} (\lambda, \mu) \\ \chi, L \end{array} \right\rangle \sum_j (-)^{L'+L''+1/2+J}$$

$$\times \sqrt{(2L+1)(2j+1)} \left\{ \begin{array}{ccc} L' & L'' & L \\ 1/2 & J & j \end{array} \right\} |(\lambda_{\mathrm B}, \mu_{\mathrm B}), \chi', L'; j; J\rangle.$$

$$(3.252)$$

The SU(3) $\supset$ O(3) isoscalar factors in (3.252) are given by Vergados (1968) for $n = 1, 2$.

3.4.2.5 Electromagnetic transitions and moments; E2   Simple expressions are obtained for the matrix elements of the E2 operator if this operator is taken as

$$T_\mu^{(E2)} = \alpha_2 \bar{Q}_{BF,\mu}^{(2)}. \qquad (3.253)$$

Since $\bar{Q}_{BF}^{(2)}$ is a generator of $\overline{SU}^{BF}(3)$, the computation of its matrix elements now proceeds exactly as in Sect. 3.4.1.5, leading to the same formal expression (3.241) and the results quoted in Tables 3.31 and 3.32.

3.4.2.6 One-nucleon transfer intensities   Expressions for one-nucleon transfer are obtained from Sect. 3.4.1.7 replacing $(n, 0)$ by $(0, n)$. Thus, we find the following predictions for the spectroscopic strengths:
(i) even–even to odd–even,

$$S_j((2N,0), \chi' = 0, L' = 0 \to (\lambda_B, \mu_B), (\lambda, \mu), \chi, L, J)$$

$$= (2j+1) \left\langle \begin{array}{cc} (2N,0) & (0,n) \\ 0 & L \end{array} \middle| \begin{array}{c} (\lambda,\mu) \\ \chi, L \end{array} \right\rangle^2 p_j^2 \delta_{\lambda_B,2N} \delta_{\mu_B,0} \delta_{j,J};$$

$$(3.254)$$

(ii) odd–even to even–even,

$$S_j((\lambda_B, \mu_B), (\lambda, \mu), \chi, L, J \to (\lambda'_B, \mu'_B), \chi', L')$$

$$= (2j+1)(2L+1) \left\{ \begin{array}{ccc} L' & L'' & L \\ 1/2 & J & j \end{array} \right\}^2$$

$$\times \left\langle \begin{array}{cc} (\lambda'_B, \mu'_B) & (0,n) \\ \chi', L' & L'' \end{array} \middle| \begin{array}{c} (\lambda,\mu) \\ \chi, L \end{array} \right\rangle^2 p_j^2 \delta_{\lambda_B,\lambda'_B} \delta_{\mu_B,\mu'_B} \delta_{\chi,\chi'}.$$

$$(3.255)$$

Table 3-34 One-nucleon transfer reaction intensities in the $\overline{SU}^{BF}(3) \otimes \overline{U}^F_s(2)$ limit

Final state[a]	Transf j	Intensity
(i) Reaction $N_B = N, \bar{N}_F = 0 \rightarrow N_B = N, \bar{N}_F = 1$ for $n = 1$		
$\|(2N,0),(0,1),(2N-1,0),0,1,j\rangle$	1/2,3/2	$\dfrac{(2j+1)N}{3(N+1)}p_j^2$
$\|(2N,0),(0,1),(2N,1),1,1,j\rangle$	1/2,3/2	$\dfrac{(2j+1)(2N+3)}{3(N+1)}p_j^2$
(ii) Reaction $N_B = N, \bar{N}_F = 0 \rightarrow N_B = N, \bar{N}_F = 1$ for $n = 2$		
$\|(2N,0),(0,2),(2N-2,0),0,0,1/2\rangle$	1/2	$\dfrac{2N}{3(N+1)}p_{1/2}^2$
$\|(2N,0),(0,2),(2N,2),0,0,1/2\rangle$	1/2	$\dfrac{2(2N+3)}{3(N+1)}p_{1/2}^2$
$\|(2N,0),(0,2),(2N-2,0),0,2,j\rangle$	3/2,5/2	$\dfrac{4(2j+1)(N-1)N}{15(N+1)(2N+1)}p_j^2$
$\|(2N,0),(0,2),(2N-1,1),1,2,j\rangle$	3/2,5/2	$\dfrac{4(2j+1)N}{5(2N+1)}p_j^2$
$\|(2N,0),(0,2),(2N,2),0,2,j\rangle$	3/2,5/2	$\dfrac{(2j+1)(2N+3)(2N+5)}{15(4N^2+8N+1)}p_j^2$
$\|(2N,0),(0,2),(2N,2),2,2,j\rangle$	3/2,5/2	$\dfrac{8(2j+1)N(N+2)^2}{5(N+1)(4N^2+8N+1)}p_j^2$

[a] Initial state is $\|[N],(2N,0),\chi = 0, L = 0\rangle$.

In Table 3.34 the explicit expressions for even–even to odd–even transfer are summarized for $n = 1$ and $n = 2$.

3.4.3 $U^{BF}(6) \otimes U^F_s(2)$ (II_3)

3.4.3.1 Lattice of algebras A special situation arises for the symmetries discussed in Sect. 3.4.1 in the case $n = 2$ and it will discussed separately here. It is special because of the occurrence of the fermion algebra $U^F_k(6)$ which can be combined with $U^B(6)$ in a way similar to the symmetries considered in Sects. 3.2.3 and 3.3.3. We thus have a situation in which the bosons have SU(3) symmetry and the fermions occupy orbits with $j = 1/2, 3/2, 5/2$.

The lattice of algebras is:

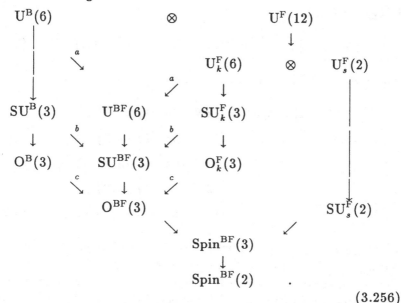

$$(3.256)$$

Only route a is not contained in the lattice of Sect. 3.4.1.1 and it is the one that will be discussed here. Generators of the coupled algebras appearing in (3.256) are defined in (3.77) and (3.219).

3.4.3.2 Basis states The basis states are labelled by

$$\left| \begin{array}{c} U^B(6) \otimes U^F(12) \supset U^B(6) \otimes U^F_k(6) \otimes U^F_s(2) \\ \downarrow \qquad \downarrow \qquad \downarrow \\ [N_B] \qquad \{N_F\} \qquad [N'_1,\ldots,N'_6] \end{array} \right.$$

$$\left. \begin{array}{c} \supset U^{BF}(6) \otimes U^F_s(2) \supset SU^{BF}(3) \otimes U^F_s(2) \\ \downarrow \qquad \downarrow \\ [N_1,\ldots,N_6] \qquad (\lambda,\mu) \end{array} \right.$$

$$\left. \begin{array}{c} \supset O^{BF}(3) \otimes SU^F_s(2) \supset Spin^{BF}(3) \supset Spin^{BF}(2) \\ \downarrow \qquad \downarrow \qquad \downarrow \qquad \downarrow \\ \chi, L \qquad S \qquad J \qquad M_J \end{array} \right\rangle .$$

$$(3.257)$$

When $N_F = 1$, one has $[N'_1,\ldots,N'_6] = [1,0,\ldots,0]$, $S = 1/2$ and $[N_1,\ldots,N_6] = [N+1-i,i,0,0,0,0], i = 0,1$, as discussed in

Sect. 3.2.3.2. We thus need the reduction from $U^{BF}(6)$ to $SU^{BF}(3)$ in two cases,

$$[N+1,0] : (\lambda,\mu) = \{(2N+2-6p-4q,2q)\}, \qquad (3.258)$$

where $p,q = 0,1,\ldots$, and $\lambda \geq 0$, and

$$
\begin{aligned}
[N,1] : (\lambda,\mu) = &\{(2N-6p-4q,2q+1)\}', \\
&\{(2N-6p-4q-2,2q+2)\}', \\
&\{(2N-6p-4q-3,2q+1)\}, \\
&\{(2N-6p-4q-4,2q)\}, \\
&\{(2N-6p-4q-5,2q+2)\}, \qquad (3.259)
\end{aligned}
$$

where $p,q = 0,1,\ldots$. The values of p and q are restricted by requiring $\lambda \geq 0$ and $\mu \geq 0$. Furthermore, a prime $\{\cdots\}'$ indicates the stronger condition $\lambda > 0$. For $N = 3r + 2$, with r an integer, the expression (3.259) must be corrected since it will give one $(0,0)$ representation too many. The remaining classification of labels is as in Sect. 3.4.1.2. To summarize, basis states in this symmetry are

$$|[N_B = N],\{N_F = 1\},[N+1-i,i],(\lambda,\mu),\chi,L,J,M_J\rangle. \quad (3.260)$$

3.4.3.3 Energy eigenvalues   Expanding the Hamiltonian as

$$
\begin{aligned}
H^{(II_{3a})} = {}&e_0 + e_1 C_1(U^B 6) + e_2 C_2(U^B 6) + e_3 C_1(U^F 12) + e_4 C_2(U^F 12) \\
&+ e_5 C_1(U^B 6) C_1(U^F 12) + \eta C_2(U^{BF} 6) + \delta'' C_2(SU^{BF} 3) \\
&+ \gamma C_2(O^{BF} 3) + \gamma' C_2(SU_s^F 2) + \gamma'' C_2(Spin^{BF} 3), \qquad (3.261)
\end{aligned}
$$

with Casimir operators as defined in Sects. 3.2.3.3 and 3.4.1.3, one finds the eigenvalues

$$
\begin{aligned}
E^{(II_{3a})}(&N_B = N, N_F = 1, [N+1-i,i],(\lambda,\mu),\chi,L,J,M_J) \\
&= E_{03}'' + \eta[(N+1-i)(N+6-i) + i(i+3)] \\
&+ \tfrac{2}{3}\delta'' C(\lambda,\mu) + 2\gamma L(L+1) + 2\gamma'' J(J+1), \quad (3.262)
\end{aligned}
$$

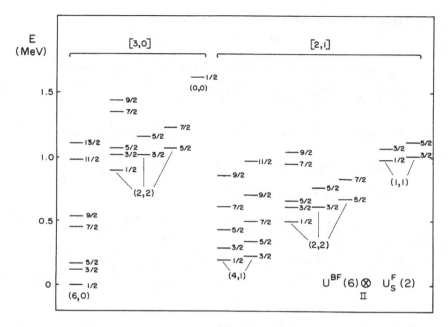

Fig. 3-24   A typical spectrum with $U^{BF}(6) \otimes U^F_s(2)$ (II) symmetry for $N_B = 2$, $N_F = 1$. The $U^{BF}(6)$ quantum numbers are shown on top (square brackets), the $SU^{BF}(3)$ quantum numbers at the bottom and the angular momentum J to the right of each level.

with

$$E''_{03} = e_0 + e_1 N + e_2 N(N + 5) + e_3 + 12e_4 + e_5 N + \tfrac{3}{2}\gamma'. \quad (3.263)$$

The excitation spectrum associated with (3.262) is shown in Fig. 3.24.

3.4.3.4 Wave functions. Isoscalar factors   The wave functions for this chain are obtained by going from a coupled to an uncoupled basis,

$$|[N_1, N_2], (\lambda, \mu), \chi, L\rangle$$
$$= \sum_{\substack{\lambda_B,\mu_B,\chi',L' \\ L''}} \theta^{[N],(\lambda_B,\mu_B),\chi',L';L''}_{[N_1,N_2],(\lambda,\mu),\chi,L} |[N], (\lambda_B, \mu_B), \chi', L'; [1], (2, 0), L''; L\rangle.$$

$$(3.264)$$

Table 3-35 A selected number of $U(6) \supset SU(3)$ isoscalar factors

$$\left\langle \begin{array}{cc} [N] & [1] \\ (\lambda_B, \mu_B) & (2,0) \end{array} \middle| \begin{array}{c} [N_1, N_2] \\ (\lambda, \mu) \end{array} \right\rangle$$

$[N_1, N_2]$	(λ, μ)	(λ_B, μ_B)	Isoscalar factor
$[N+1, 0]$	$(2N+2, 0)$	$(2N, 0)$	1
$[N+1, 0]$	$(2N-2, 2)$	$(2N, 0)$	$\left[\dfrac{2N}{(N+1)(2N-1)}\right]^{\frac{1}{2}}$
$[N+1, 0]$	$(2N-2, 2)$	$(2N-4, 2)$	$\left[\dfrac{(N-1)(2N+1)}{(N+1)(2N-1)}\right]^{\frac{1}{2}}$
$[N, 1]$	$(2N, 1)$	$(2N, 0)$	1
$[N, 1]$	$(2N-2, 2)$	$(2N, 0)$	$\left[\dfrac{(N-1)(2N+1)}{(N+1)(2N-1)}\right]^{\frac{1}{2}}$
$[N, 1]$	$(2N-2, 2)$	$(2N-4, 2)$	$-\left[\dfrac{2N}{(N+1)(2N-1)}\right]^{\frac{1}{2}}$

The θ-coefficients can be written as a product of two isoscalar factors,

$$\theta^{[N],(\lambda_B,\mu_B),\chi',L';L''}_{[N_1,N_2],(\lambda,\mu),\chi,L} = \left\langle \begin{array}{cc} [N] & [1] \\ (\lambda_B, \mu_B) & (2,0) \end{array} \middle| \begin{array}{c} [N_1, N_2] \\ (\lambda, \mu) \end{array} \right\rangle$$
$$\times \left\langle \begin{array}{cc} (\lambda_B, \mu_B) & (2,0) \\ \chi', L' & L'' \end{array} \middle| \begin{array}{c} (\lambda, \mu) \\ \chi, L \end{array} \right\rangle. \quad (3.265)$$

The second factor, associated with $SU(3) \supset O(3)$, is a special case $(n = 2)$ of the coefficient discussed in Sect. 3.4.1.4. The first isoscalar factor in (3.265), associated with $U(6) \supset SU(3)$, can be obtained in the usual way. A summary of the most important cases with $(\lambda, \mu) = (2N+2, 0)$, $(2N, 1)$ and $(2N-2, 2)$ is given in Table 3.35.

The total wave function (3.260) can now be obtained from the expansion (3.264) as

$$|[N_1, N_2], (\lambda, \mu), \chi, L\rangle$$
$$= \sum_{\substack{\lambda_B, \mu_B, \chi', L' \\ L''}} \theta^{[N],(\lambda_B,\mu_B),\chi',L';L''}_{[N_1,N_2],(\lambda,\mu),\chi,L} \sum_j (-)^{L'+L''+1/2+J}$$

$$\times \sqrt{(2L+1)(2j+1)} \begin{Bmatrix} L' & L'' & L \\ 1/2 & J & j \end{Bmatrix} |[N],(\lambda_B,\mu_B),\chi',L';j;J\rangle.$$

$$(3.266)$$

3.4.3.5 Electromagnetic transitions and moments; E2   If the E2 transition operator is taken as (3.236), it is a generator of $\mathrm{SU^{BF}}(3)$, and its matrix elements between the states (3.260) do not depend on labels associated with groups occurring before $\mathrm{SU^{BF}}(3)$ in the chain (3.257). Consequently, all results derived in Sect. 3.4.1.5 are also valid in the $\mathrm{U^{BF}}(6) \otimes \mathrm{U}_s^F(2)$ symmetry.

3.4.3.6 One-nucleon transfer intensities   Expressions for one-nucleon transfer are obtained as in Sect. 3.4.1.7 and lead to the following predictions for the spectroscopic strengths:
(i) even–even to odd–even,

$$S_j([N],(2N,0),\chi'=0,L'=0 \rightarrow [N_1,N_2],(\lambda,\mu),\chi,L,J)$$
$$= (2j+1)\left(\theta_{[N_1,N_2],(\lambda,\mu),\chi,L}^{[N],(2N,0),0,0;L}\right)^2 p_j^2 \delta_{j,J};$$

$$(3.267)$$

(ii) odd–even to even–even,

$$S_j([N_1,N_2],(\lambda,\mu),\chi,L,J \rightarrow [N],(\lambda_B,\mu_B),\chi',L')$$
$$= (2j+1)(2L+1)\begin{Bmatrix} L' & L'' & L \\ 1/2 & J & j \end{Bmatrix}^2 \left(\theta_{[N_1,N_2],(\lambda,\mu),\chi,L}^{[N],(\lambda_B,\mu_B),\chi',L';L''}\right)^2 p_j^2,$$

$$(3.268)$$

with $L'' = 0$ for $j = 1/2$ transfer and $L'' = 2$ for $j = 3/2, 5/2$ transfer.

From these expressions one derives the results shown in Table 3.36.

3.4.3.7 Examples of spectra with $\mathrm{U^{BF}}(6) \otimes \mathrm{U}_s^F(2)$ symmetry   The general situation with regard to $\mathrm{SU^{BF}}(3) \otimes \mathrm{U}_s^F(2)$ symmetries in nuclei was summarized in Sect. 3.4.1.8. It was pointed out there that such a symmetry may occur in the odd-neutron W isotopes and we now turn to a discussion of the relevant experimental data

Table 3-36 One-nucleon transfer reaction intensities in the $U^{BF}(6) \otimes U_s^F(2)$ limit for the reaction $N_B = N, N_F = 0 \to N_B = N, N_F = 1$

Final state[a]	Transf j	Intensity
$|[N+1,0],(2N+2,0),0,0,1/2\rangle$	1/2	$\dfrac{2(2N+3)}{3(2N+1)}p_{1/2}^2$
$|[N+1,0],(2N-2,2),0,0,1/2\rangle$	1/2	$\dfrac{16N^2}{3(N+1)(2N-1)(2N+1)}p_{1/2}^2$
$|[N,1],(2N-2,2),0,0,1/2\rangle$	1/2	$\dfrac{8(N-1)N}{3(N+1)(2N-1)}p_{1/2}^2$
$|[N+1,0],(2N+2,0),0,2,j\rangle$	3/2,5/2	$\dfrac{(2j+1)(2N+3)(2N+5)}{15(N+1)(2N+1)}p_j^2$
$|[N,1],(2N,1),1,2,j\rangle$	3/2,5/2	$\dfrac{(2j+1)(2N+3)}{5(N+1)}p_j^2$
$|[N+1,0],(2N-2,2),0,2,j\rangle$	3/2,5/2	$\dfrac{2(2j+1)N(2N-3)^2(2N+3)}{15(N+1)(2N-1)(2N+1)(4N^2-3)}p_j^2$
$|[N,1],(2N-2,2),0,2,j\rangle$	3/2,5/2	$\dfrac{(2j+1)(N-1)(2N-3)^2(2N+3)}{15(N+1)(2N-1)(4N^2-3)}p_j^2$
$|[N+1,0],(2N-2,2),2,2,j\rangle$	3/2,5/2	$\dfrac{16(2j+1)(N-1)N^2}{5(N+1)(2N-1)(4N^2-3)}p_j^2$
$|[N,1],(2N-2,2),2,2,j\rangle$	3/2,5/2	$\dfrac{8(2j+1)(N-1)^2N(2N+1)}{5(N+1)(2N-1)(4N^2-3)}p_j^2$

[a] Initial state is $|[N],(2N,0),\chi = 0, L = 0\rangle$.

in these nuclei. In the analysis of the experimental data, two different schemes have been used: one in which the odd neutron occupies the orbits $3p_{1/2}$, $3p_{3/2}$ and $2f_{5/2}$ (Warner, 1984) and a second one with a larger space consisting of the orbits $3p_{1/2}$, $3p_{3/2}$, $2f_{5/2}$, $2f_{7/2}$ and $1h_{9/2}$ (Kota, 1986; Bijker and Kota, 1988).

In Fig. 3.25 the experimental spectrum of $^{185}_{74}W_{111}$ is compared with the theoretical one, calculated with (3.262). The calculation describes reasonably well some of the observed bands (Warner and Bruce, 1984), although it does not reproduce all of them. On the other side, the alternative scheme using (3.231) with $n = 4$ ($j = 1/2, 3/2, 5/2, 7/2, 9/2$) fails to describe the low-lying structure of $^{185}_{74}W_{111}$. This is in part due to the neglect in this scheme of the exchange interaction of Sect. 1.4.4 (Scholten and Warner, 1984; Bijker and Scholten, 1985).

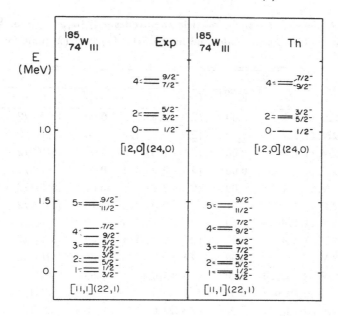

Fig. 3-25   An example of a spectrum with $U^{BF}(6) \otimes U_s^F(2)$ (II) symmetry: $^{185}_{74}W_{111}$ ($N_B = 11$, $N_F = 1$). The theoretical spectrum is calculated using (3.262) with $\eta + 3\delta'' = 43$ KeV, $2\gamma = 18$ KeV and $2\gamma'' = -1$ KeV.

Another test is provided by the one-nucleon transfer reactions or, equivalently, by the single-particle components in the wave functions. In Table 3.37 we compare the effective single-particle amplitudes, as calculated in the Nilsson model to fit the one-neutron transfer data (Casten $et~al.$, 1972), with the ones obtained in $SU^{BF}(3) \otimes U_s^F(2)$ for $n = 2$ and $n = 4$. The Nilsson model (Nilsson, 1955) corresponds to the classical limit of the interacting boson–fermion model and will be briefly discussed in Chapter 6. Here it must be viewed as representing the experimental data. In the Nilsson model the single-particle amplitudes are defined as:

$$C_{jl}^{eff}(\alpha, J = j) = \sum_i b_i(\alpha, J)C_{jl}^i, \qquad (3.269)$$

where the index i stands for the Nilsson quantum numbers $[Nn_z\Lambda]\Omega$ and $b_i(\alpha, J)$ are Coriolis mixing amplitudes for a state with angular momentum J and additional quantum numbers α. In

Table 3-37 The effective single-particle amplitudes $|C_{jl}^{\text{eff}}(\alpha, J = j)|$ for $^{185}_{74}\text{W}_{111}$ calculated in the Nilsson model and in the $\text{SU}^{\text{BF}}(3) \otimes \text{U}_s^{\text{F}}(2)$ limit for $n = 2$ and $n = 4$

	$n = 2$		$n = 4$		Nilsson	
J	$(\lambda,\mu),\chi,L$	$|C_{jl}^{\text{eff}}|$	$(\lambda,\mu),\chi,L$	$|C_{jl}^{\text{eff}}|$	$[Nn_z\Lambda]\Omega$	$|C_{jl}^{\text{eff}}|$
3/2	$(2N,1),1,1$	0.00	$(2N-2,3),1,1$	0.00	$[512]3/2$	0.14
5/2	$(2N,1),1,2$	1.12	$(2N-2,3),1,2$	0.86	$[512]3/2$	1.00
7/2	$(2N,1),1,3$	0.00	$(2N-2,3),1,3$	0.00	$[512]3/2$	0.12
9/2	$(2N,1),1,4$	0.00	$(2N-2,3),1,4$	0.60	$[512]3/2$	0.43
1/2	$(2N,1),1,1$	0.00	$(2N-2,3),1,1$	0.00	$[510]1/2$	0.03
3/2	$(2N,1),1,2$	0.91	$(2N-2,3),1,2$	0.70	$[510]1/2$	0.78
5/2	$(2N,1),1,3$	0.00	$(2N-2,3),1,3$	0.00	$[510]1/2$	0.15
7/2	$(2N,1),1,4$	0.00	$(2N-2,3),1,4$	0.53	$[510]1/2$	0.47
9/2	$(2N,1),1,5$	0.00	$(2N-2,3),1,5$	0.00	$[510]1/2$	0.03
1/2	$(2N+2,0),0,0$	0.60	$(2N,2),0,0$	0.53	$[521]1/2$	0.51
3/2	$(2N+2,0),0,2$	0.57	$(2N,2),0,2$	0.21	$[521]1/2$	0.31
5/2	$(2N+2,0),0,2$	0.70	$(2N,2),0,2$	0.25	$[521]1/2$	0.47
7/2	$(2N+2,0),0,4$	0.00	$(2N,2),0,4$	0.55	$[521]1/2$	0.38
9/2	$(2N+2,0),0,4$	0.00	$(2N,2),0,4$	0.61	$[521]1/2$	0.45

the $\text{SU}^{\text{BF}}(3) \otimes \text{U}_s^{\text{F}}(2)$ limit the effective single-particle amplitudes are given by

$$C_{jl}^{\text{eff}}(\alpha, J = j) = \sqrt{\frac{2j + 1}{2}} \left\langle \begin{array}{cc} (2N,0) & (n,0) \\ 0 & L \end{array} \middle| \begin{array}{c} (\lambda,\mu) \\ \chi, L \end{array} \right\rangle,$$
(3.270)

with $\alpha \equiv (\lambda,\mu),\chi,L$. It is seen from Table 3.37 that the single-particle amplitudes of the $\text{SU}^{\text{BF}}(3) \otimes \text{U}_s^{\text{F}}(2)$ limit for $n = 4$ are in close agreement with the Coriolis-mixed Nilsson calculation. The $n = 2$ results are realistic for $J = 1/2, 3/2$ and $5/2$, but cannot describe $J = 7/2$ and $9/2$ states, since orbits with these quantum numbers are missing from the single-particle space. In view of these difficulties, the use of dynamic symmetries in the $\text{SU}(3)$ limit of the interacting boson–fermion model must be taken with some care, since symmetries can describe the situation only partially. Numerical studies exploiting fully the model space must be done in order to obtain a more complete description of the data.

$$3.4.4 \ \overline{U}^{BF}(6) \otimes \overline{U}^{F}_{s}(2) \ (II_{4})$$

3.4.4.1 Lattice of algebras   One obtains the conjugate symmetry of (3.256) after a particle–hole transformation. The procedure is very similar to the one in Sect. 3.4.2 and will only be briefly discussed in this case. The lattice of algebras is:

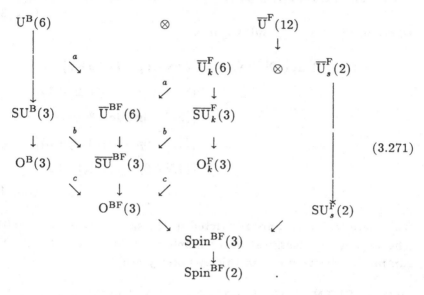

$$(3.271)$$

Generators of the coupled algebras appearing in (3.271) are defined in (3.138) and (3.245).

3.4.4.2 Basis states   The basis states are labelled by

$$\left| \begin{array}{l} U^{B}(6) \otimes \overline{U}^{F}(12) \supset U^{B}(6) \otimes \quad \overline{U}^{F}_{k}(6) \quad \otimes \overline{U}^{F}_{s}(2) \\ \quad \downarrow \qquad \quad \downarrow \qquad \qquad \qquad \qquad \downarrow \\ \quad [N_{B}] \qquad \{\tilde{N}_{F}\} \qquad \qquad \qquad [N'_{1},\ldots,N'_{6}] \\[4pt] \supset \quad \overline{U}^{BF}(6) \quad \otimes \overline{U}^{F}_{s}(2) \supset \overline{SU}^{BF}(3) \otimes \overline{U}^{F}_{s}(2) \\ \qquad \quad \downarrow \qquad \qquad \qquad \qquad \qquad \downarrow \\ \quad [N_{1},\ldots,N_{6}] \qquad \qquad \qquad (\lambda,\mu) \\[4pt] \supset O^{BF}(3) \otimes SU^{F}_{s}(2) \supset Spin^{BF}(3) \supset Spin^{BF}(2) \\ \quad \downarrow \qquad \qquad \downarrow \qquad \qquad \downarrow \qquad \qquad \downarrow \\ \quad \chi, L \qquad \quad S \qquad \qquad J \qquad \qquad M_{J} \end{array} \right\rangle .$$

$$(3.272)$$

When $\bar{N}_F = 1$, one has $[N_1', \ldots, N_6'] = [1, \ldots, 1, 0]$, $S = 1/2$ and $[N_1, \ldots, N_6] = [N + 1 - i, 1, 1, 1, 1, i], i = 0, 1$, as discussed in Sect. 3.2.4.2. The reduction from $\overline{U}^{BF}(6)$ to $\overline{SU}^{BF}(3)$ is thus needed in two cases,

$$[N, 1, 1, 1, 1, 1] \equiv [N - 1] : (\lambda, \mu) = \{(2N - 2 - 6p - 4q, 2q)\}, \tag{3.273}$$

where $p, q = 0, 1, \ldots$, and $\lambda \geq 0$, and

$$
\begin{aligned}
[N + 1, 1, 1, 1, 1, 0] : (\lambda, \mu) &= \{(2N - 6p - 4q, 2q + 2)\}, \\
&\quad \{(2N - 6p - 4q - 1, 2q + 1)\}, \\
&\quad \{(2N - 6p - 4q - 2, 2q)\}', \\
&\quad \{(2N - 6p - 4q - 3, 2q + 2)\}, \\
&\quad \{(2N - 6p - 4q - 4, 2q + 1)\}',
\end{aligned}
\tag{3.274}
$$

where $p, q = 0, 1, \ldots$, are restricted in the same way as in (3.259). The remaining classification of labels is as in Sect. 3.4.1.2. To summarize, basis states in this symmetry are

$$|[N_B = N], \{\bar{N}_F = 1\}, [N + 1 - i, 1, 1, 1, 1, i], (\lambda, \mu), \chi, L, J, M_J). \tag{3.275}$$

3.4.4.3 Energy eigenvalues   Writing the Hamiltonian as

$$
\begin{aligned}
H^{(II_{4a})} &= e_0 + e_1 C_1(U^B 6) + e_2 C_2(U^B 6) + e_3 C_1(\overline{U}^F 12) + e_4 C_2(\overline{U}^F 12) \\
&\quad + e_5 C_1(U^B 6) C_1(\overline{U}^F 12) + \eta C_2(\overline{U}^{BF} 6) + \delta'' C_2(\overline{SU}^{BF} 3) \\
&\quad + \gamma C_2(O^{BF} 3) + \gamma' C_2(SU_s^F 2) + \gamma'' C_2(Spin^{BF} 3), \tag{3.276}
\end{aligned}
$$

one finds the eigenvalues

$$
\begin{aligned}
E^{(II_{4a})}&(N_B = N, \bar{N}_F = 1, [N + 1 - i, 1, 1, 1, 1, i], (\lambda, \mu), \chi, L, J, M_J) \\
&= E_{04}'' + \eta[(N + 1 - i)(N + 6 - i) + (i - 4)(i - 1)] \\
&\quad + \tfrac{2}{3}\delta'' C(\lambda, \mu) + 2\gamma L(L + 1) + 2\gamma'' J(J + 1), \tag{3.277}
\end{aligned}
$$

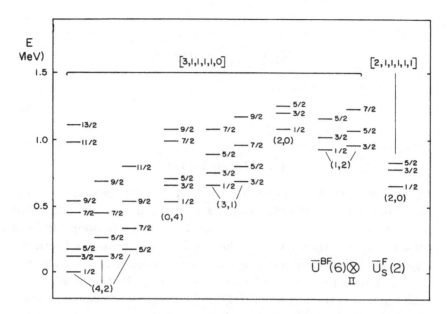

Fig. 3-26   A typical spectrum with $\overline{U}^{BF}(6) \otimes \overline{U}_s^F(2)$ (II) symmetry for $N_B = 2, N_F = 1$. The $\overline{U}^{BF}(6)$ quantum numbers are shown on top (square brackets), the $\overline{SU}^{BF}(3)$ quantum numbers at the bottom and the angular momentum J to the right of each level.

with

$$E_{03}'' = e_0 + e_1 N + e_2 N(N+5) + 11e_3 + 22e_4 + 11e_5 N + \tfrac{3}{2}\gamma'. \quad (3.278)$$

The excitation spectrum of (3.277) is shown in Fig. 3.26.

This concludes our description of coupled Bose–Fermi algebras and dynamic symmetries. In view of the diversity of single-particle orbits that one encounters in nuclei, several other cases can be (and have been) constructed. The methods developed in this chapter provide a way to derive results of interest for all cases.

4

Superalgebras

4.1 Introduction

In the previous chapters we have considered the fermion algebras constructed from bilinear products of fermion creation and annihilation operators, and the combined boson plus fermion algebras obtained by taking the sum of the generators of boson and fermion algebras. These combined algebras are particularly useful in the description of properties of odd–even nuclei. In this description each nucleus is treated separately and the values of the parameters are obtained by comparison with individual nuclei. In view of the large number of nuclear species that are observed experimentally, it is of interest to attempt a classification in which several nuclei are described simultaneously. Since some of these contain an even number of protons and neutrons and some an odd number, the mathematical construct one needs for a simultaneous description is more complicated than that of Lie algebras. This mathematical construction (super or graded Lie algebras) was originally developed for applications to elementary particle physics (Miyazawa, 1966; Ramond, 1971), but has found in nuclear physics its most extensive use. In this chapter, the basic ingredients of the approach will be discussed and some applications presented. Overviews of the theory of superalgebras and its application to the interacting boson model have been given by Balantekin (1982), Jolie (1986) and Vervier (1987).

4.2 Graded Lie algebras

Consider a set of operators X and Y, called bosonic and fermionic respectively, satisfying the commutation relations

$$[X_a, X_b] = \sum_c c^c_{ab} X_c,$$

$$[X_a, Y_b] = \sum_c d^c_{ab} Y_c,$$

$$\{Y_a, Y_b\} = \sum_c f^c_{ab} X_c, \qquad (4.1)$$

together with the Jacobi identities

$$[[X_a, X_b], X_c] + [[X_b, X_c], X_a] + [[X_c, X_a], X_b] = 0,$$
$$[X_a, [X_b, Y_c]] + [X_b, [Y_c, X_a]] + [Y_c, [X_a, X_b]] = 0,$$
$$[X_a, \{Y_b, Y_c\}] + \{Y_b, [Y_c, X_a]\} - \{Y_c, [X_a, Y_b]\} = 0,$$
$$[Y_a, \{Y_b, Y_c\}] + [Y_b, \{Y_c, Y_a\}] + [Y_c, \{Y_a, Y_b\}] = 0. \qquad (4.2)$$

In these equations, the curly brackets denote anticommutation,

$$\{A, B\} = AB + BA. \qquad (4.3)$$

The set X, Y is said to form a graded Lie algebra g^*. The quantities c^c_{ab}, d^c_{ab} and f^c_{ab} in (4.1) are called graded structure constants.

Associated with each graded algebra, g^*, there is a graded Lie group, G^*. In applications in nuclear physics, one makes use only of graded algebras and not of groups. Following common use, we shall not distinguish between a graded algebra and its corresponding group and shall denote both of them by capital letters. All 'classical' graded Lie algebras have been classified by Kac (1975; 1977) in a manner similar to that used by Cartan (1894) in the classification of 'normal' Lie algebras. The classification is given in Table 4.1, where both the commonly used labels and the Kac labels are given. This, together with Table 2.1 of Volume 1, provides a complete classification of algebraic constructs that can be used in nuclear structure physics.

Graded Lie algebras can be realized in terms of creation and annihilation operators for bosons $b^\dagger_\alpha, b_\alpha$ ($\alpha = 1, \ldots, n$) and fermions $a^\dagger_i, a_i$ ($i = 1, \ldots, m$). Consider, for example, the bilinear products

$$\begin{aligned}
G^B_{\alpha\beta} &= b^\dagger_\alpha b_\beta, & (\alpha, \beta &= 1, \ldots, n), \\
G^F_{ij} &= a^\dagger_i a_j, & (i, j &= 1, \ldots, m), \\
F^\dagger_{i\alpha} &= a^\dagger_i b_\alpha, & (i &= 1, \ldots, m; \alpha = 1, \ldots, n), \\
F_{\alpha i} &= b^\dagger_\alpha a_i, & (\alpha &= 1, \ldots, n; i = 1, \ldots, m). \quad (4.4)
\end{aligned}$$

Table 4-1 Admissible graded Lie algebras

Name	Label	Kac label
[Special]Unitary	[S]U($n|m$)	A(n, m)
Orthosymplectic	Osp($n|2m$), n = odd	B(($n-1$)/2, m)
Orthosymplectic	Osp($n|2m$), n = even	B($n/2, m$)
Orthosymplectic	Osp($1|2n$)	C[n]
Others	A[n],P[n],F[4]	A[n],P[n],F[4]
Exceptional	G[3]	G[3]
Other	D[1,2;α]	D[1,2;α]

There are n^2 operators G^{B}, m^2 operators G^{F}, mn operators $F^\dagger$ and mn operators F. The $(m+n)^2$ operators together satisfy commutation relations appropriate to the graded Lie algebra U($n|m$). The operators G are of bosonic nature (X-type), while the operators F are of fermionic nature (Y-type). A superscript B or F has been attached to the operators G in order to indicate that they are constructed with boson and fermion creation or annihilation operators. It is customary to write the operators G and F in a matrix form,

$$G^* = \left(\begin{array}{c|c} b_\alpha^\dagger b_\beta & b_\alpha^\dagger a_i \\ \hline a_i^\dagger b_\alpha & a_i^\dagger a_j \end{array} \right). \tag{4.5}$$

This form distinguishes clearly the Bose sector, $b_\alpha^\dagger b_\beta$ and $a_i^\dagger a_j$, from the Fermi sector, $b_\alpha^\dagger a_i$ and $a_i^\dagger b_\alpha$.

4.3 Subalgebras

A subset of operators of G^*, closed with respect to commutation, forms a subalgebra of G^*. For applications to nuclear physics, there are two types of subalgebras

$$G^* \supset G'^* \supset G''^* \supset \cdots \tag{4.6}$$

of particular importance. In the first type, the subalgebra G'^* is already a 'normal' Lie algebra formed by the Bose sector of G^*.

In particular, for the unitary superalgebra $U(n|m)$,

$$U(n|m) \supset U^B(n) \otimes U^F(m), \tag{4.7}$$

where again a superscript B and F has been attached to the letter U to indicate that the first group refers to bosons and the second to fermions. Furthermore, as discussed in Sect. 2.5.2, the product sign $\otimes$ is used in (4.7) instead of the plus sign $\oplus$, to conform with common practice.

The second type is that in which the subalgebra G'^* is still a graded algebra. A particularly interesting case here is (Morrison and Jarvis, 1985)

$$U(n|m) \supset OSp(n|m), \qquad m = \text{even}. \tag{4.8}$$

As in the case of normal Lie algebras, the search for appropriate subalgebras is simplified by the classification of Table 4.1 and Table 2.1 of Volume 1. Here again, the letter S, denoting special transformations, will be deleted except for the unitary algebras for which it is essential.

4.4 Representations of superalgebras

Superalgebras have different classes of representations. The representations of graded Lie unitary algebras can be characterized by a set of integers. It is convenient to introduce a Young supertableau,

$$[\nu_1, \nu_2, \ldots, \nu_{n+m}\} = \overbrace{\boxtimes \boxtimes \boxtimes \cdots \boxtimes}^{\nu_1}$$
$$\overbrace{\boxtimes \boxtimes \cdots \boxtimes}^{\nu_2}$$
$$\vdots$$
$$\overbrace{\boxtimes \cdots \boxtimes}^{\nu_{n+m}}, \tag{4.9}$$

where boxes have been crossed to distinguish (4.9) from the ordinary Young tableau. Particularly important is the totally supersymmetric tableau,

$$[\mathcal{N}\} = \overbrace{\boxtimes \boxtimes \cdots \boxtimes}^{\mathcal{N}}. \tag{4.10}$$

This tableau looks similar to the totally symmetric tableau of normal Lie algebras but has a different meaning. It is characterized by the total number of bosons plus fermions, $\mathcal{N} = N_{\mathrm{B}} + N_{\mathrm{F}}$, and it implies that all the bosonic indices are symmetrized and the fermionic indices are antisymmetrized.

A crucial problem in applications is to construct the representations of the subalgebra G''^* contained in a given representation of the algebra G^*. For the interesting case (4.7), the decomposition is given by Balantekin and Bars (1981):

$$
\overbrace{\boxtimes\boxtimes\cdots\boxtimes}^{\mathcal{N}} \; : \; \left(\overbrace{\square\square\cdots\square}^{\mathcal{N}} \otimes 1 \right)
$$

$$
\oplus \left(\overbrace{\square\square\cdots\square}^{\mathcal{N}-1} \otimes \square \right)
$$

$$
\vdots
$$

$$
\oplus \left(\overbrace{\square\square\cdots\square}^{\mathcal{N}-k} \otimes \left.\begin{array}{c}\square\\\square\\\vdots\\\square\end{array}\right\} k \right)
$$

$$
\oplus \cdots , \tag{4.11}
$$

where the series stops either when $k = m$ or $\mathcal{N} - k = 0$, whichever comes first.

4.5 Dynamic supersymmetries

Dynamic supersymmetries arise when the Hamiltonian operator can be written in terms only of Casimir operators, $\mathcal{C}$, of a chain of superalgebras (4.6),

$$
H = \alpha^* \mathcal{C}(G^*) + \alpha'^* \mathcal{C}(G'^*) + \cdots . \tag{4.12}
$$

The importance of dynamic supersymmetries is that, since the Casimir operators $\mathcal{C}$ are diagonal in the basis given by chain (4.6), they provide a solution to the eigenvalues of H,

$$
E = \alpha^* \langle \mathcal{C}(G^*) \rangle + \alpha'^* \langle \mathcal{C}(G'^*) \rangle + \cdots , \tag{4.13}
$$

Table 4-2 Eigenvalues of some Casimir operators of superalgebras

Algebra	Representation	order	$\langle \mathcal{C} \rangle$
$U(n\|m)$	$[\mathcal{N}\}$	1	$\mathcal{N}$
		2	$\mathcal{N}(\mathcal{N} + n - m - 1)$
$SU(n\|m)$	$[\mathcal{N}\}$	2	$\dfrac{\mathcal{N}(n - m - 1)(n - m + \mathcal{N})}{n - m}$

where $\langle \mathcal{C}(G^*) \rangle$ denotes the expectation value of $\mathcal{C}$ in the appropriate representation of G^*. The eigenvalues of some Casimir operators of Lie algebras are given in Table 2.8 of Volume 1. For graded algebras, not all eigenvalues have been evaluated. In analogy with normal Lie algebras, graded algebras have Casimir invariants of various orders, p, denoted here by $C_p(G^*)$. For applications in nuclear physics one is interested in the eigenvalues of the linear and quadratic invariants of the unitary superalgebras in the totally supersymmetric representation $[\mathcal{N}\}$. These are given in Table 4.2. Note that the expression for the eigenvalue of $C_2(SU(n|m))$ fails for $n = m$. The expressions in Table 4.2 reduce to those given in Volume 1 in the limit in which the graded algebra becomes a normal Lie algebra.

4.6   Classification of dynamic supersymmetries

In nuclear structure physics a dynamic supersymmetry occurs whenever a set of nuclei can be assigned to single representation of a superalgebra. As discussed in Chapter 1, the collective degrees of freedom span a six-dimensional bosonic space, while the single-particle degrees of freedom span a fermionic space of dimension $\Omega = \sum_i (2j_i + 1)$. The superalgebras appropriate to describe nuclear states built from these degrees of freedom are the unitary superalgebras $U(6|\Omega)$. Since the wave functions must be totally symmetric in the bosonic indices and totally antisymmetric in the fermionic indices, one needs to consider the totally supersymmetric representations $[\mathcal{N}\}$ of $U(6|\Omega)$. The superalgebra

Table 4-3 Partial classification of supersymmetries used in nuclear physics

Supersymmetry	Chain	Bose–Fermi symmetry	Experimental examples
(i) Supersymmetries associated with $O^B(6)$			
$U(6|4)$	(III_1)	$Spin^{BF}(6)$	Os–Ir–Pt–Au
$U(6|2)$	(III_2)	$Spin^{BF}(3)$	
$U(6|12)$	(III_3)	$U^{BF}(6) \otimes U_s^F(2)$	Pt
$U(6|\overline{12})$	(III_4)	$\overline{U}^{BF}(6) \otimes \overline{U}_s^F(2)$	
$U(6|12)$	(III_5)	$Spin'^{BF}(5) \otimes SU_{k'}^F(3)$	
(ii) Supersymmetries associated with $U^B(5)$			
$U(6|4)$	(I_1)	$Spin^{BF}(5)$	Cu–Zn
$U(6|2)$	(I_2)	$Spin^{BF}(3)$	Ru–Rh–Pd–Ag
$U(6|12)$	(I_3)	$U^{BF}(6) \otimes U_s^F(2)$	Se–As
$U(6|\overline{12})$	(I_4)	$\overline{U}^{BF}(6) \otimes \overline{U}_s^F(2)$	
(iii) Supersymmetries associated with $SU^B(3)$			
$U(6|2m)$	(II_1)	$SU^{BF}(3) \otimes U_s^F(2)$	Er–Tm
$U(6|\overline{2m})$	(II_2)	$\overline{SU}^{BF}(3) \otimes \overline{U}_s^F(2)$	
$U(6|12)$	(II_3)	$U^{BF}(6) \otimes U_s^F(2)$	W
$U(6|\overline{12})$	(II_4)	$\overline{U}^{BF}(6) \otimes \overline{U}_s^F(2)$	

$U(6|\Omega)$ is broken down either directly, Eq. (4.7), or at a subsequent step, into normal Lie algebras which are then combined as in the previous chapters. Consequently, the study of supersymmetry in nuclei has two parts. The first one is identical to the study of Bose–Fermi symmetries. In the second part, one attempts to classify a set of nuclei within a single representation $[\mathcal{N}\}$ of $U(6|\Omega)$. This presupposes that each individual nucleus is well described by a Bose–Fermi symmetry and, in addition, parameters appearing in the Hamiltonian and other operators must be identical for all nuclei belonging to the representation $[\mathcal{N}\}$. Several supersymmetric schemes have been employed in the classification of nuclear states. A partial list is shown in Table 4.3 (Vervier, 1987). Others will be mentioned at the end of this chapter. Three examples of supersymmetric schemes will be discussed in detail.

4.7   U(6|4) (III₁)

A simple (and much studied) case is that in which the fermionic space is composed by a single level with $j = 3/2$. In this case, $\Omega = 4$, and one considers the superalgebra U(6|4).

4.7.1 Supermultiplets

In order to study the occurrence of supersymmetries, one must first identify the set of nuclei belonging to the representation $[\mathcal{N}\}$. When $G^* \equiv U(6|4)$, Eq. (4.11) shows that for $\mathcal{N} \geq 4$ there are five nuclei belonging to this set. Denoting the quantum numbers associated with the groups in (4.7) by

$$\left| \begin{array}{ccc} U(6|4) & \supset \quad U^B(6) & \otimes \quad U^F(4) \\ \downarrow & \downarrow & \downarrow \\ [\mathcal{N}\} & [N_B] & \{N_F\} \end{array} \right\rangle , \qquad (4.14)$$

the five nuclei are characterized by

$$N_B = \mathcal{N}, N_F = 0,$$
$$N_B = \mathcal{N} - 1, N_F = 1,$$
$$N_B = \mathcal{N} - 2, N_F = 2,$$
$$N_B = \mathcal{N} - 3, N_F = 3,$$
$$N_B = \mathcal{N} - 4, N_F = 4, \qquad (\mathcal{N} \geq 4). \qquad (4.15)$$

These nuclei are alternately even–even and odd–even and have an increasing number of unpaired fermions. States with $N_F = 0$ and $N_F = 1$ are the lowest states of the corresponding nuclei, while states with $N_F \geq 2$ are at higher energies. The set (4.15) forms a supermultiplet. Possible supermultiplets in the Os–Pt–Hg region are shown in Fig. 4.1. One such family comprises $^{190}_{76}\text{Os}_{114}$ ($N_B = 9, N_F = 0$), $^{191}_{77}\text{Ir}_{114}$ ($N_B = 8, N_F = 1$), $^{192}_{78}\text{Pt}^*_{114}$ ($N_B = 7, N_F = 2$), $^{193}_{79}\text{Au}^*_{114}$ ($N_B = 6, N_F = 3$) and $^{194}_{80}\text{Hg}^{**}_{114}$ ($N_B = 5, N_F = 4$). The star indicates that the states are excited configurations in the corresponding nucleus. The boson number N_B for each nucleus is counted as discussed in Volume 1.

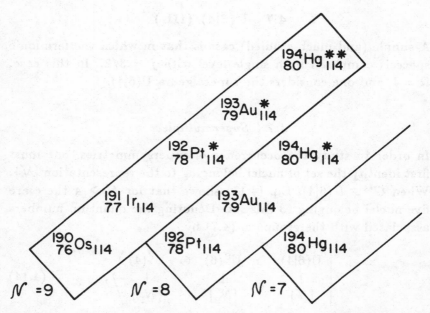

Fig. 4-1   U(6|4) supermultiplets in the Os–Pt–Hg region. Each multiplet is identified by the total number of bosons plus fermions, $\mathcal{N}$.

4.7.2 Energy eigenvalues

The complete classification scheme for the nuclei in the supermultiplets of Fig. 4.1 depends on the further breaking of $U^B(6) \otimes U^F(4)$ into subalgebras. Individual nuclei in the neutron-deficient Os–Ir region are well described by the chain III_1 of Chapter 3. The complete classification scheme can thus be obtained by embedding the chain (3.3) into U(6|4),

$$
\left.
\begin{array}{l}
\text{U}(6|4) \supset \text{U}^B(6) \ \otimes \ \text{U}^F(4) \supset \text{O}^B(6) \ \otimes \ \text{SU}^F(4) \\
\quad \downarrow \qquad\qquad \downarrow \qquad\qquad \downarrow \qquad\qquad \downarrow \\
\quad [\mathcal{N}] \qquad\quad [N_B] \qquad\quad \{N_F\} \qquad\quad \Sigma \\
\ \\
\supset \text{Spin}^{BF}(6) \supset \text{Spin}^{BF}(5) \supset \text{Spin}^{BF}(3) \supset \text{Spin}^{BF}(2) \\
\qquad \downarrow \qquad\qquad\quad \downarrow \qquad\qquad\quad \downarrow \qquad\qquad\quad \downarrow \\
\ (\sigma_1, \sigma_2, \sigma_3) \qquad (\tau_1, \tau_2) \qquad\quad \nu_\Delta, J \qquad\qquad M_J
\end{array}
\right\} . \quad (4.16)
$$

The values of N_B and N_F contained in a given $\mathcal{N}$ are obtained from (4.15). The values of Σ are given by Eq. (2.50) of Volume 1,

$$[N]: \Sigma = N_B - 2t; \quad t = 0, 1, \ldots, \frac{N_B}{2} \text{ or } \frac{N_B - 1}{2}, \tag{4.17}$$

$(N_B = \text{even or odd}).$

The next step involves the combination of $O^B(6)$ and $SU^F(4)$. This can be done as in Sect. 3.2.1.2 where it was explicitly carried out for $N_F = 1$. When $N_F = 0$, the result is trivial since in that case $\sigma_1 = \Sigma$, $\sigma_2 = 0$ and $\sigma_3 = 0$. The complete result in Spin(6) notation is

$$
\begin{aligned}
N_F = 0 &: (\sigma_1 = \mathcal{N} - 2t, \sigma_2 = 0, \sigma_3 = 0); \\
N_F = 1 &: (\sigma_1 = \mathcal{N} - 2t - \tfrac{1}{2}, \sigma_2 = \tfrac{1}{2}, \sigma_3 = \tfrac{1}{2}), \\
&\quad (\sigma_1 = \mathcal{N} - 2t - \tfrac{3}{2}, \sigma_2 = \tfrac{1}{2}, \sigma_3 = -\tfrac{1}{2}); \\
N_F = 2 &: (\sigma_1 = \mathcal{N} - 2t - 1, \sigma_2 = 0, \sigma_3 = 0), \\
&\quad (\sigma_1 = \mathcal{N} - 2t - 2, \sigma_2 = 1, \sigma_3 = 0), \\
&\quad (\sigma_1 = \mathcal{N} - 2t - 3, \sigma_2 = 0, \sigma_3 = 0); \\
N_F = 3 &: (\sigma_1 = \mathcal{N} - 2t - \tfrac{5}{2}, \sigma_2 = \tfrac{1}{2}, \sigma_3 = -\tfrac{1}{2}), \\
&\quad (\sigma_1 = \mathcal{N} - 2t - \tfrac{7}{2}, \sigma_2 = \tfrac{1}{2}, \sigma_3 = \tfrac{1}{2}); \\
N_F = 4 &: (\sigma_1 = \mathcal{N} - 2t - 4, \sigma_2 = 0, \sigma_3 = 0). \tag{4.18}
\end{aligned}
$$

The representations $(\sigma_1, \tfrac{1}{2}, \tfrac{1}{2})$ and $(\sigma_1, \tfrac{1}{2}, -\tfrac{1}{2})$ are, as mentioned in Sect. 2.6, equivalent and it is sufficient to label them by $(\sigma_1, \tfrac{1}{2}, \tfrac{1}{2})$. The further reduction from $\text{Spin}^{BF}(6)$ is as in Sect. 3.2.1.2.

Energy eigenvalues can be obtained by adding to (3.15) the Casimir operators of the superalgebra $U(6|4)$,

$$
\begin{aligned}
H'^{(III_{1a})} = \ &e_0 + e_6 \mathcal{C}_1(U(6|4)) + e_7 \mathcal{C}_2(U(6|4)) + e_1 \mathcal{C}_1(U^B 6) \\
&+ e_2 \mathcal{C}_2(U^B 6) + e_3 \mathcal{C}_1(U^F 4) + e_4 \mathcal{C}_2(U^F 4) \\
&+ e_5 \mathcal{C}_1(U^B 6)\mathcal{C}_1(U^F 4) + \eta \mathcal{C}_2(O^B 6) + \eta' \mathcal{C}_2(\text{Spin}^{BF} 6) \\
&+ \beta \mathcal{C}_2(\text{Spin}^{BF} 5) + \gamma \mathcal{C}_2(\text{Spin}^{BF} 3). \tag{4.19}
\end{aligned}
$$

This Hamiltonian is diagonal in the basis (4.16) with eigenvalues

$$
\begin{aligned}
E'^{(\mathrm{III}_{1a})}&(\mathcal{N}, N_B, N_F, \Sigma, (\sigma_1, \sigma_2, \sigma_3), (\tau_1, \tau_2), \nu_\Delta, J, M_J) \\
&= e_0 + e_6\mathcal{N} + e_7\mathcal{N}(\mathcal{N}+1) + e_1 N_B + e_2 N_B(N_B+5) \\
&\quad + e_3 N_F + e_4 N_F(5 - N_F) + e_5 N_B N_F \\
&\quad + 2\eta\Sigma(\Sigma+4) + 2\eta'[\sigma_1(\sigma_1+4) + \sigma_2(\sigma_2+2) + \sigma_3^2] \\
&\quad + 2\beta[\tau_1(\tau_1+3) + \tau_2(\tau_2+1)] + 2\gamma J(J+1). \qquad (4.20)
\end{aligned}
$$

The expression (4.20) is obtained using Table 4.2 and Table 2.8 of Volume 1.

4.7.3 Tests of U(6|4) supersymmetry

4.7.3.1 Excitation energies Excitations energies are given by the last four terms in (4.20)

$$
\begin{aligned}
E_{\mathrm{exc}} &= 2\eta\Sigma(\Sigma+4) + 2\eta'[\sigma_1(\sigma_1+4) + \sigma_2(\sigma_2+2) + \sigma_3^2] \\
&\quad + 2\beta[\tau_1(\tau_1+3) + \tau_2(\tau_2+1)] + 2\gamma J(J+1). \qquad (4.21)
\end{aligned}
$$

The extent to which excitation energies in all nuclei belonging to a supermultiplet can be described by this single formula is a test of supersymmetry. In order for the test to be meaningful, at least two nuclei, one even–even and one odd–even, must be measured. Experimental evidence shows that at least ten nuclei in the Os–Pt region, five even and five odd, have experimental spectra that can be described by (4.21). These are given in Table 4.4. The experimental spectra of the first pair of nuclei are shown in Fig. 4.2. The corresponding spectra expected on the basis of the supersymmetry are shown in Fig. 4.3. States in the even–even and odd–even nuclei can be approximately described with the same parameters, $2\eta = -15$ KeV, $2\eta' = -28$ KeV, $2\beta = 30$ KeV and $2\gamma = 10$ KeV. Note that all states in Figs. 4.2 and 4.3 belong to the maximum allowed representations of Spin(6). In odd–even nuclei, according to the classification scheme of the previous section, one expects a splitting of the representation $(\Sigma, 0, 0)$, observed in even–even nuclei, into two representations with $(\Sigma - \frac{1}{2}, \frac{1}{2}, \frac{1}{2})$ and $(\Sigma - \frac{3}{2}, \frac{1}{2}, -\frac{1}{2})$.

Table 4-4 U(6|4) multiplets

$\mathcal{N}$	N_B	N_F	Nucleus
9	9	0	$^{190}_{76}\text{Os}_{114}$
	8	1	$^{191}_{77}\text{Ir}_{114}$
8	8	0	$^{192}_{76}\text{Os}_{116}$
	7	1	$^{193}_{77}\text{Ir}_{116}$
8	8	0	$^{192}_{78}\text{Pt}_{114}$
	7	1	$^{193}_{79}\text{Au}_{114}$
7	7	0	$^{194}_{78}\text{Pt}_{116}$
	6	1	$^{195}_{79}\text{Au}_{116}$
6	6	0	$^{196}_{78}\text{Pt}_{118}$
	5	1	$^{197}_{79}\text{Au}_{118}$

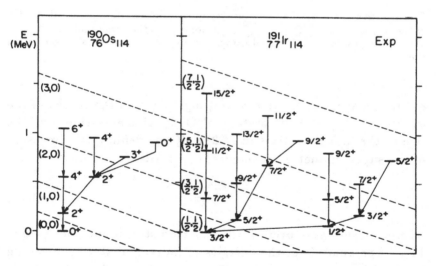

Fig. 4-2 An example of U(6|4) supersymmetry in nuclei: experimental spectra of the pair of nuclei $^{190}_{76}\text{Os}_{114}$ – $^{191}_{77}\text{Ir}_{114}$ belonging to the supermultiplet $\mathcal{N} = 9$.

This splitting is similar to the spin–orbit splitting obtained by coupling spin $s = 1/2$ to an orbital angular momentum, L. Some of the states in $^{191}_{77}\text{Ir}_{114}$ at energies above 500 KeV might belong to the excited configuration with $(\sigma_1, \sigma_2, |\sigma_3|) = (\frac{15}{2}, \frac{1}{2}, \frac{1}{2})$. The

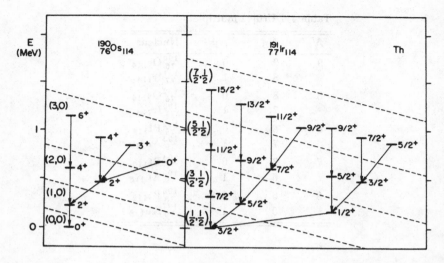

Fig. 4-3 An example of U(6|4) supersymmetry in nuclei: theoretical spectra of the pair of nuclei $^{190}_{76}\text{Os}_{114} - ^{191}_{77}\text{Ir}_{114}$ belonging to the supermultiplet $\mathcal{N} = 9$. The energy levels are calculated using (4.20) with parameters as described in the text.

excitation energies of these states, together with the energy of the $\sigma_1 = 9$, $\tau_1 = 0$, $J^P = 0^+$ state in $^{190}_{76}\text{Os}_{114}$, give an estimate of the size of the first two terms in (4.21). If one defines, as a measure of the supersymmetry breaking, the quantity

$$\phi = \frac{\sum_i |E_i^{\text{th}} - E_i^{\text{exp}}|}{\sum_i E_i^{\text{exp}}}, \qquad (4.22)$$

where the sum goes over all the observed states, one obtains $\phi = 14\%$. The analysis of the energies of individual states, including those with $\sigma_1 = \frac{15}{2}$, is shown in Table 4.5, adapted from Balantekin et al. (1981). A similar analysis for other supermultiplets shows that the breaking of supersymmetry is of the order of $\phi \approx 20\%$ in this region.

An important question is the location of the remaining members of the supermultiplet. From Fig. 4.1 one observes that the same multiplet to which $^{190}_{76}\text{Os}_{114}$ and $^{191}_{77}\text{Ir}_{114}$ belong, also contains $^{192}_{78}\text{Pt}^*_{114}$, $^{193}_{79}\text{Au}^*_{114}$ and $^{194}_{80}\text{Hg}^{**}_{114}$. Excited configurations with two or more unpaired fermions are very difficult to measure in view

Table 4-5 Comparison between experimental and calculated energies in $^{190}_{76}$Os$_{114}$ and $^{191}_{77}$Ir$_{114}$

Nucleus	σ_1	τ_1	J^P	E_{exp}(KeV)	E_{th}(KeV)	Δ(KeV)
$^{190}_{76}$Os$_{114}$	9	0	0^+	0	0	0
		1	2^+	187	180	-7
		2	2^+	558	360	-198
			4^+	548	500	-48
		3	0^+	912	540	-372
			3^+	756	660	-96
			4^+	955	740	-215
			6^+	1050	960	-90
		4	2^+	1115	900	-215
			4^+	1163	1040	-123
			5^+	1204	1140	-64
			6^+	1474	1260	-214
			8^+	1667	1560	-107
	7	0	0^+	1733	1720	-13
$^{191}_{77}$Ir$_{114}$	$\frac{17}{2}$	$\frac{1}{2}$ $\frac{3}{2}$	$3/2^+$	0	0	0
			$1/2^+$	82	120	$+38$
			$5/2^+$	129	200	$+71$
			$7/2^+$	343	270	-73
		$\frac{5}{2}$	$3/2^+$	179	360	$+181$
			$5/2^+$	351	410	$+59$
			$7/2^+$	686	480	-206
			$9/2^+$	502	570	$+68$
			$11/2^+$	832	680	-152
		$\frac{7}{2}$	$5/2^+$	588	680	$+92$
			$7/2^+$	504	750	$+246$
			$9/2^+$	812	840	$+28$
			$9/2^+$	946	840	-106
			$11/2^+$	1207	950	-257
			$13/2^+$	1004	1080	$+76$
			$15/2^+$	1418	1230	-188
		$\frac{9}{2}$	$11/2^+$	991	1280	$+289$
			$13/2^+$	1397	1410	$+13$
			$17/2^+$	1599	1730	$+131$
			$19/2^+$	2112	1920	-192
		$\frac{11}{2}$	$21/2^+$	2311	2520	$+209$
	$\frac{15}{2}$	$\frac{1}{2}$ $\frac{3}{2}$	$3/2^+$	539	560	$+21$
			$1/2^+$	624	680	$+56$
			$5/2^+$	748	760	$+12$

of the high density of states in that energy region. Preliminary experimental results indicate that the state with quantum numbers $\Sigma = \mathcal{N} - 2$, $(\sigma_1, \sigma_2, \sigma_3) = (\mathcal{N} - 1, 0, 0)$, $(\tau_1, \tau_2) = (0, 0)$ and $J = 0$ of the configuration with $N_B = \mathcal{N} - 2$ bosons and $N_F = 2$ fermions is located at $E^* = 1542$ KeV in $^{192}_{78}\text{Pt}_{114}$ and $E^* = 1479$ KeV in $^{194}_{78}\text{Pt}_{116}$.

4.7.3.2 Binding energies

A direct test of supersymmetry for energies would require a measurement of the binding energies of all five members of the supermultiplet. In fact, one can see from (4.20) that even after eliminating two coefficients by using the condition $\mathcal{N} = N_B + N_F$, there remain, for fixed $\mathcal{N}$, four independent parameters, in addition to 2η, $2\eta'$, 2β and 2γ, previously determined from the excitation energies. Since the binding energies of the excited configurations, E^*, have not been measured, this direct test is, at present, not possible. A relatively simple test of the supersymmetry scheme for binding energies is provided by the study of two-neutron separation energies. For even–even nuclei these were defined and discussed in Sect. 2.7 of Volume 1. Here, one can define the two-nucleon separation energies for even–even and odd–even nuclei as

$$S_2(\mathcal{N}, N_F = 0) = E_B(\mathcal{N} + 1, N_F = 0) - E_B(\mathcal{N}, N_F = 0),$$
$$S_2(\mathcal{N}, N_F = 1) = E_B(\mathcal{N} + 1, N_F = 1) - E_B(\mathcal{N}, N_F = 1), \quad (4.23)$$

where E_B denotes the binding energy. Using (4.20), one can show that, if supersymmetry applies and furthermore, if the coefficients 2η, $2\eta'$, 2β and 2γ are independent of $\mathcal{N}$, the separation energies in even–even and odd–even nuclei should be linear functions of $\mathcal{N}$ with identical slope, given by $2(e_2 + e_7 + 2\eta + 2\eta')$. The experimental situation in the Os and Ir isotopes is shown in Fig. 4.4. The quantity $(e_2 + e_7)$ extracted from this figure is approximately 325 KeV.

4.7.3.3 Electromagnetic transitions and moments; E2

Besides energies, dynamic supersymmetries make also definite predictions for other quantities, for example rates of electromagnetic transitions and moments. The study of these quantities within a

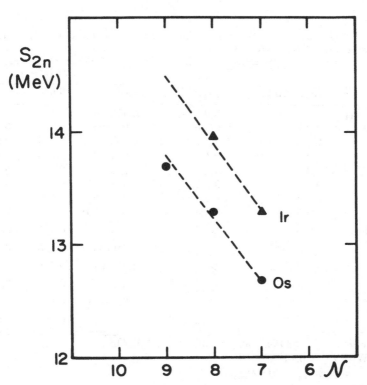

Fig. 4-4   Two-neutron separation energies in $_{76}$Os and $_{77}$Ir as a function of $\mathcal{N}$.

supersymmetric scheme is similar to that of the energies. Supersymmetry implies that all transitions in nuclei belonging to the same supermultiplet be described by the same operator. A particularly important case is provided by electric quadrupole properties. The general E2 transition operator was given in Sect. 3.2.1.5 and can, to a good approximation, be taken as

$$T_\mu^{(E2)} = \alpha_2 G_\mu^{(2)}, \tag{4.24}$$

where $G_\mu^{(2)}$ is a generator of the Spin^{BF}(6) algebra. Supersymmetry implies that even–even and odd–even nuclei belonging to the same supermultiplet, can be described with the same value of α_2. A comparison between the E2 transition rates in $_{76}^{190}$Os$_{114}$ and

174   4 *Superalgebras*

Table 4-6 Comparison between experimental and calculated B(E2) values in $^{190}_{76}\text{Os}_{114}$ and $^{191}_{77}\text{Ir}_{114}$ ($\mathcal{N} = 9$)

Nucleus	$(\sigma_1, \tau_1, J)_i$	$\rightarrow$	$(\sigma_1, \tau_1, J)_f$	B(E2) $(e^2 b^2)$		
				Exp[a]	Th[b]	Δ[c]
$^{190}_{76}\text{Os}_{114}$	9,1,2		9,0,0	0.478(12)	0.478	0
	9,2,2		9,0,0	0.046(2)	0	-0.046
	9,2,2		9,1,2	0.259(15)	0.654	$+0.395$
	9,2,4		9,1,2	0.622(44)	0.654	$+0.032$
	9,3,4		9,1,2	0.010(2)	0	-0.010
	9,3,4		9,2,2	0.488(100)	0.375	-0.113
	9,3,4		9,2,4	0.362(72)	0.340	-0.022
	9,3,6		9,2,4	1.038(330)	0.715	-0.323
$^{191}_{77}\text{Ir}_{114}$	$\frac{17}{2}, \frac{3}{2}, 1/2$		$\frac{17}{2}, \frac{1}{2}, 3/2$	0.130(3)	0.425	$+0.295$
	$\frac{17}{2}, \frac{3}{2}, 5/2$		$\frac{17}{2}, \frac{1}{2}, 3/2$	0.640(30)	0.425	-0.215
	$\frac{17}{2}, \frac{3}{2}, 7/2$		$\frac{17}{2}, \frac{1}{2}, 3/2$	0.293(6)	0.425	$+0.132$
	$\frac{17}{2}, \frac{5}{2}, 3/2$		$\frac{17}{2}, \frac{1}{2}, 3/2$	0.073(13)	0	-0.073
	$\frac{17}{2}, \frac{5}{2}, 5/2$		$\frac{17}{2}, \frac{1}{2}, 3/2$	0.0111(4)	0	-0.011
	$\frac{17}{2}, \frac{5}{2}, 7/2$		$\frac{17}{2}, \frac{1}{2}, 3/2$	0.065(6)	0	-0.065

[a] From Lederer (1982) and Mundy *et al.* (1984).
[b] With $\alpha_2 = 0.143\ eb$.
[c] $\Delta = B(\text{E2})_{\text{th}} - B(\text{E2})_{\text{exp}}$.

$^{191}_{77}\text{Ir}_{114}$ is shown in Table 4.6. The value of α_2 for both nuclei is adjusted so as to reproduce the $2^+_1 \rightarrow 0^+_1$ transition in $^{190}_{76}\text{Os}_{114}$ and only the most important transitions in $^{191}_{77}\text{Ir}_{114}$ are shown. Defining, once more, as a measure of symmetry breaking, the average deviation divided by the average value,

$$\phi' = \frac{\sum_i |B(\text{E2})_i^{\text{th}} - B(\text{E2})_i^{\text{exp}}|}{\sum_i B(\text{E2})_i^{\text{exp}}}, \qquad (4.25)$$

where the sum goes over all observed values, one obtains $\phi' = 39\%$. The difference between the observed and calculated values is thus rather large in this case, but this might be due in part to the uncertainties in the experimental data. Similar tests have been done for other pairs of nuclei in the same region. The situation is summarized in Fig. 4.5, where the observed $B(\text{E2}; 2^+_1 \rightarrow 0^+_1)$

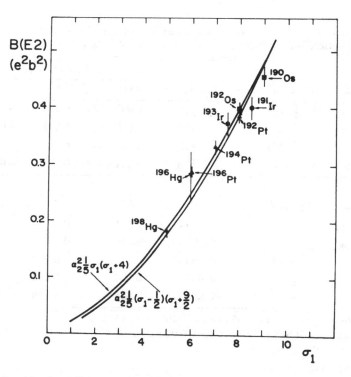

Fig. 4-5   Absolute electromagnetic transition rates in the Os–Pt region as a function of the Spin^{BF}(6) quantum number σ_1. The lines are the predictions of the supersymmetry scheme, (4.26), with the same overall normalization α_2^2 for all nuclei. The circles, squares and triangles are the experimental points.

values in even–even nuclei and the values of

$$\tfrac{1}{3}[B(E2; 1/2_1^+ \to 3/2_1^+) + B(E2; 5/2_1^+ \to 3/2_1^+)$$
$$+ B(E2; 7/2_1^+ \to 3/2_1^+)]$$

in odd–even nuclei are plotted as a function of the Spin^{BF}(6) quantum number σ_1. The supersymmetric predictions, obtained by making use of Eq. (2.136) of Volume 1 and the results of Table 3.4 are

$$B(E2; 2_1^+ \to 0_1^+) = \tfrac{1}{5}\sigma_1(\sigma_1 + 4)\alpha_2^2,$$
$$\tfrac{1}{3}[B(E2; 1/2_1^+ \to 3/2_1^+) + B(E2; 5/2_1^+ \to 3/2_1^+) + B(E2; 7/2_1^+ \to 3/2_1^+)]$$
$$= \tfrac{1}{5}(\sigma_1 - \tfrac{1}{2})(\sigma_1 + \tfrac{9}{2})\alpha_2^2. \tag{4.26}$$

Table 4-7 Comparison between experimental and calculated $B(M1)$ values and magnetic dipole moments in $^{190}_{76}\mathrm{Os}_{114}$ and $^{191}_{77}\mathrm{Ir}_{114}$ ($\mathcal{N} = 9$)

Nucleus	$(\sigma_1, \tau_1, J)_i$	$\rightarrow$	$(\sigma_1, \tau_1, J)_f$	$B(M1)$ (μ_N^2)		
				Exp[a]	Th[b]	Δ[c]
$^{191}_{77}\mathrm{Ir}_{114}$	$\frac{17}{2},\frac{3}{2},1/2$		$\frac{17}{2},\frac{1}{2},3/2$	0.00086(2)	0.023	+0.022
	$\frac{17}{2},\frac{3}{2},5/2$		$\frac{17}{2},\frac{1}{2},3/2$	0.0461(6)	0.011	−0.035
	$\frac{17}{2},\frac{5}{2},3/2$		$\frac{17}{2},\frac{1}{2},3/2$	0.0029(6)	0	−0.003
	$\frac{17}{2},\frac{5}{2},5/2$		$\frac{17}{2},\frac{1}{2},3/2$	0.0143(12)	0	−0.014

Nucleus	$(\sigma_1, \tau_1, J)_i$	μ (μ_N)		
		Exp[a]	Th[b]	Δ[d]
$^{190}_{76}\mathrm{Os}_{114}$	9,1,2	0.671(24)	0.671	0
$^{191}_{77}\mathrm{Ir}_{114}$	$\frac{17}{2},\frac{1}{2},3/2$	0.146(1)	0.146	0
	$\frac{17}{2},\frac{3}{2},1/2$	0.540(5)	0.343	−0.197
	$\frac{17}{2},\frac{3}{2},5/2$	0.450(23)	0.634	+0.184
	$\frac{17}{2},\frac{3}{2},7/2$	1.680(245)	0.824	−0.856

[a] From Mundy *et al.* (1984) and Kölbl *et al.* (1986).
[b] With $\beta_1 = 0.518$ μ_N and $t_1 = -0.023$ μ_N.
[c] $\Delta = B(M1)_{th} - B(M1)_{exp}$.
[d] $\Delta = \mu_{N\,th} - \mu_{N\,exp}$.

It appears that the scale factor α_2 is the same not only for nuclei in the same multiplet but for the entire region.

4.7.3.4 Electromagnetic transitions and moments; M1 The general form of the M1 transition operator is given in (3.41). A test of supersymmetry here is to see the extent to which this operator describes magnetic dipole properties of even–even and odd–even nuclei with the same coefficients β_1 and t_1. The results of this study for the pair of nuclei $^{190}_{76}\mathrm{Os}_{114} - ^{191}_{77}\mathrm{Ir}_{114}$ is shown in Table 4.7, adapted from Vervier (1987). As in the corresponding study of Bose–Fermi symmetry schemes, M1 transitions are poorly described by supersymmetry, while magnetic moments are described somewhat better.

Table 4-8 Comparison between experimental intensities
of two-nucleon transfer reactions and those calculated
using supersymmetry

$\mathcal{N}_i$	$\mathcal{N}_f$	Reaction	$I_{exp}{}^a$	$I_{th}{}^b$
9	8	$^{191}_{77}\text{Ir}_{114} \rightarrow {}^{193}_{77}\text{Ir}_{116}$	1.00(10)	1.00
		$^{190}_{76}\text{Os}_{114} \rightarrow {}^{192}_{76}\text{Os}_{116}$	0.64(12)	1.00
8	7	$^{192}_{78}\text{Pt}_{114} \rightarrow {}^{194}_{78}\text{Pt}_{116}$	0.97(13)	1.02
8	7	$^{193}_{77}\text{Ir}_{116} \rightarrow {}^{195}_{77}\text{Ir}_{118}$	0.94(18)	0.93
		$^{192}_{76}\text{Os}_{116} \rightarrow {}^{194}_{76}\text{Os}_{118}$	0.64(12)	0.93
7	6	$^{194}_{78}\text{Pt}_{116} \rightarrow {}^{196}_{78}\text{Pt}_{118}$	0.97(12)	0.96

^{a} From Cizewski *et al.* (1981).
^{b} Normalized to the $^{191}_{77}\text{Ir}_{114} \rightarrow {}^{193}_{77}\text{Ir}_{116}$ reaction.

4.7.3.5 Two-nucleon transfer intensities The intensities of transfer reactions provide in part direct and in part indirect tests of
supersymmetry. In the case of two-nucleon transfer, they are
indirect tests because the transfer operators (3.65) connect different supermultiplets. They are thus similar to the tests provided
by the separation energies. The experimental situation for reactions in the Os–Ir–Pt region is shown in Table 4.8. While there
is agreement between the experimental and calculated intensities
in Pt and Ir, the observed intensities in Os are well below the
supersymmetry predictions.

4.7.3.6 One-nucleon transfer intensities The form of one-nucleon
transfer operators was specified in Chapter 3. These operators
connect in part nuclei belonging to different supermultiplets and
in part nuclei belonging to the same supermultiplet, as shown in
Fig. 4.6. In particular, the operators (3.62) and their conjugates
are generators of the superalgebra G^*, belonging to its Fermi sector, $F^\dagger$ and F, in (4.4). A test of the selection rules provided by
supersymmetry is thus particularly important. This test was already given in Table 3.9, parts (i) and (ii). The extent to which the
experimental intensities agree with the column denoted by 'Th' in
Table 3.9 is also a measure of the goodness of the U(6|4) symmetry.

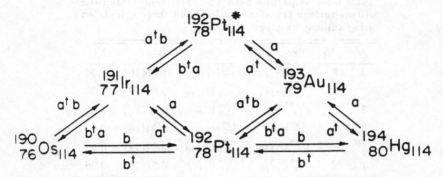

Fig. 4-6   Schematic illustration of the operators inducing transfer reactions.

In fact, the values given in this column are calculated assuming that states of both the intial and final nucleus are described by Spin(6) wave functions, an assumption satisfied automatically if the two nuclei belong to the same supermultiplet.

The other, indirect test is provided by transitions connecting nuclei in different supermultiplets. The appropriate transfer operator is given by (3.57). This operator is outside the $U(6|4)$ algebra. A test of the goodness of supersymmetry here is given in Table 3.9, parts (iii) and (iv), under the column denoted by 'Th'. Extensive tests of the predictions of $U(6|4)$ with regard to one-nucleon transfer have been reported by Blasi (1984).

It is of interest to note that, if absolute intensities of one-nucleon and two-nucleon transfer were measurable, one could test another important aspect of supersymmetry. The set of transfer operators $a_i^\dagger$ and $b_\alpha^\dagger$ transforms as the representation $[1]$ of $U(6|4)$ and, consequently, their matrix elements can be calculated simultaneously by considering the matrix element of a one-box representation between $[\mathcal{N}]$ and $[\mathcal{N}+1]$,

$$ (\overbrace{\boxtimes\boxtimes\cdots\boxtimes}^{\mathcal{N}}\,|\,\boxtimes\,|\,\overbrace{\boxtimes\boxtimes\cdots\boxtimes}^{\mathcal{N}+1}). \qquad (4.27) $$

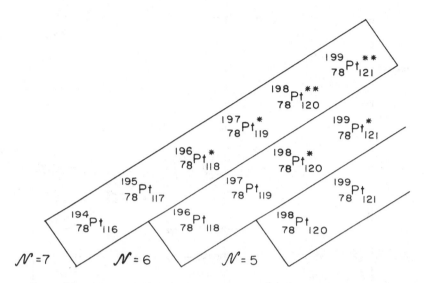

Fig. 4-7   U(6|4) supermultiplets in the Pt region. Each multiplet is identified by the total number of bosons plus fermions, $\mathcal{N}$.

4.8   U(6|12) (III₃)

4.8.1 Supermultiplets

This supersymmetry corresponds to the embedding of the Bose–Fermi algebra of Sect. 3.2.3 (fermions with $j = 1/2, 3/2, 5/2$) into the superalgebra U(6|12),

$$\left| \begin{array}{ccccc} U(6|12) & \supset & U^B(6) & \otimes & U^F(12) \\ \downarrow & & \downarrow & & \downarrow \\ [\mathcal{N}] & & [N_B] & & \{N_F\} \end{array} \right\rangle. \qquad (4.28)$$

If $\mathcal{N} \geq 12$ the supermultiplet contains 13 nuclei, if $\mathcal{N} < 12$ it contains $\mathcal{N}+1$ nuclei. Possible U(6|12) supersymmetric multiplets are the even–even and odd–even Pt isotopes shown in Fig. 4.7.

4.8.2 Tests of U(6|12) supersymmetry

A study of the goodness of the U(6|12) supersymmetry can be done in the same way as for U(6|4).

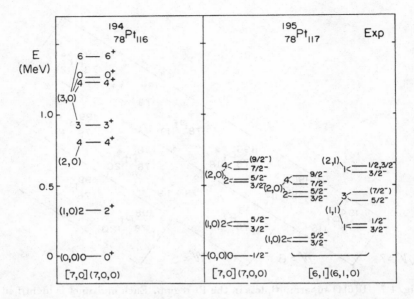

Fig. 4-8   An example of U(6|12) supersymmetry in nuclei: experimental spectra of the pair of nuclei $^{194}_{78}\text{Pt}_{116} - ^{195}_{78}\text{Pt}_{117}$ belonging to the supermultiplet $\mathcal{N} = 7$.

4.8.2.1 Energies   The complete classification and eigenvalue expression can be obtained by attaching to (3.79) the part described in (4.28). The eigenvalue expression is:

$$E'^{(\text{III}3a)} = e_6 \mathcal{N} + e_7 \mathcal{N}(\mathcal{N} - 7) + E^{(\text{III}3a)}, \qquad (4.29)$$

where $E^{(\text{III}3a)}$ is given by (3.90). A simple test of supersymmetry is provided, as usual, by a comparison of the excitation energies of even–even and odd–even nuclei. This is shown in Figs. 4.8 and 4.9 for the pair $^{194}_{78}\text{Pt}_{116} - ^{195}_{78}\text{Pt}_{117}$. The quality of the agreement between the experimental and calculated spectra is somewhat better than that in the case of U(6|4), with a deviation, as defined in (4.22), of $\phi = 11\%$.

4.8.2.2 Electromagnetic transitions and moments; E2   Electromagnetic transitions and moments can be obtained by using the results of Sect. 3.2.3.7 for odd–even nuclei and those of Volume 1

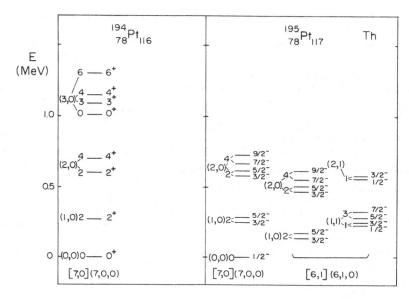

Fig. 4-9 An example of U(6|12) supersymmetry in nuclei: theoretical spectra of the pair of nuclei $^{194}_{78}$Pt$_{116}$–$^{195}_{78}$Pt$_{117}$ belonging to the supermultiplet $\mathcal{N} = 7$. The energy levels are calculated using (3.90) with $\eta + 2\eta' = 8$ KeV, $2\beta = 56$ KeV, $2\gamma = 0$ and $2\gamma'' = 7$ KeV.

for even–even nuclei. If, in first approximation, the operator for electric quadrupole transitions is written as

$$T^{(E2)}_\mu = \alpha_2 G^{(2)}_\mu, \qquad (4.30)$$

where $G^{(2)}_\mu$ is a generator of $\mathrm{U^{BF}}(6)$, all transitions are given in terms of α_2. A test of supersymmetry is to see to what extent the same value of α_2 describes both even–even and odd–even nuclei belonging to the same supermultiplet $\mathcal{N}$. This comparison is shown in Table 4.9 for the pair $^{194}_{78}$Pt$_{116}$ – $^{195}_{78}$Pt$_{117}$. Only a subset of the E2 transitions, known in $^{195}_{78}$Pt$_{117}$, is shown. Notice that, if α_2 is adjusted to reproduce the $2^+_1 \rightarrow 0^+_1$ transition strength in $^{194}_{78}$Pt$_{116}$, the non-zero transitions in $^{195}_{78}$Pt$_{117}$ are overpredicted by a factor of about two. This discrepancy is probably due to use of an oversimplified transition operator (4.30). If the more general form (3.119) is taken with $\alpha_2 \neq f_2$, it is possible to simultaneously reproduce the E2 transitions in even–even and odd–even nuclei.

Table 4-9 Comparison between experimental and calculated $B(E2)$ values in $^{194}_{78}\text{Pt}_{116}$ and $^{195}_{78}\text{Pt}_{117}$ ($\mathcal{N} = 7$)

Nucleus	$((\sigma_1,\sigma_2,\sigma_3),$ $(\tau_1,\tau_2),L,J)_i$	$((\sigma_1,\sigma_2,\sigma_3),$ $(\tau_1,\tau_2),L,J)_f$	$B(E2)$ (e^2b^2)		
			Exp[a]	Th[b]	Δ[c]
$^{194}_{78}\text{Pt}_{116}$	(7,0,0),(1,0),2,2	(7,0,0),(0,0),0,0	0.374(16)	0.374	0
	(7,0,0),(2,0),2,2	(7,0,0),(0,0),0,0	0.0014(2)	0	−0.001
	(7,0,0),(2,0),2,2	(7,0,0),(1,0),2,2	0.580(70)	0.500	−0.080
	(7,0,0),(2,0),4,4	(7,0,0),(1,0),2,2	0.470(30)	0.500	+0.030
$^{195}_{77}\text{Pt}_{117}$	(7,0,0),(1,0),2,3/2	(7,0,0),(0,0),0,1/2	0.190(10)	0.374	+0.184
	(7,0,0),(1,0),2,5/2	(7,0,0),(0,0),0,1/2	0.170(10)	0.374	+0.204
	(7,0,0),(2,0),2,3/2	(7,0,0),(0,0),0,1/2	0.017(1)	0	−0.017
	(7,0,0),(2,0),2,5/2	(7,0,0),(0,0),0,1/2	0.008(4)	0	−0.008
	(6,1,0),(1,0),2,3/2	(7,0,0),(0,0),0,1/2	0.038(6)	0	−0.038
	(6,1,0),(1,0),2,5/2	(7,0,0),(0,0),0,1/2	0.066(4)	0	−0.066

[a] From Harmatz (1977) and Mauthofer *et al.* (1986).
[b] With $\alpha_2 = 0.156\ eb$.
[c] $\Delta = B(E2)_{th} - B(E2)_{exp}$.

For example, the values quoted in Table 3.20 ($\alpha_2 = -f_2 = 0.151$ eb) will lead to $B(E2; 2_1^+ \to 0_1^+) = 0.351\ e^2b^2$, in good agreement with the value observed in $^{194}_{78}\text{Pt}_{116}$.

4.9   U(6|2) (I$_2$)

As a final example of supersymmetric schemes used in the classification of nuclei, we treat here the simple case of bosons with U(5) symmetry and fermions with $j = 1/2$.

4.9.1 Supermultiplets

This supersymmetry corresponds to the embedding of the algebra discussed in Sect. 3.3.2 into U(6|2),

$$\left| \begin{array}{ccc} \text{U}(6|2) & \supset & \text{U}^{\text{B}}(6) \otimes \text{U}^{\text{F}}(2) \\ \downarrow & & \downarrow \quad\quad \downarrow \\ [\mathcal{N}] & & [N_{\text{B}}] \quad \{N_{\text{F}}\} \end{array} \right\rangle . \quad (4.31)$$

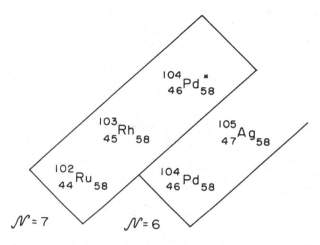

Fig. 4-10 U(6|2) supermultiplets in the Ru–Pd region. Each multiplet is identified by the total number of bosons plus fermions, $\mathcal{N}$.

Supermultiplets here contain 3 nuclei, if $\mathcal{N} \geq 2$. Examples of U(6|2) supermultiplets are shown in Fig. 4.10 (Vervier, 1987).

4.9.2 Tests of U(6|2) *supersymmetry*

Energy eigenvalues are obtained by adding to (3.177) the part arising from (4.31),

$$E'^{(\text{I}_2)} = e_6\mathcal{N} + e_7\mathcal{N}(\mathcal{N}+3) + E^{(\text{I}_2)}, \qquad (4.32)$$

where $E^{(\text{I}_2)}$ is given by (3.179). An analysis of the experimental spectra for the pair of nuclei $^{102}_{44}\text{Ru}_{58} - {}^{103}_{45}\text{Rh}_{58}$ shows that supersymmetry here is an excellent approximation, Figs. 4.11 and 4.12. Similar conclusions are reached by a study of electromagnetic transitions. Table 4.10 shows the results of analysis of the experimental $B(\text{E2})$ values in $^{102}_{44}\text{Ru}_{58}$ and $^{103}_{45}\text{Rh}_{58}$. The overall agreement of this type of supersymmetry is excellent, with an average deviation ϕ or ϕ' of the order of 10% or less.

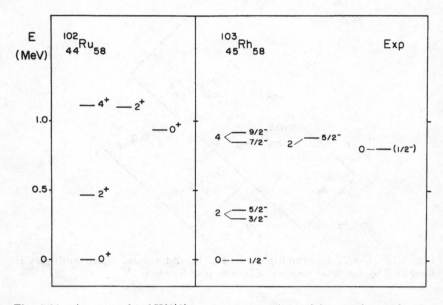

Fig. 4-11   An example of U(6|2) supersymmetry in nuclei: experimental spectra of the pair of nuclei $^{102}_{44}\text{Ru}_{58}$ – $^{103}_{45}\text{Rh}_{58}$ belonging to the supermultiplet $\mathcal{N} = 7$.

4.10   Supersymmetries associated with SU(3)

Several cases of supersymmetries associated with SU(3) have been investigated (Vervier, 1987). Of particular interest are the Er–Tm and the W isotopes. However, in view of the fact that all single-particle orbits in a major shell are mixed when the bosons are described by an SU(3) symmetry, a detailed study of supersymmetries associated with SU(3) requires a fermionic space Ω of large dimensions. The appropriate formalism for these large dimensions has been developed recently (Bijker and Kota, 1988) and was described in Chapter 3. The consequences of this formalism with respect to supersymmetry still need to be investigated.

4.11   General supersymmetry schemes

In the supersymmetries classified in Table 4.3 and described in the previous sections, fermionic spaces of various dimensions Ω

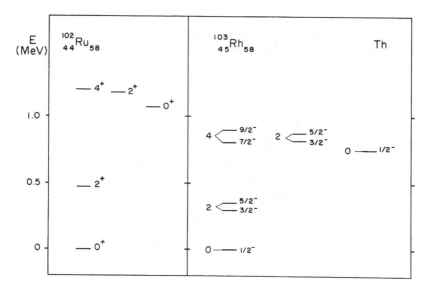

Fig. 4-12   An example of U(6|2) supersymmetry in nuclei: theoretical spectra of the pair of nuclei $^{102}_{44}\mathrm{Ru}_{58} - ^{103}_{45}\mathrm{Rh}_{58}$ belonging to the supermultiplet $\mathcal{N} = 7$. The energy levels are calculated using (3.179) with $\epsilon = 480$ KeV, $\epsilon' = -140$ KeV, $\alpha = 0$, $2\beta = 13$ KeV, $2\gamma = -8$ KeV and $2\gamma' = 11$ KeV.

Table 4-10 Comparison between experimental and calculated $B(E2)$ values in $^{102}_{44}\mathrm{Ru}_{58}$ and $^{103}_{45}\mathrm{Rh}_{58}$ $(\mathcal{N} = 7)$

Nucleus	$(n_d, v, L, J)_i$	$\rightarrow$	$(n_d, v, L, J)_f$	$B(E2)$ $(e^2 b^2)$		
				Exp[a]	Th[b]	Δ[c]
$^{102}_{44}\mathrm{Ru}_{58}$	1,1,2,2		0,0,0,0	0.123(1)	0.123	0
	2,2,2,2		0,0,0,0	0.013(2)	0	−0.013
$^{103}_{45}\mathrm{Rh}_{58}$	1,1,2,3/2		0,0,0,1/2	0.109(8)	0.105	−0.004
	1,1,2,5/2		0,0,0,1/2	0.111(3)	0.105	−0.006

[a] From De Gelder *et al.* (1982), Vervier and Janssens (1982) and Vervier (1987).
[b] With $\alpha_2 = 0.133$ *eb*.
[c] $\Delta = B(E2)_{th} - B(E2)_{exp}$.

Table 4-11 Some algebras that can be constructed in the space $\Omega = 20$ $(j = 1/2, 3/2, 5/2, 7/2)$

Boson algebra	Bose–Fermi algebra	Representation
(i) Pseudo-spin algebras		
$U^B(5)$	$U^{BF}(5) \otimes U_s^F(2)$	$(k = 1, 3) \otimes (s = 1/2)$
	$U'^{BF}(5) \otimes U_{s'}^F(4)$	$(k' = 2) \otimes (s' = 3/2)$
$SU^B(3)$	$SU^{BF}(3) \otimes U_s^F(2)$	$(k = 1, 3) \otimes (s = 1/2)$
$O^B(6)$	$O^{BF}(6) \otimes U_s^F(2)$	$(k = 1, 3) \otimes (s = 1/2)$
	$\text{Spin}'^{BF}(6) \otimes U_{k'}^F(5)$	$(s' = 3/2) \otimes (k' = 2)$
(ii) Spinor algebras		
$O^B(6)$	$\text{Spin}^{BF}(6)$	$j = 1/2, 3/2, 5/2, 7/2$

are coupled to a fixed bosonic space of dimension six. This procedure yields supersymmetries appropriate to particular regions of the periodic table, much in the same way in which each individual subalgebra U(5), SU(3) and O(6) of the bosonic space yields symmetries appropriate to certain regions. One may attempt an even more ambitious scheme in which all states of a major shell can be classified in terms of representations of a large superalgebra of fixed Ω given by the largest possible value within that major shell. These enlarged schemes are indeed possible, at least partially. Consider, for example, the 50–82 shell. The single-particle orbits in this shell are $2d_{5/2}$, $1g_{7/2}$, $1h_{11/2}$, $2d_{3/2}$ and $3s_{1/2}$. If one treats separately the so-called unnatural-parity orbit $1h_{11/2}$, one has a space composed of $j = 1/2, 3/2, 5/2, 7/2$, with $\Omega = 20$. This space is particularly interesting since one can accommodate within it all three algebras U(5), SU(3) and O(6). A supersymmetric description of nuclei in this major shell, excluding the $1h_{11/2}$ orbit, can thus be obtained by means of the superalgebra U(6|20). Different chains can be reached by breaking the angular momenta in different combinations of pseudo-orbital and pseudo-spin angular momenta, as shown in Table 4.11. Many of these schemes have been investigated and for some of them approximate experimental examples have been found (Ling *et al.*, 1984; Jolie *et al.*, 1987).

Another general supersymmetric scheme can be constructed for nuclei in the 28–50 shell. If one excludes the unnatural-parity orbit $1g_{9/2}$, the remaining orbits, $1p_{3/2}$, $1p_{1/2}$ and $1f_{5/2}$, span a 12-dimensional space. The resulting superalgebra $U(6|12)$ admits classifications corresponding to the three chains $U(5)$, $SU(3)$ and $O(6)$. Due to the similarities between the three lattices that can be constructed, it is possible in this case to consider *transitional* Hamiltonians which simultaneously describe even–even and odd–even nuclei, and hence to formulate a supersymmetric model for transitional nuclei (Frank *et al.*, 1987). Although these schemes might provide an overall classification of observed states, they are of limited importance in a detailed description of states in individual nuclei.

5

Numerical studies

5.1 Introduction

Dynamic symmetries and supersymmetries provide a convenient framework within which spectra of nuclei, either individually (symmetries) or in a set (supersymmetries) can be analyzed. However, they usually provide only a first approximation to the observed properties and furthermore there are nuclei for which they cannot be used. In these cases, one needs to do numerical studies. The starting point for these studies is the diagonalization of the Hamiltonian for the combined system of bosons and fermions written in one of its forms. Computer programs are available for the numerical solution of this problem (Scholten, 1979). The structure of the Hamiltonian is as in (1.16). Numerical studies are done by first analyzing the spectra of even–even nuclei as in Volume 1. This analysis determines the parameters appearing in H_B. In a second step, the spectra of odd–even nuclei are studied. For odd–even nuclei with one unpaired particle, H_F contains only the single-particle energies, η_j. If states originating from only one single-particle level are studied, there is only one single-particle energy, η, which can be chosen as zero on the energy scale. If states originating from m single-particle orbits are included, the number of input parameters for the calculation is $m - 1$, since the lowest level can be chosen as zero on the energy scale. The crucial property that determines the structure of spectra of odd–even nuclei is the coupling between the collective degrees of freedom (bosons) and the single-particle degrees of freedom (fermions). This coupling is written as V_{BF} in (1.18). In general, V_{BF} contains many parameters. However, it has been found that special forms of the interaction contain

188

most of the features of the coupling expected from microscopic considerations and yet are simple enough to make a phenomeno-logical study of nuclear spectra possible. One such a form is given in Sect. 1.4.4 in terms of a monopole, quadrupole and exchange interaction. The interplay between these three interactions and the structure of the collective part of H determine most of the properties of odd–even nuclei. It is therefore of interest to study this interplay in detail in order to be able to identify typical fea-tures of the spectra that can be used in understanding nuclear properties.

The simplest situation is provided by the case of one single-particle level of large j (usually called the unnatural-parity orbit). Examples of this orbit are the $1g_{9/2}$ level in the 28–50 shell, the $1h_{11/2}$ level in the 50–82 shell and the $1i_{13/2}$ level in the 82–126 shell. In the following sections the case of $j = 9/2$ will be dis-cussed. When only one single-particle level is included, there are three parameters in the boson–fermion interaction, A_j, Γ_{jj} and Λ^j_{jj}. Results will be obtained as a function of these parameters. As discussed in Volume 1, boson structures have three different and characteristic limits corresponding to the subalgebra chains U(5), SU(3) and O(6) of U(6). These will provide three different benchmarks for structures in odd–even nuclei and will be dis-cussed in the subsequent sections. Since the three limits have a geometric analogue (see Chapter 3 of Volume 1), the correspond-ing situations in odd–even nuclei can also be denoted with these analogues and can be called the particle–vibration, particle–rotor and particle–γ-unstable-rotor limits. The connection between al-gebras and geometry in odd–even nuclei will be further explored in Chapter 6.

5.2   Features of the U(5) limit

We consider here the main features of the energy spectra arising from the coupling of a single particle with angular momentum j to bosons with $U^B(5)$ symmetry. The boson wave functions are characterized by the quantum numbers $[N_B], n_d, v, \tilde{n}_\Delta, L, M_L$

discussed in Volume 1 and the eigenvalues of H_B are given by

$$E_B^{(I)}(N_B, n_d, v, \tilde{n}_\Delta, L, M_L)$$
$$= E_{0B} + \epsilon n_d + \alpha n_d(n_d + 4) + 2\beta v(v + 3) + 2\gamma L(L + 1). \quad (5.1)$$

The fermion wave functions are characterized simply by the angular momentum j and its projection m_j. A basis for the diagonalization of the Hamiltonian is provided by the product of the boson and the fermion wave functions. Since no symmetry is assumed, except rotational invariance, boson and fermion wave functions are only coupled at the level of $\text{Spin}^{BF}(3)$ with ordinary Clebsch–Gordan coefficients,

$$|[N_B], n_d, v, \tilde{n}_\Delta, L; j; J, M_J\rangle$$
$$= \sum_{M_L m_j} (LM_L\, jm_j | JM_J)\, |[N_B], n_d, v, \tilde{n}_\Delta, L, M_L\rangle\, |j, m_j\rangle. \quad (5.2)$$

Note that both H_B and H_F are diagonal. The eigenvalues of H_B are given in (5.1), while the eigenvalue of H_F simply is η_j. One thus needs to diagonalize only V_{BF} for which we take the special form of Sect. 1.4.4. It is instructive to consider separately the effect of the three terms, monopole, quadrupole and exchange. The monopole term can be rewritten as

$$V_{BF}^{MON} = -A_j \hat{n}_d \hat{n}_j / \sqrt{5(2j + 1)}, \quad (5.3)$$

where $\hat{n}_d$ and $\hat{n}_j$ represent the number operators for d bosons and j fermions and only one term is kept in the sum since we are considering only one single-particle level. The operator (5.3) is diagonal in the basis (5.2) with eigenvalues

$$\Delta E_{n_d} = -A_j n_d / \sqrt{5(2j + 1)}, \quad (5.4)$$

since $\langle \hat{n}_j \rangle = 1$ here. This term can be added to the term ϵn_d in (5.1) and thus only causes a renormalization of the boson energy, as shown in Fig. 5.1(a).

Consider next the quadrupole interaction which can be written in the present case as

$$V_{BF}^{QUAD} = \Gamma_{jj} \hat{Q}_B^\chi \cdot \hat{q}_j / \sqrt{5}, \quad (5.5)$$

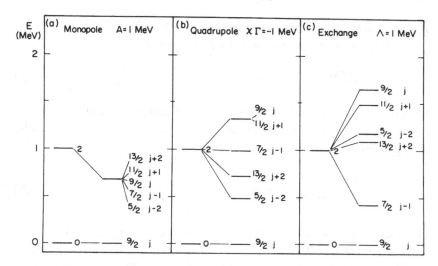

Fig. 5-1   Schematic illustration of the effects of the boson–fermion interaction $V_{\rm BF}$ in the spectra of odd–even nuclei with $j = 9/2$ and $U^{\rm B}(5)$ symmetry: (a) monopole interaction; (b) quadrupole interaction; (c) exchange interaction.

where $\hat{Q}_{\rm B}^{\chi}$ and $\hat{q}_j$ represent the quadrupole operators of s,d bosons and j fermions respectively. The quadrupole interaction has a diagonal part in the basis (5.2) given by

$$V_{\rm BF}^{\rm QUAD'} = \chi\Gamma_{jj}[[d^{\dagger} \times \tilde{d}]^{(2)} \times [a_j^{\dagger} \times \tilde{a}_j]^{(2)}]^{(0)}, \qquad (5.6)$$

and an off-diagonal part

$$V_{\rm BF}^{\rm QUAD''} = \Gamma_{jj}[[s^{\dagger} \times \tilde{d} + d^{\dagger} \times \tilde{s}]^{(2)} \times [a_j^{\dagger} \times \tilde{a}_j]^{(2)}]^{(0)}. \qquad (5.7)$$

When ϵ is large the off-diagonal part gives a negligible contribution and the result is dominated by the diagonal part. This results in a splitting of the multiplets obtained by coupling the boson states $0^+; 2^+; 4^+, 2^+, 0^+; \ldots$; to the single particle j with a very definite structure, Fig. 5.1(b). For example, the splitting of the multiplet with $n_{\rm d} = 1$ is given by

$$\Delta E'_{n_{\rm d}=1}(J) = (\chi\Gamma_{jj})\sqrt{5}(-)^{J-j} \left\{ \begin{array}{ccc} 2 & j & J \\ j & 2 & 2 \end{array} \right\}, \qquad (5.8)$$

obtained from (5.6) by an appropriate recoupling.

Finally, the exchange interaction is given by

$$V_{\mathrm{BF}}^{\mathrm{EXC}} = \Lambda_{jj}^j : [[d^\dagger \times \tilde{a}_j]^{(j)} \times [\tilde{d} \times a_j^\dagger]^{(j)}]^{(0)} : . \qquad (5.9)$$

This interaction is also diagonal in the basis (5.2) and it causes a splitting of the multiplets as shown in Fig. 5.1(c). For example, the splitting of the $n_{\mathrm{d}} = 1$ multiplet is given by

$$\Delta E_{n_{\mathrm{d}}=1}''(J) = \Lambda_{jj}^j \sqrt{2j+1} \left\{ \begin{array}{ccc} j & 2 & J \\ j & 2 & j \end{array} \right\}. \qquad (5.10)$$

This result can again be obtained by a recoupling of (5.9).

As discussed in Chapter 3 of Volume 1, the dynamic symmetry I corresponds to a spherical geometry. The coupling of an odd particle to the bosons has thus the geometric analogue in the particle–vibration model (Bohr and Mottelson, 1975). This aspect will be further discussed in Chapter 6. It is worth mentioning here that in the particle–vibration model only the off-diagonal quadrupole term is usually kept. The resulting spectra are therefore somewhat different from those obtained here. Figures 5.2 and 5.3 summarize the situation when the single-particle orbit has $j = 9/2$. Figure 5.4 instead shows an experimental example of a spectrum built on the $1h_{11/2}$ proton orbit coupled to bosons with U(5) symmetry.

5.3   Features of the SU(3) limit

In this case the boson wave functions are characterized by the quantum numbers $[N_{\mathrm{B}}], (\lambda, \mu), \tilde{\chi}, L, M_L$ and the eigenvalues of H_{B} are given by

$$E_{\mathrm{B}}^{(\mathrm{II})}(N_{\mathrm{B}}, (\lambda, \mu), \tilde{\chi}, L, M_L)$$
$$= E_{0\mathrm{B}} + (\tfrac{3}{4}\kappa - \kappa')L(L+1) - \kappa(\lambda^2 + \mu^2 + \lambda\mu + 3\lambda + 3\mu). \qquad (5.11)$$

The fermion wave functions are characterized by the angular momentum j and its projection m_j. A basis for the diagonalization

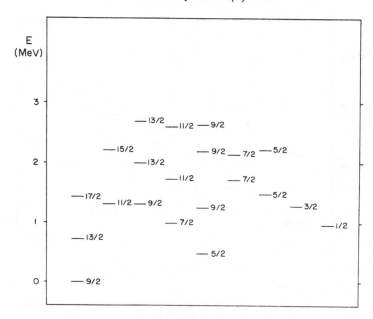

Fig. 5-2   Schematic illustration of the spectrum of states obtained by coupling bosons with $U^B(5)$ symmetry ($N_B = 2$) to a fermion with $j = 9/2$ through a quadrupole interaction.

of H is provided by the product

$$|[N_B], (\lambda, \mu), \tilde{\chi}, L; j; J, M_J\rangle$$
$$= \sum_{M_L m_j} (L M_L \, j m_j | J M_J) \, |[N_B], (\lambda, \mu), \tilde{\chi}, L, M_L\rangle \, |j, m_j\rangle. \quad (5.12)$$

The boson–fermion interaction has the same form as in the previous section. In this limit, the boson quadrupole operator takes its SU(3) form

$$\hat{Q}_{B,\mu} = [s^\dagger \times \tilde{d} + d^\dagger \times \tilde{s}]_\mu^{(2)} - \frac{\sqrt{7}}{2} [d^\dagger \times \tilde{d}]_\mu^{(2)}, \quad (5.13)$$

where the minus sign has been chosen in front of the coefficient $\sqrt{7}/2$ corresponding to prolate deformation. In order to study the features of the spectra obtained in this limit, we consider first the case in which there is only a quadrupole boson–fermion interaction

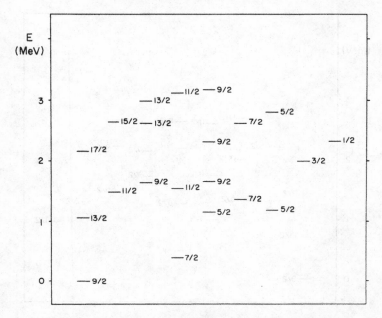

Fig. 5-3   Schematic illustration of the spectrum of states obtained by coupling bosons with $U^B(5)$ symmetry ($N_B = 2$) to a fermion with $j = 9/2$ through an exchange interaction.

($A_j = 0$, $\Lambda^j_{jj} = 0$). The resulting spectrum is shown in Fig. 5.5. It consists of a series of rotational bands starting at some value K_0 of angular momentum. In the case shown in Fig. 5.5 where $j = 9/2$, the values of K_0 are $9/2, 7/2, 5/2, 3/2, 1/2$. Above these bands are other rotational bands corresponding to the coupling to the boson representations ($\lambda = 2N_B - 4, \mu = 2$), the β- and γ-vibrations discussed in Volume 1.

If one now introduces the exchange interaction , $\Lambda^j_{jj} \neq 0$, one obtains spectra as shown in Fig. 5.6. The structure of the bands remains unchanged. However, the relative position of the band heads changes. In the example shown in the figure, the lowest band is now $K_0 = 5/2$ followed by $K_0 = 7/2, 3/2, 1/2$ and $9/2$. The further introduction of the monopole interaction does not change the structure of the spectrum nor the relative location of the bands. It only expands or compresses the energy scale.

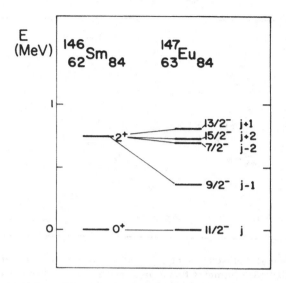

Fig. 5-4   An experimental example of a spectrum built on a $j = 11/2$ fermion coupled to bosons with $U^B(5)$ symmetry: $^{147}_{63}\text{Eu}_{84}$.

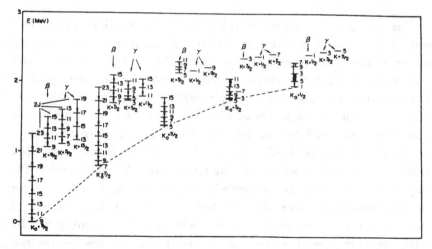

Fig. 5-5   Schematic illustration of the spectrum of states obtained by coupling bosons with $SU^B(3)$ symmetry ($N_B = 7$) to a fermion with $j = 9/2$ through a quadrupole interaction.

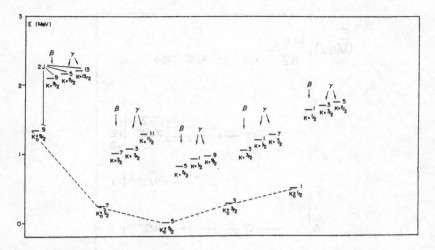

Fig. 5-6   Schematic illustration of the spectrum of states obtained by coupling bosons with $SU^B(3)$ symmetry to a fermion with $j = 9/2$ through a quadrupole and exchange interaction. Only the band heads are shown. On top of each of them there is built a rotational-like band as in Fig. 5.5.

The geometric analogue of this limit is the particle–axial-rotor model (Bohr and Mottelson, 1975). A simpler version of this model is the Nilsson model (Nilsson, 1955). In the Nilsson model one assumes that the single particle moves in an average field with axial deformation. The single-particle levels in this field correspond to the band heads of Fig. 5.5. Band heads of odd–even nuclei are obtained by successively filling the single-particle levels in the deformed field. States above or below the last filled states are described as particle or hole states. The successive filling of the states has its analogue in the interacting boson–fermion in the exchange interaction, as has been shown by Semmes *et al.* (1986) by comparing the interacting boson–fermion model with core–quasiparticle coupled calculations. As this interaction increases the ordering of the bands heads changes in a manner similar to that of the Nilsson model. However, in the interacting boson–fermion model one simultaneously obtains the band heads, the rotational bands built on them and the β- and γ-vibrations arising from the coupling of the single particle j with the boson

SU(3) representation $(2N - 4, 2)$. The latter features must be put *ad hoc* in the Nilsson model.

5.4 Features of the O(6) limit

Here the boson wave functions are characterized by the quantum numbers $[N_{\rm B}], \sigma, \tau, \nu_\Delta, L, M_L$ and the eigenvalues of $H_{\rm B}$ are given by

$$E_{\rm B}^{\rm (III)}(N_{\rm B}, \sigma, \tau, \nu_\Delta, L, M_L)$$
$$= E_{\rm 0B} + 2\eta\sigma(\sigma + 4) + 2\beta\tau(\tau + 3) + 2\gamma L(L + 1). \quad (5.14)$$

The fermion wave functions are again characterized by j and m_j. A basis for the diagonalization of H is thus provided by the product

$$|[N_{\rm B}], \sigma, \tau, \nu_\Delta, L; j; J, M_J\rangle$$
$$= \sum_{M_L m_j} (L M_L \, j m_j | J M_J) \, |[N_{\rm B}], \sigma, \tau, \nu_\Delta, L, M_L\rangle \, |j, m_j\rangle. \quad (5.15)$$

The boson–fermion interaction again is given by the terms discussed in Sect. 1.4.4 but with a boson quadrupole operator which is appropriate in this limit,

$$\hat{Q}_{{\rm B},\mu} = [s^\dagger \times \tilde{d} + d^\dagger \times \tilde{s}]_\mu^{(2)}. \quad (5.16)$$

In order to study the features of the spectrum obtained in this case we start again with $A_j = 0$ and $\Lambda_{jj}^j = 0$. The resulting spectrum is shown in Fig. 5.7. Its structure is very regular and simple. Several bands appear with a triangular-like structure which does not resemble either of the two previous coupling schemes. Nonetheless, as in the previous two cases, the qualitative features of the low-lying spectrum can be constructed by means of some simple rules:

(i) There are first three bands, denoted by T_0, T_2 and T_4 in Fig. 5.7; the lowest angular momenta in these bands are $9/2$, $5/2$ and $1/2$ respectively. (The corresponding rule for the coupling

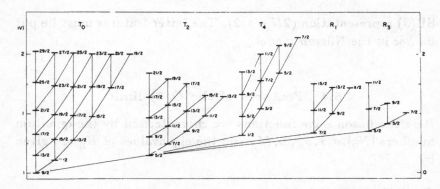

Fig. 5-7   Schematic illustration of the spectrum of states obtained by coupling bosons with $O^B(6)$ symmetry ($N_B = 5$) to a fermion with $j = 9/2$ through a quadrupole interaction.

of a single-particle level with $j \neq 9/2$ is that the lowest angular momenta are $K_0 = j - k$, $k = 0, 2, 4, \ldots$; $K_0 \geq 1/2$.)

(ii) At higher excitation energies there are two additional bands, denoted by R_1 and R_3 in Fig. 5.7; the lowest angular momenta in these bands are 7/2 and 3/2 respectively. (The corresponding rule when $j \neq 9/2$ is that the lowest angular momenta are $K_0 = j - k$, $k = 1, 3, 5, \ldots$; $K_0 \geq 1/2$.)

(iii) Within each band, states can be classified by a quantum number $\tau' = 0, 1, 2, \ldots$; the bands stop at some value $\tau' = \tau'_{\max}$ related to N_B. The angular momenta J contained in each τ'-multiplet are given by the rule

$$J = K_0 + 2\tau', K_0 + 2\tau' - 1, \ldots, K_0 + \tau'. \qquad (5.17)$$

(iv) The energy levels are approximately given by the formula

$$E(K_0, \tau', J) = A'(K_0) + B'\tau'(\tau' + 3) + C'J(J + 1), \qquad (5.18)$$

where A' depends only on K_0 and B' and C' are appropriate constants; large deviations from this formula occur only in the band with $K_0 = 1/2$.

If one now introduces the exchange interaction, $\Lambda^j_{jj} \neq 0$, one observes that the structure of the bands remains the same but

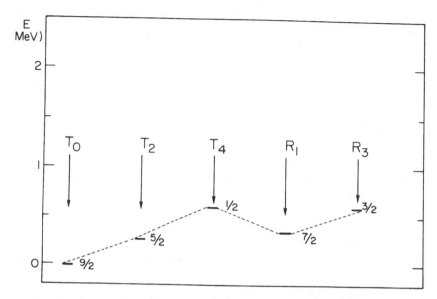

Fig. 5-8   Schematic illustration of the spectrum of states obtained by coupling bosons with $O^B(6)$ symmetry to a fermion with $j = 9/2$ through a quadrupole and exchange interaction. Only the band heads are shown. On top of each of them there is built a triangular-like structure as in Fig. 5.7.

that the relative location of the band heads, K_0, changes. Figure 5.8 shows the situation. This result is similar to that obtained in the previous section. Finally, the effect of the monopole term, A_j, is that of expanding or compressing the spectrum. An experimental spectrum in which the triangular-like pattern discussed here is observed, is shown in Fig. 5.9. One may remark that the particularly simple structure of the spectrum in this limit may be a consequence of some degree of symmetry. Indeed, the Bose–Fermi symmetry discussed in Sect. 3.2 produces exactly this pattern. The symmetry there is obtained by coupling bosons with O(6) symmetry to a single fermion with $j = 3/2$. It appears thus that although the value of j here ($j = 9/2$) does not match the value required for constructing the $Spin^{BF}(6)$ algebra, the nature of the coupling is still such that a high degree of regularity remains. This observation leads to the concept of a quasi-symmetry (Paar

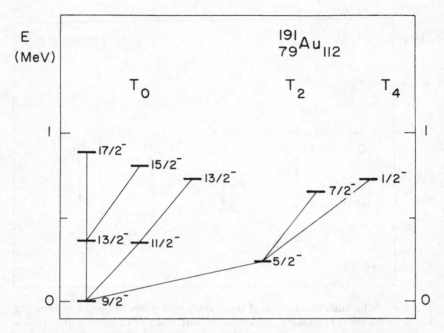

Fig. 5-9   An experimental example of a spectrum built on a $j = 9/2$ fermion coupled to bosons with $O^B(6)$ symmetry: $^{191}_{79}$Au$_{112}$.

et al., 1982). Quasi-symmetries have been exploited in detail to construct simple schemes appropriate to odd–even nuclei when no exact symmetry can be constructed. These quasi-symmetries are useful in providing an understanding of the observed features.

The geometric analogue of the limit discussed in this section is the particle–γ-unstable-rotor model (Leander, 1976). The limit also shares some properties with the particle–triaxial-rotor model (Meyer-ter-Vehn, 1975).

5.5   Transitional classes

After having studied the features of the limiting U(5), SU(3) and O(6) cases, the next step is that of studying the basic features of the transitional classes A, B, C and D, discussed in Volume 1.

We describe here only some results obtained for the transitional classes A and C.

The transitional class A (Sect. 2.10 of Volume 1) corresponds to nuclei intermediate between the U(5) and SU(3) limits. Typical examples of even–even nuclei in this transitional class are the $_{62}$Sm isotopes. A study of odd–even nuclei in this transitional class can be done by using the approach discussed in Volume 1 for the even–even nuclei and by subsequently coupling to them the odd particle. The coupling of the odd particle is described by three parameters A_j, Γ_{jj} and Λ^j_{jj}. In a way similar to that described in Volume 1 one can expand the coupling parameters as a function of the boson number $N_{\rm B}$ and thus study several nuclei with changing $N_{\rm B}$. Retaining the first term in the expansion, one has:

$$A_j(N_{\rm B}) = A_j(N_{\rm B0}) + \left.\frac{\partial A_j}{\partial N_{\rm B}}\right|_{N_{\rm B}=N_{\rm B0}} (N_{\rm B} - N_{\rm B0}) + \cdots,$$

$$\Gamma_{jj}(N_{\rm B}) = \Gamma_{jj}(N_{\rm B0}) + \left.\frac{\partial \Gamma_{jj}}{\partial N_{\rm B}}\right|_{N_{\rm B}=N_{\rm B0}} (N_{\rm B} - N_{\rm B0}) + \cdots,$$

$$\Lambda^j_{jj}(N_{\rm B}) = \Lambda^j_{jj}(N_{\rm B0}) + \left.\frac{\partial \Lambda^j_{jj}}{\partial N_{\rm B}}\right|_{N_{\rm B}=N_{\rm B0}} (N_{\rm B} - N_{\rm B0}) + \cdots. \quad (5.19)$$

In some cases, the lowest, constant, term suffices and one has an even simpler situation. This occurs, for example, for odd-proton nuclei in the Sm region when studied as a function of neutron number. To a good approximation, the parameters of the boson–fermion interaction do not change appreciably with neutron number and one can attempt a calculation with constant parameters. An example of this calculation is shown in Fig. 5.10 (Scholten, 1980). The single-particle considered here is in the $1h_{11/2}$ proton orbit. As one can see from the figure, the structure of both the observed and the calculated spectrum changes from a situation typical of the coupling scheme discussed in Sect. 5.2 to a situation typical of that of Sect. 5.3.

In Fig. 5.11 an example of the transitional class C is shown (Bucurescu *et al.*, 1985). In this case the $1g_{9/2}$ proton orbit is coupled to the $_{44}$Ru core. The resulting odd–even spectrum changes

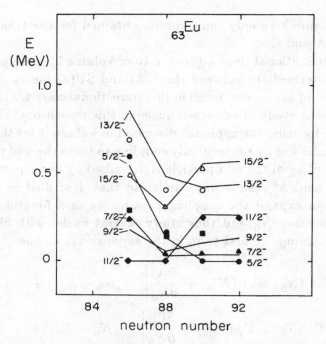

Fig. 5-10   An example of an interacting boson–fermion model-1 calculation for transitional class A: the $_{63}$Eu isotopes. The calculated (lines) and experimental (circles, triangles, squares) energies of negative-parity levels are plotted against neutron number.

smoothly with neutron number, reflecting the smooth transition from U(5) to O(6) of the underlying even–even core.

5.6   Full numerical studies

The interacting boson–fermion model-1 has been used extensively in the last decade to compute spectra of odd–even nuclei in a variety of situations covering most of the periodic table. A summary of available calculations is given in Table 5.1. These studies include both single-j and multiple-j situations. When several single-particle orbits are considered, the use of all coefficients A_j, $\Gamma_{jj'}$ and $\Lambda_{jj'}^{j''}$ as independent parameters is not possible since this leads to too many free parameters. For those situations, a further

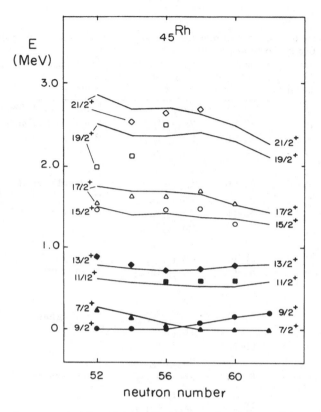

Fig. 5-11 An example of an interacting boson–fermion model-1 calculation for transitional class C: the $_{45}$Rh isotopes. The calculated (lines) and experimental (circles, triangles, squares) energies of positive-parity levels are plotted against neutron number.

input in the calculation is introduced by making models for the j dependence of the coefficients A_j, $\Gamma_{jj'}$ and $\Lambda_{jj'}^{j''}$. These models are based on the microscopic theory of the interacting boson model. The use of the generalized seniority scheme, for example, leads to the following dependence of the coefficients on j (Scholten, 1985):

$$A_j = -\sqrt{5(2j+1)}A_0,$$
$$\Gamma_{jj'} = \sqrt{5}\gamma_{jj'}\Gamma_0,$$
$$\Lambda_{jj'}^{j''} = -2\sqrt{\frac{5}{2j''+1}}\beta_{jj''}\beta_{j'j''}\Lambda_0, \qquad (5.20)$$

Table 5-1 List of available numerical calculations of nuclear properties using the interacting boson–fermion model-1

Proton number	Odd particle	Neutron number	Single-particle orbits	Reference
37 Rb	π	42–46	$1g_{9/2}$	Kaup *et al.* (1980)
38 Sr	ν	43,45	$2d_{5/2},1g_{9/2}$	Bucurescu *et al.* (1983)
39 Y	π	40–46	$2d_{5/2},1g_{9/2}$ $2p_{1/2},2p_{3/2},$ $1f_{5/2}$	Bucurescu *et al.* (1988)
43 Tc	π	54–60	$1g_{9/2}$	De Gelder *et al.* (1983)
45 Rh	π	52–62	$2d_{5/2},1g_{9/2}$	Bucurescu *et al.* (1985)
		56-64	$2d_{5/2},1g_{9/2}$ $2p_{1/2},2p_{3/2},$ $1f_{5/2}$	Jolie *et al.* (1985b)
46 Pd	ν	59,63	$1h_{11/2}$	Casten and Smith (1979)
47 Ag	π	58–62	$2d_{5/2},1g_{9/2}$ $2p_{1/2},2p_{3/2},$ $1f_{5/2}$	Maino and Mengoni (1988)
54 Xe	ν	63–77	$2f_{7/2},1h_{9/2},$ $1h_{11/2}$	Cunningham (1982)
		69-77	$3s_{1/2},2d_{3/2},$ $2d_{5/2},1g_{7/2}$	
56 Ba	ν	67–77	$2f_{7/2},1h_{9/2},$ $1h_{11/2}$	Cunningham (1982)
		73-77	$3s_{1/2},2d_{3/2},$ $2d_{5/2},1g_{7/2}$	
61 Pm	π	86–92	$1h_{11/2}$ $2d_{5/2},1g_{7/2}$	Scholten and Ozzello (1984)
63 Eu	π	84–92	$1h_{11/2}$ $2d_{5/2},1g_{7/2}$	Scholten and Blasi (1982)
64 Gd	ν	89–93	$3p_{1/2},3p_{3/2},$ $2f_{5/2},2f_{7/2},$ $1h_{9/2}$	Chuu and Hsieh (1989)
77 Ir	π	108–114	$1h_{11/2}$	Bijker and Dieperink (1982)
78 Pt	ν	107–115	$1i_{13/2}$	Bijker and Dieperink (1982)
79 Au	π	106–116	$1h_{9/2}$	Bijker and Dieperink (1982)
		106-116	$1h_{9/2}$	Wood (1981b)

with

$$\gamma_{jj'} = (u_j u_{j'} - v_j v_{j'}) Q_{jj'},$$
$$\beta_{jj'} = (u_j v_{j'} + v_j u_{j'}) Q_{jj'}. \tag{5.21}$$

The $Q_{jj'}$ in (5.21) are matrix elements of the quadrupole operator in the single-particle basis,

$$Q_{jj'} = \langle j \parallel Y^{(2)} \parallel j' \rangle. \tag{5.22}$$

The coefficients u_j and v_j describe the occupation probabilities of the states. They are obtained either by solving the BCS equations (see, for example, Ring and Schuck, 1980) or by other procedures (Otsuka and Arima, 1978).

When several j values are considered, one also needs the single-particle energies ϵ_j. If a BCS calculation is used to generate the occupation probabilities u_j and v_j, the values of the single-particle energies ϵ_j are taken from the same calculation and given in terms of the Fermi energy λ_F and the energy gap Δ_F as

$$\epsilon_j = \sqrt{(E_j - \lambda_F)^2 + \Delta_F^2}. \tag{5.23}$$

The Fermi energy λ_F and the gap Δ_F are in turn obtained in terms of the strength of the pairing interaction, G. The values of E_j in (5.23) are the single-particle levels in the absence of the pairing interaction.

The generalized seniority scheme is appropriate for nuclei in the neighborhood of closed shells and described by the U(5) limit. It becomes increasingly poor as one moves away from that limit. An improvement to the formulas (5.20) and (5.21) that allows one to study nuclei away from the U(5) limit is that in which, in addition to the exchange interaction (1.33), one adds the exchange interactions given by (1.34) and (1.35). Particularly important is the interaction (1.34). The inclusion of this term considerably affects the results in regions where the SU(3) limit is appropriate.

6

Geometry

6.1 Introduction

Every algebraic structure has associated with it geometric structures. The choice of the geometric structure with which it is most convenient to visualize the situation depends on the physics that one wishes to expose. For boson systems of the type discussed in Volume 1 and also for those used in the description of molecules (Iachello and Levine, 1982), there is a very natural geometric structure provided by the coset space $U(n)/U(n-1) \otimes U(1)$. This leads in the case of nuclei to a description in terms of five variables, α_μ ($\mu = 0, \pm 1, \pm 2$), which can then be associated with the shape of a liquid drop with quadrupole deformation (Bohr and Mottelson, 1975). Similarly, in molecules, use of the coset space mentioned above leads to a description in terms of three variables, r_μ ($\mu = 0, \pm 1$), which can be associated with the vector distance between the two atoms in the molecule.

For fermionic systems or when bosons and fermions coexist, the introduction of a geometric space is not so obvious. One can, if one wishes, introduce coset spaces, as briefly discussed in the following section, but even with this introduction, the geometric structure of the problem remains as abstract as before. A simpler situation arises if one is interested only in the case of a single fermion coupled to a system of bosons. In this case one can analyze the algebraic structure in terms of the motion of a single particle in a potential well generated by the bosons. This is not a geometric structure in the true sense of the word but nonetheless it provides a very convenient and simple scheme for studying the problem. This scheme leads to the most common 'geometric' model for odd–even nuclei, the Nilsson model (Nilsson, 1955). In this chapter, after a

206

Table 6-1 Coset spaces for fermions

Space	Dimension
$\dfrac{U(n)}{USp(n)}, n = \text{even}$	$n(n-1)/2$

brief discussion of the general problem of geometric structures of fermion systems, the connection between the interacting boson–fermion model and the Nilsson model will be analyzed.

6.2 Coset spaces

The geometric spaces that can be associated in a straigthforward way with algebraic structures are the coset spaces. These were briefly discussed in Chapter 3 of Volume 1 and are extensively discussed by Gilmore (1974). Cosets are constructed by decomposing an algebra g into two parts:

$$g = h \oplus p, \tag{6.1}$$

where h is a subalgebra of g,

$$g \supset h, \tag{6.2}$$

and p is not closed with respect to commutation. For fermion systems the construction of cosets is somewhat more involved than that of boson systems. Consider the case of fermions with angular momentum j. As discussed in Chapter 2, the algebraic structure is that of $u(2j + 1)$. Subalgebras can be formed as discussed in Sect. 2.3. The stability algebra h can be taken as $sp(2j + 1)$. The group associated with $u(2j + 1)$ is $U(n, C)$ defined over the complex numbers C and with $n = 2j + 1$. The group associated with $sp(2j + 1)$ is $USp(n, C)$, where the unitary label U has been written explicitly. The factor algebra g/h then produces the coset $U(n)/USp(n)$. The dimension of the associated space is shown in Table 6.1. Geometric variables can be associated with the generators in p and intrinsic states can be constructed appropriately.

As an example, consider the case of $j = 3/2$ discussed in Sect. 2.3.2. Here, $g \equiv u(4)$ and $h \equiv sp(4)$, with generators

$$
\begin{aligned}
g : \; & [a^\dagger_{3/2} \times \tilde{a}_{3/2}]^{(3)}_\mu, \\
& [a^\dagger_{3/2} \times \tilde{a}_{3/2}]^{(2)}_\mu, \\
& [a^\dagger_{3/2} \times \tilde{a}_{3/2}]^{(1)}_\mu, \\
& [a^\dagger_{3/2} \times \tilde{a}_{3/2}]^{(0)}_0; \\[4pt]
h : \; & [a^\dagger_{3/2} \times \tilde{a}_{3/2}]^{(3)}_\mu, \\
& [a^\dagger_{3/2} \times \tilde{a}_{3/2}]^{(1)}_\mu; \\[4pt]
p : \; & [a^\dagger_{3/2} \times \tilde{a}_{3/2}]^{(2)}_\mu, \\
& [a^\dagger_{3/2} \times \tilde{a}_{3/2}]^{(0)}_0.
\end{aligned}
\tag{6.3}
$$

The algebra g has 16 generators, the algebra h has 10 generators and the remainder p has 6 generators. The coset space, according to Table 6.1, has dimension 6, the same as the number of generators in p.

When several single-particle levels are included, the breaking of the algebra $u(\Omega)$, $\Omega = \sum_j (2j+1)$, can occur in several ways. The construction of the associated geometric spaces is more complex and will not be discussed here.

6.3   Classical limit of bosons

As mentioned in the introduction, the use of coset spaces is not particularly illuminating here. It is more convenient to analyze the situation in terms of the motion of a single particle in an external field. To this end, we return to the boson–fermion Hamiltonian (1.18) of Chapter 1 and consider the case of a single particle with angular momentum j. The corresponding Hamiltonian is

$$
H = H_{\mathrm{B}} + H_{\mathrm{F}} + V_{\mathrm{BF}},
\tag{6.4}
$$

where H_B is the interacting boson Hamiltonian, Eq. (1.21) of Volume 1, and

$$H_F = \epsilon_j \hat{n}_j = -\epsilon_j \sqrt{2j+1}[a_j^\dagger \times \tilde{a}_j]_0^{(0)},$$

$$
\begin{aligned}
V_{BF} = & -k_0 \sqrt{2j+1}[[s^\dagger \times \tilde{s}]^{(0)} \times [a_j^\dagger \times \tilde{a}_j]^{(0)}]_0^{(0)} \\
& + k_2 \sqrt{5}[[s^\dagger \times \tilde{d} + d^\dagger \times \tilde{s}]^{(2)} \times [a_j^\dagger \times \tilde{a}_j]^{(2)}]_0^{(0)} \\
& - \sum_L x_L (-)^L \sqrt{2L+1}[[d^\dagger \times \tilde{d}]^{(L)} \times [a_j^\dagger \times \tilde{a}_j]^{(L)}]_0^{(0)} \quad (6.5)
\end{aligned}
$$

where k_0, k_2 and x_L are related to the coefficients in (1.20) through

$$
\begin{aligned}
k_0 &= -w_{00jj}^{\prime(0)}/\sqrt{2j+1}, \\
k_2 &= w_{20jj}^{\prime(2)} = w_{02jj}^{\prime(2)}, \\
x_L &= -w_{22jj}^{\prime(L)}.
\end{aligned}
\quad (6.6)
$$

The first step in analyzing the situation from the point of view of the motion of a single particle in an external field is that of freezing the dynamics of the bosons by taking their static classical limit. This is discussed in Chapter 3 of Volume 1. One first introduces intrinsic states in terms of Bohr variables, β and γ,

$$|N_B; \beta, \gamma\rangle = (N_B!)^{-\frac{1}{2}} [b_c^\dagger(\beta, \gamma)]^{N_B} |0\rangle,$$

$$b_c^\dagger(\beta, \gamma) = (1+\beta^2)^{-\frac{1}{2}} \left[\beta \cos\gamma \, d_0^\dagger + \tfrac{1}{\sqrt{2}} \beta \sin\gamma \, (d_{+2}^\dagger + d_{-2}^\dagger) + s^\dagger \right].$$

$$(6.7)$$

The intrinsic states (6.7) are those used by Leviatan (1987) and contain an additional normalization factor $(1 + \beta^2)^{-\frac{1}{2}}$ relative to those discussed in Sect. 3.3 of Volume 1. By considering the matrix element of H between the intrinsic state (6.7) and integrating out the boson degrees of freedom one obtains

$$H'(N_B; \beta, \gamma) = E^B(N_B; \beta, \gamma) + \sum_{m_1, m_2} g_{m_1, m_2}^j(\beta, \gamma) a_{j, m_1}^\dagger a_{j, m_2}. \quad (6.8)$$

In this equation, $E^B(N_B; \beta, \gamma)$ is the expectation value of the boson Hamiltonian given explicitly by

$$
\begin{aligned}
E^B(N_B; \beta, \gamma) = & E_0 + \epsilon_s N_B + \tfrac{1}{2} u_0 N_B(N_B - 1) \\
& + N_B(N_B - 1) \frac{\beta^2}{(1+\beta^2)^2} (f_1'\beta^2 + f_2'\beta \cos 3\gamma + f_3'),
\end{aligned}
$$

$$(6.9)$$

with

$$f_1' = \tfrac{1}{10}c_0 + \tfrac{1}{7}c_2 + \tfrac{9}{35}c_4 - \tfrac{1}{2}u_0 + \frac{\epsilon_d - \epsilon_s}{N_B - 1},$$

$$f_2' = -\sqrt{\tfrac{4}{35}}v_2,$$

$$f_3' = \frac{v_0 + u_2}{\sqrt{5}} - u_0 + \frac{\epsilon_d - \epsilon_s}{N_B - 1}. \tag{6.10}$$

This result is identical to that of Eq. (3.20) of Volume 1, but written in a slightly different form. The factor $g_{m_1,m_2}^j(\beta,\gamma)$ depends on β and γ and the parameters in H_F and V_{BF}. Since it is real and symmetric in m_1 and m_2, H' is Hermitian. For fixed β and γ and given parameters in H_F and V_{BF}, one can compute g_{m_1,m_2}^j. A diagonalization of this matrix produces the single-particle energies in the presence of the boson condensate (deformed single-particle energies). As a specific example, consider the case of prolate axial symmetry ($\gamma = 0°$). The geometry of the bosons is in this case that depicted in Fig. 3.6 of Volume 1. In this special case, g_{m_1,m_2}^j can be brought to diagonal form by simply choosing the ms as the projections of the angular momenta along the boson symmetry axis. Calling this projection K, one has (Leviatan, 1988):

$$H'(N_B;\beta,\gamma = 0°) = E^B(N_B;\beta,\gamma = 0°) + \sum_K \lambda_{jK}(\beta)a_{j,K}^\dagger a_{j,K},$$

$$\tag{6.11}$$

with

$$\lambda_{jK}(\beta) = \epsilon_j + B_j(\beta) + C_j(\beta)K^2 + D_j(\beta)K^4. \tag{6.12}$$

The coefficients B_j, C_j and D_j are given in terms of the parameters in V_{BF} by:

$$B_j(\beta) = k_0 N_B + N_B \frac{\beta}{1+\beta^2}\left[\beta\left(-k_0 + \frac{x_0}{\sqrt{5(2j+1)}}\right)\right.$$

$$+ j(j+1)\left(k_2 2\sqrt{5}P_j + \beta x_2\sqrt{\tfrac{10}{7}}P_j\right)$$

$$\left.+ 54\beta x_4\left(\frac{2(2j-4)!}{35(2j+5)!}\right)^{\frac{1}{2}}\frac{(j+2)!}{(j-2)!}\right],$$

$$C_j(\beta) = -3N_{\mathrm{B}}\frac{\beta}{1+\beta^2}\left[k_2 2\sqrt{5}P_j + \beta x_2\sqrt{\tfrac{10}{7}}P_j\right.$$

$$\left. + 30(6j(j+1)-5)\beta x_4\left(\frac{2(2j-4)!}{35(2j+5)!}\right)^{\frac{1}{2}}\right],$$

$$D_j(\beta) = 630N_{\mathrm{B}}\frac{\beta^2}{1+\beta^2}x_4\left(\frac{2(2j-4)!}{35(2j+5)!}\right)^{\frac{1}{2}},$$

$$P_j = [(2j-1)j(2j+1)(j+1)(2j+3)]^{-\frac{1}{2}}. \tag{6.13}$$

The values $\lambda_{jK}(\beta)$ represent the single-particle energies in the deformed field due to bosons with axial symmetry ($\gamma = 0°$) and deformation β. These single-particle energies are doubly degenerate with levels $\pm K$ belonging to the same eigenvalue. For small β, each λ_{jK} is linear in β. For a given β, the order of the levels depends on the values of the coefficients B_j, C_j and D_j.

It is instructive to consider the special form of the boson–fermion interaction discussed in Sect. 1.4.4. For a single-j level this form is specified by three parameters, $A \equiv A_j$, $\Gamma \equiv \Gamma_{jj}$ and $\Lambda \equiv \Lambda_{jj}^j$. The monopole interaction does not introduce K-dependent terms and will be neglected. The remaining interactions produce single-particle energies:

$$\lambda_{jK}(\beta) = \epsilon_j + N_{\mathrm{B}}\frac{\beta}{1+\beta^2}P_j\left(3K^2 - j(j+1)\right)$$

$$\times \left[\Gamma\left(\beta\chi\sqrt{\tfrac{2}{7}} - 2\right) - P_j\beta\Lambda\sqrt{2j+1}\left(3K^2 - j(j+1)\right)\right]. \tag{6.14}$$

The behavior of the energies with β is then dictated by the interplay between the strengths Γ of the quadrupole interaction and Λ of the exchange interaction. A typical situation is shown in Fig. 6.1.

6.4   The Nilsson model

Spectra of odd–even deformed nuclei are often described in terms of the Nilsson model. Single-particle states are calculated by solving the Schrödinger equation for a particle moving in a deformed

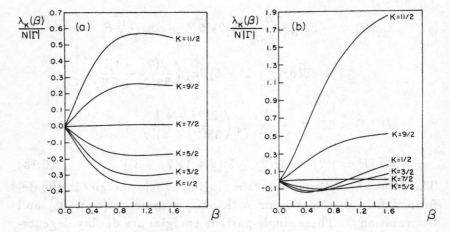

Fig. 6-1 Schematic representation of the single-particle energies as a function of deformation β: (a) quadrupole interaction; (b) quadrupole and exchange interaction. The energy levels are labelled with the values of K (Leviatan, 1988).

potential. The single-particle Hamiltonian is of the form

$$H = \frac{\vec{p}^{\,2}}{2m} + v_c(r)\left(1 + \hat{\beta}Y_0^{(2)}(\theta)\right) + v_{so}(r)\vec{l}\cdot\vec{s}, \qquad (6.15)$$

where $\hat{\beta}$ is the deformation parameter introduced by Bohr (1952), $v_c(r)$ is the potential energy of a particle moving in the axially deformed field generated by the core and $v_{so}(r)$ is the spin–orbit interaction. The single-particle energy levels can be written in perturbation theory as

$$\lambda_{jK} = \epsilon_j + \hat{\beta}P_j\left(3K^2 - j(j+1)\right)\langle j \parallel v_c(r)Y^{(2)} \parallel j\rangle, \qquad (6.16)$$

where ϵ_j are the eigenvalues of

$$H_0 = \frac{\vec{p}^{\,2}}{2m} + v_c(r) + v_{so}(r)\vec{l}\cdot\vec{s}. \qquad (6.17)$$

In general, the solution will involve a diagonalization since states of different j are coupled by the $Y^{(2)}$ term. This general solution as well as the corresponding one for the interacting boson–fermion model will be discussed below.

By comparing (6.14) and (6.16) one can see that the Nilsson model corresponds to the classical limit of the interacting boson–fermion model with a pure quadrupole boson–fermion interaction. The strength of the interaction Γ is related to the parameters of the Nilsson model by

$$\Gamma = \frac{\hat{\beta}\langle j \parallel v_{\mathrm{c}}(r)Y^{(2)} \parallel j\rangle}{N_{\mathrm{B}}\dfrac{\beta}{1+\beta^2}\left(\beta\chi\sqrt{\tfrac{2}{7}} - 2\right)}. \tag{6.18}$$

As discussed in Volume 1, Sect. 3.3, the Bohr variable $\hat{\beta}$ and the interacting boson variable β are related by a scale transformation. Using the relations given there, one can eliminate $\hat{\beta}$ from (6.18) and obtain:

$$\Gamma = -\frac{\langle j \parallel v_{\mathrm{c}}(r)Y^{(2)} \parallel j\rangle}{0.76\left(\dfrac{eZR_0^2}{e_{\mathrm{B}}}\right)}. \tag{6.19}$$

6.5   The Nilsson model plus BCS

The Nilsson model by itself is not sufficient to determine the correct ordering of levels in odd–even nuclei. In order to do that one needs either to place particles by hand in each K-level or to perform a calculation with a residual interaction. If the residual interaction is taken as a pairing interaction in the deformed basis, with strength G_{def}, single-particle levels can be obtained by performing a BCS (Bardeen, Cooper and Schrieffer, 1957) calculation. The new single-particle energies are given by

$$\epsilon_{jK} = \sqrt{(\lambda_{jK} - \lambda_{\mathrm{F}})^2 + \Delta_{\mathrm{F}}^2}, \tag{6.20}$$

where the Fermi energy, λ_{F}, and the pairing gap, Δ_{F}, are given in terms of G_{def} and by imposing conservation of particle number

$$n = \sum_{K} 2v_K^2, \tag{6.21}$$

where v_K^2 is the occupation probability of the K-level. The relation between the Nilsson plus BCS model and the classical limit of the interacting boson–fermion model is more complex than that described in Sect. 6.4 for the Nilsson model. A single-particle spectrum similar to that given by (6.20) can be obtained by including in the interacting boson–fermion model both the quadrupole and the exchange interaction. By equating the values of λ_{jK} reported in Fig. 6.1(b) to those of ϵ_{jK}, one can then obtain a relation between Λ and λ_F which can be used either way to relate the two models.

Although the Nilsson model appears as a special case of the classical limit of the interacting boson–fermion model, there are, however, differences. The major difference occurs for large deformations. In the interacting boson–fermion model, the energy levels converge, for large β, to a constant. This is due to the finiteness of the boson number N_B, related to the finiteness of the number of particles. In the Nilsson model, the energy levels originating from a single-j orbit diverge for large β, as shown in Fig. 6.2. There is, in this model, no cut-off effect since the deformed field is not generated microscopically by the other particles but it is a given external field. Detailed comparisons of the properties of the Nilsson model and the interacting boson–fermion model have been reported by Frank et al. (1986) and Arima et al. (1987).

6.6 Classical limit of bosons and fermions

The treatment of Sect. 6.3 can be improved by considering the classical limit of the combined system of bosons plus one fermion. As discussed in Sect. 1.3, the wave functions for the interacting boson–fermion model can be written as the product functions for bosons and fermions. In the case in which only one fermion is considered, one can then introduce intrinsic wave functions of the type (Pittel and Frank, 1986):

$$|N_B; \beta, \gamma, \alpha_{j,m}\rangle = \Theta[b_c^\dagger(\beta, \gamma)]^{N_B}|o_B\rangle \left[\sum_{jm} \alpha_{j,m} a_{j,m}^\dagger\right]|o_F\rangle, \quad (6.22)$$

where $|o_B\rangle$ and $|o_F\rangle$ denote the vacua of bosons and fermions and Θ is a normalization constant. By taking the expectation value of

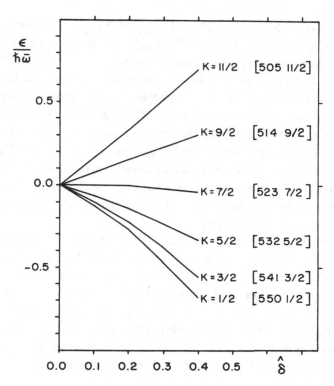

Fig. 6-2 Energy levels of a single particle with $l = 5, j = 11/2$ in a deformed potential as a function of deformation. The energy levels are given in units of the average oscillator frequency $\hbar\bar{\omega}$ (Nilsson, 1955).

H in the state (6.22) one obtains:

$$E^{\mathrm{BF}}(N_{\mathrm{B}}; \beta, \gamma, \alpha_{j,m}) = E^{\mathrm{B}}(N_{\mathrm{B}}; \beta, \gamma) + \sum_{m_1, m_2} g^j_{m_1, m_2}(\beta, \gamma) \alpha_{j, m_1} \alpha_{j, m_2}.$$

(6.23)

The amplitudes $\alpha_{j,m}$ and the values of β and γ can be taken as variational parameters. The equilibrium deformations β^{e}, γ^{e} and $\alpha^{\mathrm{e}}_{j,m}$ are defined by the global minimum of the combined energy surface. Since the N_{B}-dependence is quadratic for the boson part and linear for the boson–fermion part, terms in V_{BF} can cause a shift in the equilibrium position of the order $1/N_{\mathrm{B}}$ from the value in the adjacent even–even nucleus. In the case of prolate axial symmetry, one can look for solutions of the extremum equations of the type $\beta > 0$, $\gamma = 0°$ and $\alpha_{j,m} = \delta_{m,K}$. This amounts to finding the minimum in β of $E^B(N_B; \beta, \gamma = 0°) + \lambda_{jK}(\beta)$ for each K and then selecting the equilibrium set β^e, $\gamma^e = 0°$, K^e which gives the lowest value for the energy surface. This equilibrium set determines the intrinsic wave function of the ground-state band $|N_B; \beta^e, \gamma^e = 0°, K^e\rangle$, while states with $K \neq K^e$ represent intrinsic states of excited bands. In the limit of large N_B, the value of β^e becomes independent of K^e and is governed only by $E^B(N_B; \beta, \gamma)$. This method can be used to derive approximate analytic formulas for a variety of nuclear properties such as the moment of inertia or the decoupling parameter of rotational bands in deformed nuclei (Dukelsky and Lima, 1986).

The study of the classical limit of bosons and fermions can be developed further by using techniques similar to those introduced in the case of bosons only. The analysis is carried out by introducing operators that create β- and γ-vibrations (Leviatan, 1987). The role of β- and γ-vibrations in odd–even nuclei can, in principle, be studied by resolving the combined boson–fermion Hamiltonian into an intrinsic and a collective part or, alternatively, the approach proposed by Pittel and Frank (1986) can be followed.

6.7   Multiple j

The results of the previous sections can be generalized to the case of many-j shells. For example, in the presence of many-j shells, the Hamiltonian (6.8) becomes

$$H'(N_B; \beta, \gamma) = E^B(N_B; \beta, \gamma) + \sum_{\substack{j_1, m_1 \\ j_2, m_2}} g_{j_1 m_1, j_2 m_2}(\beta, \gamma) a^\dagger_{j_1, m_1} a_{j_2, m_2}.$$

$$(6.24)$$

The single-particle energies for a fixed deformation β, γ are obtained by diagonalizing the matrix $g_{j_1 m_1, j_2 m_2}(\beta, \gamma)$. In nuclei with axial symmetry, $\gamma = 0°$, the result obtained in this way is similar to that obtained in the multiple-j Nilsson model. An example is shown in Fig. 6.3. Deviations from the Nilsson model occur only for large deformations, as before. The parameters $\Gamma_{jj'}$ of

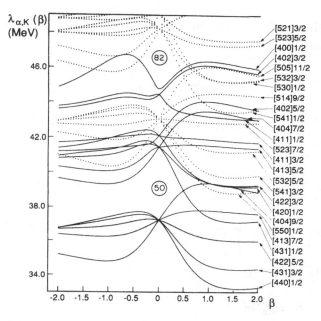

Fig. 6-3   Single-particle levels $\lambda_{\alpha,K}$ as a function of deformation β. The levels are labelled by the corresponding Nilsson quantum numbers (Leviatan and Shao, 1990).

the boson–fermion quadrupole interaction (1.32) can be obtained from the matrix elements of the operator $v_{\rm c}(r)Y^{(2)}$ in the same way as they were obtained in (6.19) for a single-j shell.

6.8   Bose–Fermi condensates

The Bose–Fermi condensates (6.22) can be used to obtain a variety of results. Particularly important are the matrix elements of electromagnetic transition operators. As an example, consider the matrix elements of the M1 operator

$$T_{\mu}^{(\rm M1)} = T_{\rm B,\mu}^{(\rm M1)} + T_{\rm F,\mu}^{(\rm M1)} = \sqrt{\frac{3}{4\pi}} \left(g_{\rm B}\hat{L}_{\rm B,\mu} + g_{\rm F}\hat{L}_{\rm F,\mu} \right). \qquad (6.25)$$

Using the states (6.22) in the limit of large $N_{\rm B}$, one obtains
(Leviatan, 1988)

$$B(\text{M1}; K, J_{\rm i} \to K, J_{\rm f}) = (g_{\rm B} - g_{\rm F})^2 \frac{3}{4\pi}$$

$$\times \left[K \, (J_{\rm i} K \, 10 | J_{\rm f} K) + \frac{(-)^{J_{\rm i}+j}}{\sqrt{2}} (j + \tfrac{1}{2})(J_{\rm i} - \tfrac{1}{2} \, 11 | J_{\rm f} \tfrac{1}{2}) \delta_{K, \frac{1}{2}} \right]^2,$$

(6.26)

and the magnetic moments

$$\mu_J = g_{\rm B} J + (g_{\rm F} - g_{\rm B}) \frac{J}{J+1} \left[K^2 - \frac{(-)^{J+j}}{2} (J + \tfrac{1}{2})(j + \tfrac{1}{2}) \delta_{K, \frac{1}{2}} \right].$$

(6.27)

Similarly, for the E2 operator

$$T_\mu^{(\text{E2})} = T_{{\rm B},\mu}^{(\text{E2})} + T_{{\rm F},\mu}^{(\text{E2})}$$

$$= e_{\rm B} \left([s^\dagger \times \tilde{d} + d^\dagger \times \tilde{s}]_\mu^{(2)} + \chi [d^\dagger \times \tilde{d}]_\mu^{(2)} \right) + e_{\rm F} [a_j^\dagger \times \tilde{a}_j]_\mu^{(2)},$$

(6.28)

one obtains

$$B(\text{E2}; K, J_{\rm i} \to K, J_{\rm f}) = (J_{\rm i} K \, 20 | J_{\rm f} K)^2$$

$$\times \left[e_{\rm B} \frac{N_{\rm B} \beta}{1 + \beta^2} \left(2 - \beta \chi \sqrt{\tfrac{2}{7}} \right) - e_{\rm F} \sqrt{\frac{5}{2j+1}} (jK \, 20 | jK) \right]^2. \quad (6.29)$$

Bose–Fermi condensates can thus be used as realistic alterna-
tives to the diagonalization of the boson–fermion Hamiltonian,
especially in those cases in which $N_{\rm B}$ is large. They produce results
that are similar to those of Nilsson-like models.

Part II

THE INTERACTING
BOSON–FERMION MODEL-2

7

Operators

7.1 Introduction

The interacting boson–fermion model-1 describes properties of odd–even nuclei by coupling collective and single-particle degrees of freedom much in the same way this is done in the collective model (Bohr and Mottelson, 1975). The collective degrees of freedom are described either by shape variables α_μ ($\mu = 0, \pm1, \pm2$) or by boson operators s, d_μ ($\mu = 0, \pm1, \pm2$), with no direct link to the underlying microscopic structure. A microscopic description of nuclei is provided by the spherical shell model. Collective features in this model can be obtained by introducing the concept of correlated pairs with angular momentum and parity $J^P = 0^+$ and $J^P = 2^+$. A treatment of these pairs as bosons leads to the interacting boson model. However, since there are protons and neutrons, one has the possibility of forming proton and neutron pairs. In heavy nuclei, the neutron excess prevents the formation of correlated proton–neutron pairs and one thus is led to consider only proton–proton and neutron–neutron pairs. The corresponding model is the interacting boson model-2 (Arima et al., 1977; Otsuka et al., 1978). The introduction of fermions in this models leads to the interacting boson–fermion model-2. In addition to a more direct connection with the spherical shell model, the interacting boson–fermion model-2 has features that cannot be obtained in the interacting boson–fermion model-1.

The structure of model-2 is very similar to that of model-1. In order to avoid repetitions, the discussion here and in the following chapters will therefore be kept short and will concentrate mostly on numerical studies.

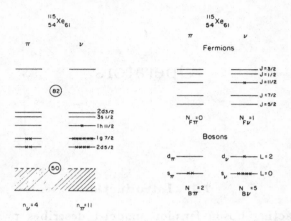

Fig. 7-1 Schematic representation of the shell-model problem for $^{115}_{54}\mathrm{Xe}_{61}$ (left) and the boson–fermion problem that replaces it (right).

7.2   Bosons and fermions

Consider an odd–even nucleus in the spherical shell model, Fig. 7.1. Single-particle levels here are denoted by nlj with n being the principal quantum number, l the orbital angular momentum and j the total angular momentum, $j = l \pm 1/2$. When many particles occupy the valence shells, the diagonalization of the residual interaction in the shell model space is unmanageable. A truncation can be obtained first by assuming that the closed shells are inert and second by considering only those configurations arising from pairing together particles to states with angular momentum and parity $J^{\mathrm{P}} = 0^+$ and $J^{\mathrm{P}} = 2^+$. In odd–even nuclei at least one particle remains unpaired. In odd-proton nuclei it is a proton, in odd-neutron nuclei it is a neutron. One can also consider situations in which both one proton and one neutron are unpaired or cases in which two neutrons or two protons are unpaired. The former situation will arise in odd–odd nuclei while the latter will correspond to excited states in even–even nuclei. (These excited states are often called two-quasi-particle states.) The general situation is thus described by proton (neutron) bosons with $J^{\mathrm{P}} = 0^+$, denoted by s_π (s_ν), and proton (neutron) bosons with $J^{\mathrm{P}} = 2^+$, denoted by d_π (d_ν). This is identical to the situation in even–even nuclei. In addition, there are unpaired protons, denoted by a_π, and neutrons, a_ν. As in the case of even–even nuclei, in order to take into account the particle–hole conjugation in particle space, the number of proton and neutron bosons, $N_{B\pi}$ and $N_{B\nu}$, and of proton and neutron fermions, $N_{F\pi}$ and $N_{F\nu}$, is counted from the nearest closed shell, i.e. if more than half of the shell is full, $N_{B\pi(\nu)}$ and $N_{F\pi(\nu)}$ are taken as holes. Thus, for example, for $^{119}_{54}\text{Xe}_{65}$, $N_{B\pi} = (54 - 50)/2 = 2$, $N_{B\nu} = (64 - 50)/2 = 7$ and $N_{F\nu} = 65 - 64 = 1$ while for $^{127}_{54}\text{Xe}_{73}$, $N_{B\pi} = (54 - 50)/2 = 2$, $\bar{N}_{B\nu} = (82 - 74)/2 = \bar{4}$ and $\bar{N}_{F\nu} = 74 - 73 = \bar{1}$. A bar is sometimes placed over the numbers $N_{B\pi(\nu)}$ and $N_{F\pi(\nu)}$ to indicate that these are hole states. The total number of bosons and fermions is then:

$$N_B = N_{B\pi} + N_{B\nu},$$
$$N_F = N_{F\pi} + N_{F\nu}. \tag{7.1}$$

Properties of this model, with protons and neutrons explicitly introduced, will now be discussed.

7.3 Boson and fermion operators

The building blocks of the interacting boson–fermion model-2 are boson and fermion operators for protons and neutrons. The boson operators are identical to those introduced in Volume 1,

$$s^\dagger_\pi, d^\dagger_{\pi,\mu}, s^\dagger_\nu, d^\dagger_{\nu,\mu}, \qquad (\mu = 0, \pm 1, \pm 2),$$
$$s_\pi, d_{\pi,\mu}, s_\nu, d_{\nu,\mu}, \qquad (\mu = 0, \pm 1, \pm 2), \tag{7.2}$$

or, in a more compact notation,

$$b^\dagger_{\rho,l,m}; \qquad b_{\rho,l,m}; \qquad (\rho = \pi, \nu; l = 0, 2; -l \le m \le l). \tag{7.3}$$

These operators satisfy Bose commutation relations,

$$[b_{\rho,l,m}, b^\dagger_{\rho',l',m'}] = \delta_{\rho\rho'}\delta_{ll'}\delta_{mm'},$$
$$[b_{\rho,l,m}, b_{\rho',l',m'}] = [b^\dagger_{\rho,l,m}, b^\dagger_{\rho',l',m'}] = 0. \tag{7.4}$$

In addition, there are now fermion creation and annihilation operators,

$$a^\dagger_{\rho,j,m};\ a_{\rho,j,m};\ (\rho = \pi, \nu; j = j_1, j_2, \ldots, j_n; m = \pm\tfrac{1}{2}, \pm\tfrac{3}{2}, \ldots, \pm j).$$
$$(7.5)$$

The fermion operators satisfy Fermi anticommutation relations,

$$\{a_{\rho,j,m}, a^\dagger_{\rho',j',m'}\} = \delta_{\rho\rho'}\delta_{jj'}\delta_{mm'},$$
$$\{a_{\rho,j,m}, a_{\rho',j',m'}\} = \{a^\dagger_{\rho,j,m}, a^\dagger_{\rho',j',m'}\} = 0, \qquad (7.6)$$

These are the same as in (1.6) except for the extra index $\rho = \pi, \nu$. The values over which the index j runs are now determined by the single-particle levels in the valence shell, Fig. 7.1. For example, for $^{127}_{54}\text{Xe}_{73}$, the values of j_ν are 5/2, 7/2, 11/2, 1/2 and 3/2. The principal quantum number, n, is as usual not written as an index on the fermion operators, unless one considers large spaces in which there are two single-particle states with the same j. If only valence shells are included, this never occurs.

Boson and fermion operators are assumed to commute,

$$[b_{\rho,l,m}, a_{\rho',j',m'}] = [b_{\rho,l,m}, a^\dagger_{\rho',j',m'}] = [b^\dagger_{\rho,l,m}, a_{\rho',j',m'}]$$
$$= [b^\dagger_{\rho,l,m}, a^\dagger_{\rho',j',m'}] = 0. \qquad (7.7)$$

Spherical tensors are constructed in the usual way, as shown, for example, in (1.10).

7.3.1 Isospin

Instead of the label $\rho = \pi, \nu$, it is possible, for bosons as well as for fermions, to introduce another, equivalent label. For bosons it is called F-spin and was defined in Eq. (4.8) of Volume 1. For fermions the label is precisely identical to isospin. Protons can be characterized by $T = 1/2$ and projection $T_z = +1/2$, while neutrons are characterized by $T = 1/2$ and $T_z = -1/2$, i.e.

$$|\pi\rangle = |\tfrac{1}{2}, +\tfrac{1}{2}\rangle,$$
$$|\nu\rangle = |\tfrac{1}{2}, -\tfrac{1}{2}\rangle. \qquad (7.8)$$

Using the isospin label, fermion creation and annihilation operators are denoted by

$$a^\dagger_{\frac{1}{2},m_t,j,m}; \qquad a_{\frac{1}{2},m_t,j,m}; \qquad (m_t = \pm\tfrac{1}{2}). \qquad (7.9)$$

When the isospin label is used, spherical tensors are built from creation and annihilation operators of the type

$$\tilde{a}_{\frac{1}{2},m_t,j,m} = (-)^{1/2-m_t+j-m} a^\dagger_{\frac{1}{2},-m_t,j,-m}. \qquad (7.10)$$

Isospin for fermions does not play an important role in the interacting boson–fermion model-2 since protons and neutrons occupy different single-particle states. It plays instead an important role in more elaborate versions of the model, such as those discussed in Part III.

7.4 Basis states

Basis states in the interacting boson–fermion model-2 are rather complex. Denoting the indices l, m as α and the indices j, m as i, basis states can be written as

$$\mathcal{BF}: \qquad a^\dagger_{\pi,i} a^\dagger_{\pi,i'} \cdots a^\dagger_{\nu,i} a^\dagger_{\nu,i'} \cdots b^\dagger_{\pi,\alpha} b^\dagger_{\pi,\alpha'} \cdots b^\dagger_{\nu,\alpha} b^\dagger_{\nu,\alpha'} \cdots |o\rangle. \quad (7.11)$$

If no fermion creation operators are present, (7.11) describes a state in an even–even nucleus, if one fermion operator is present, (7.11) describes a state in an odd–even nucleus, if one proton and one neutron creation operator is present, the state is in an odd–odd nucleus, etc. Angular momentum couplings are chosen in such a way that bosons and fermions are first coupled among themselves, followed by the final coupling,

$$\mathcal{BF}: [[[a^\dagger_{\pi,j} \times a^\dagger_{\pi,j'} \times \cdots]^{(J_\pi)} \times [a^\dagger_{\nu,j} \times a^\dagger_{\nu,j'} \times \cdots]^{(J_\nu)}]^{(J_F)}$$
$$\times [[b^\dagger_{\pi,l} \times b^\dagger_{\pi,l'} \times \cdots]^{(L_\pi)} \times [b^\dagger_{\nu,l} \times b^\dagger_{\nu,l'} \times \cdots]^{(L_\nu)}]^{(L_B)}]^{(J)}_M |o\rangle.$$
$$(7.12)$$

7.5   Physical operators

7.5.1 The Hamiltonian operator

The Hamiltonian operator has now the general form

$$H = H_B + H_F + V_{BF}, \qquad (7.13)$$

with

$$
\begin{aligned}
H_B &= H_{\pi B} + H_{\nu B} + V_{\pi\nu B}, \\
H_F &= H_{\pi F} + H_{\nu F} + V_{\pi\nu F}, \\
V_{BF} &= V_{\pi\pi BF} + V_{\pi\nu BF} + V_{\nu\pi BF} + V_{\nu\nu BF}.
\end{aligned}
\qquad (7.14)
$$

The various parts have the same structure as those discussed in Chapter 1, except that indices π, ν appear everywhere.

7.5.2 Special forms of the interaction

The most general Hamiltonian (7.13)–(7.14) contains many parameters. A phenomenological study using all the parameters is nearly impossible. In the analysis of experimental data simpler Hamiltonians are quite often used which contain the essential features of the interaction. The part describing the bosons is usually treated in terms of the Talmi Hamiltonian, Eq. (4.44) of Volume 1, which contains the basic features of the effective nucleon–nucleon interaction that emerge from pairing, quadrupole and symmetry energy. In addition, in some calculations a d-boson number-conserving interaction arising from a seniority-conserving nucleon–nucleon interaction between like particles is introduced. The adopted boson Hamiltonian is then

$$H = E_0 + \epsilon_\pi \hat{n}_{d_\pi} + \epsilon_\nu \hat{n}_{d_\nu} + \kappa \hat{Q}_\pi^\chi \cdot \hat{Q}_\nu^\chi + \lambda' \hat{M}_{\pi\nu} + V_{\pi\pi} + V_{\nu\nu}. \quad (7.15)$$

The operators $\hat{n}_{d_\pi}$, $\hat{n}_{d_\nu}$, $\hat{Q}_\pi^\chi$ and $\hat{Q}_\nu^\chi$ have the same meaning as in Volume 1. In terms of the boson operators they are given by:

$$
\begin{aligned}
\hat{n}_{d_\rho} &= \sum_\mu d^\dagger_{\rho,\mu} d_{\rho,\mu}, \\
\hat{Q}^\chi_{\rho,\mu} &= [s^\dagger_\rho \times \tilde{d}_\rho + d^\dagger_\rho \times \tilde{s}_\rho]^{(2)}_\mu + \chi_\rho [d^\dagger_\rho \times \tilde{d}_\rho]^{(2)}_\mu, \quad \rho = \pi, \nu.
\end{aligned}
\qquad (7.16)
$$

The Majorana operator $\hat{M}_{\pi\nu}$ is given by:

$$\hat{M}_{\pi\nu} = [s_\nu^\dagger \times d_\pi^\dagger - s_\pi^\dagger \times d_\nu^\dagger]^{(2)} \cdot [\tilde{s}_\nu \times \tilde{d}_\pi - \tilde{s}_\pi \times \tilde{d}_\nu]^{(2)}$$
$$- 2 \sum_{k=1,3} \xi_k [d_\nu^\dagger \times d_\pi^\dagger]^{(k)} \cdot [\tilde{d}_\nu \times \tilde{d}_\pi]^{(k)}. \qquad (7.17)$$

The coefficients ξ_k have been introduced in (7.17) relative to Eq. (4.43) of Volume 1, in order to allow for different strengths of the last two terms relative to the first one. This result arises from microscopic calculations of the coefficients. The d-boson number-conserving interaction is:

$$V_{\rho\rho} = \sum_{L=0,2,4} \tfrac{1}{2} c_L^{(\rho)} [d_\rho^\dagger \times d_\rho^\dagger]^{(L)} \cdot [\tilde{d}_\rho \times \tilde{d}_\rho]^{(L)}, \qquad \rho = \pi, \nu. \quad (7.18)$$

The part related to the fermions is described in terms of an effective nucleon–nucleon interaction. This interaction can be taken either as a schematic interaction (such as a surface δ-function interaction), as often used in shell-model calculations (Brussaard and Glaudemans, 1977), or as the effective interaction arising from the free nucleon–nucleon interaction. In most calculations only one proton or one neutron is unpaired. In these cases only the one-body part of H_F matters. This is just the single-particle energy

$$H_F' = E_0 + \sum_{j_\pi} \epsilon_{j_\pi} \hat{n}_{j_\pi} + \sum_{j_\nu} \epsilon_{j_\nu} \hat{n}_{j_\nu}, \qquad (7.19)$$

where

$$\hat{n}_{j_\rho} = \sum_{m_\rho} a_{\rho,j_\rho,m_\rho}^\dagger a_{\rho,j_\rho,m_\rho}, \qquad \rho = \pi, \nu. \qquad (7.20)$$

In odd–odd nuclei there is one unpaired proton and one unpaired neutron. In these cases one needs also the proton–neutron interaction. This can be taken in the form of a quadrupole interaction

$$H_F'' = \sum_{j_\pi j_\pi' j_\nu j_\nu'} \tfrac{1}{2} v_{j_\pi j_\pi' j_\nu j_\nu'} [[a_{j_\pi}^\dagger \times \tilde{a}_{j_\pi'}]^{(2)} \times [a_{j_\nu}^\dagger \times \tilde{a}_{j_\nu'}]^{(2)}]_0^{(0)}, \quad (7.21)$$

or, alternatively, a surface δ-function interaction is used.

The most important part of the Hamiltonian for odd–even nuclei is the boson–fermion interaction. The microscopic theory of the interacting boson–fermion model suggests specific forms for this interaction. The three important terms are, as in Chapter 1, the monopole, the quadrupole and the exchange interaction. The monopole and quadrupole terms are written in the same form as in Chapter 1,

$$V_{\rm BF}^{\rm MON} = \sum_{j_\pi} A_{j_\pi}\left(\hat{n}_{d_\pi}\hat{n}_{j_\pi}\right) + \sum_{j_\nu} A_{j_\nu}\left(\hat{n}_{d_\nu}\hat{n}_{j_\nu}\right) \qquad (7.22)$$

$$V_{\rm BF}^{\rm QUAD} = \sum_{j_\pi j_\pi'} \Gamma_{j_\pi j_\pi'}\hat{Q}_\nu^{\rm X}\cdot\hat{q}_{j_\pi j_\pi'} + \sum_{j_\nu j_\nu'} \Gamma_{j_\nu j_\nu'}\hat{Q}_\pi^{\rm X}\cdot\hat{q}_{j_\nu j_\nu'}, \qquad (7.23)$$

where $\hat{n}_{j_\pi}$ and $\hat{n}_{j_\nu}$ are defined in (7.20) and the fermion quadrupole operators $\hat{q}_{j_\rho j_\rho'}$ are given by

$$\hat{q}_{j_\rho j_\rho',\mu} = [a_{j_\rho}^\dagger \times \tilde{a}_{j_\rho'}]_\mu^{(2)}, \qquad \rho = \pi,\nu. \qquad (7.24)$$

The microscopic structure of the interacting boson model suggests that the monopole interaction acts predominantly between like particles (proton fermions with proton bosons and neutron fermions with neutron bosons), while the quadrupole interaction acts predominantly between unlike particles (protons with neutrons) (Iachello and Talmi, 1987). These considerations are built in the special forms (7.22) and (7.23). The last term in the boson–fermion interaction is the exchange term. In the interacting boson–fermion model-2 this term has a form somewhat different from the corresponding term in Chapter 1,

$$V_{\rm BF}^{\rm EXC} = [s_\pi^\dagger \times \tilde{d}_\pi]^{(2)} \cdot \left\{ \sum_{j_\nu j_\nu' j_\nu''} \Lambda_{j_\nu j_\nu'}^{j_\nu''} : [[d_\nu^\dagger \times \tilde{a}_{j_\nu''}]^{(j_\nu)} \times [\tilde{s}_\nu \times a_{j_\nu'}^\dagger]^{(j_\nu')}]^{(2)} : \right\}$$

$$+ [s_\nu^\dagger \times \tilde{d}_\nu]^{(2)} \cdot \left\{ \sum_{j_\pi j_\pi' j_\pi''} \Lambda_{j_\pi j_\pi'}^{j_\pi''} : [[d_\pi^\dagger \times \tilde{a}_{j_\pi''}]^{(j_\pi)} \times [\tilde{s}_\pi \times a_{j_\pi'}^\dagger]^{(j_\pi')}]^{(2)} : \right\}$$

$$+ \text{Hermitian conjugate.} \qquad (7.25)$$

This form again is suggested by the microscopic structure of the model. It should be noted that, if no distinction is made between protons and neutrons, the form (7.25) can approximately be rewritten as (1.33) by appropriately contracting the s-boson operators.

7.5.3 Transition operators

Transition operators can be written in the same way as in Chapter 1. There are now four terms describing proton and neutron bosons and fermions,

$$T_\mu^{(L)} = T_{\pi B,\mu}^{(L)} + T_{\nu B,\mu}^{(L)} + T_{\pi F,\mu}^{(L)} + T_{\nu F,\mu}^{(L)}. \qquad (7.26)$$

The boson terms are given in Part II of Volume 1. The fermion terms can, to the lowest order, be written as:

$$T_{\pi F,\mu}^{(L)} = f_{\pi,0}^{(0)}\delta_{L0} + \sum_{j_\pi j'_\pi} f_{j_\pi j'_\pi}^{(L)} [a_{j_\pi}^\dagger \times \tilde{a}_{j'_\pi}]_\mu^{(L)},$$

$$T_{\nu F,\mu}^{(L)} = f_{\nu,0}^{(0)}\delta_{L0} + \sum_{j_\nu j'_\nu} f_{j_\nu j'_\nu}^{(L)} [a_{j_\nu}^\dagger \times \tilde{a}_{j'_\nu}]_\mu^{(L)}. \qquad (7.27)$$

Particularly important in odd–even nuclei are the transition operators which induce E2 and M1 transitions. It is customary in the operators to separate the dependence on the angular momenta j_π and j_ν from the coefficients that determine the strengths of the transitions. This is done by introducing effective charges and moments. For E2 transitions, one has:

$$f_{j_\rho j'_\rho}^{(2)} = -e_\rho^{\mathrm{F}} \langle n_\rho, l_\rho | r^2 | n'_\rho, l'_\rho \rangle \langle l_\rho, \tfrac{1}{2}, j_\rho \parallel Y^{(2)} \parallel l'_\rho, \tfrac{1}{2}, j'_\rho \rangle / \sqrt{5}, \quad \rho = \pi, \nu, \qquad (7.28)$$

where now the single-particle indices $n, l, s = 1/2, j$ are written explicitly. The quantities e_π^{F} and e_ν^{F} are the fermion effective charges. The free values of these charges are 1 and 0 respectively, in units of the electron charge. Shell model calculations indicate that $e_\pi^{\mathrm{F}} \approx 1.5e$ and $e_\nu^{\mathrm{F}} \approx 0.5e$ (Brussaard and Glaudemans, 1977). Following Volume 1, the boson part is written as

$$T_{\rho B,\mu}^{(\mathrm{E2})} = e_\rho^{\mathrm{B}} \hat{Q}_{\rho,\mu}^\chi, \qquad \rho = \pi, \nu. \qquad (7.29)$$

A superscript B has been added to e_ρ in order to distinguish it from the fermion charges. The units of e_ρ^{B} are different from those of e_ρ^F since the radial integral is already included in (7.29). The boson effective charges e_ρ^B have the same units as the product

$$e_\rho^{'F} = e_\rho^F \langle n_\rho, l_\rho | r^2 | n_\rho', l_\rho' \rangle, \qquad \rho = \pi, \nu, \tag{7.30}$$

that is, the units are $e\ fm^2$.

For M1 transitions, the fermion part of the operator is written in the form

$$f_{j_\rho j_\rho'}^{(1)} = -\sqrt{\frac{3}{4\pi}} \langle l_\rho, \tfrac{1}{2}, j_\rho \parallel g_{l,\rho}^F \vec{l} + g_{s,\rho}^F \vec{s} \parallel l_\rho', \tfrac{1}{2}, j_\rho' \rangle \delta_{l_\rho l_\rho'} / \sqrt{3}, \quad \rho = \pi, \nu. \tag{7.31}$$

The quantities $g_{l,\pi}^F$, $g_{l,\nu}^F$, $g_{s,\pi}^F$ and $g_{s,\nu}^F$ are the single-particle g-factors. The free values are $g_{l,\pi}^F = 1$, $g_{l,\nu}^F = 0$, $g_{s,\pi}^F = 5.58$ and $g_{s,\nu}^F = -3.82$ in units of nuclear magnetons, μ_N. In actual calculations, the spin factors g_s are renormalized. Typical values in shell-model calculations are $g_s^{\mathrm{renorm}} \approx 0.7 g_s^{\mathrm{free}}$. Following Volume 1, the boson part of the M1 operator is usually written as

$$T_{\rho B, \mu}^{(M1)} = \sqrt{\frac{3}{4\pi}} g_\rho^B \hat{L}_{\rho,\mu}, \qquad \rho = \pi, \nu, \tag{7.32}$$

where $\hat{L}_\rho$ is the angular momentum operator of the ρ bosons. The boson g-factors have the same units as the fermion g-factors since no radial integrals are involved in M1 transitions.

7.5.4 Transfer operators

Transfer operators assume a particularly important role in the interacting boson–fermion model-2. This is because the transferred particle is either a proton or a neutron (or a pair of protons or neutrons) and it is thus natural to compute matrix elements of transfer operators within a framework of a model that explicitly treats proton and neutron degrees of freedom. The form of the transfer operators employed in calculations is identical to that used in Chapter 1. There are two types of one-nucleon transfer operators, those that change the boson number by one unit and those that do not. When expanded in terms of creation and annihilation operators, the transfer operators of the second kind can be written as:

$$P_{+,m_\rho}^{(j_\rho)} = p_{j_\rho} a_{j_\rho,m_\rho}^\dagger + \sum_{j_\rho'} q_{j_\rho'}^{(j_\rho)} [[s_\rho^\dagger \times \tilde{d}_\rho]^{(2)} \times a_{j_\rho'}^\dagger]_{m_\rho}^{(j_\rho)}$$

$$+ \sum_{j_\rho'} q_{j_\rho'}^{'(j_\rho)} [[d_\rho^\dagger \times \tilde{s}_\rho]^{(2)} \times a_{j_\rho'}^\dagger]_{m_\rho}^{(j_\rho)}$$

$$+ \sum_{k,j_\rho'} q_{k,j_\rho'}^{''(j_\rho)} [[d_\rho^\dagger \times \tilde{d}_\rho]^{(k)} \times a_{j_\rho'}^\dagger]_{m_\rho}^{(j_\rho)} + \cdots. \qquad (7.33)$$

Those of the first kind can be written as

$$P_{+,m_\rho}^{'(j_\rho)} = p_{j_\rho}' [s_\rho^\dagger \times \tilde{a}_{j_\rho}]_{m_\rho}^{(j_\rho)} + \sum_{j_\rho'} q_{j_\rho'}^{'''(j_\rho)} [d_\rho^\dagger \times \tilde{a}_{j_\rho'}]_{m_\rho}^{(j_\rho)} + \cdots. \qquad (7.34)$$

The substraction operators, P_-, are obtained by taking the Hermitian conjugate of (7.33) and (7.34).

Two-neutron addition and subtraction operators are written in terms of boson operators alone, at least if one considers only states with at most one unpaired particle. The explicit expression is given in Volume 1, Sect. 4.5.

7.5.5 Beta-decay operators

With the introduction of proton and neutron degrees of freedom one can also compute probabilities of beta decay. In this process, a proton is transformed into a neutron (or *vice versa*) with the emission of an electron and an antineutrino. In the ground state of an even–even nucleus where all particles are paired, one must break a pair in order to have a beta-decay process. Operators connecting even–even and odd–odd nuclei therefore have a complex structure. In contrast, the decay of odd–even nuclei proceeds predominantly through the conversion of the odd (unpaired) particle from proton to neutron (β^+-decay) or from neutron to proton (β^--decay). In this case, one can therefore write the decay operator in a relatively simple form. There are two types of decay possible, Fermi decay with no change of angular momentum and Gamow–Teller decay with a change of one unit of angular momentum. Fermi (F) and

Gamow–Teller (GT) β^--decay operators can be written as

$$\hat{T}_0^{\beta\mathrm{F}} = \sum_{j_\pi,j_\nu} \eta_{j_\pi,j_\nu}^{\mathrm{F}} [P_{j_\pi}^\dagger \times \tilde{P}_{j_\nu}]_0^{(0)},$$

$$\hat{T}_\mu^{\beta\mathrm{GT}} = \sum_{j_\pi,j_\nu} \eta_{j_\pi,j_\nu}^{\mathrm{GT}} [P_{j_\pi}^\dagger \times \tilde{P}_{j_\nu}]_\mu^{(1)}. \qquad (7.35)$$

The operators $P^\dagger$ and $\tilde{P}$ in (7.35) are particle creation and annihilation operators in the space of bosons and fermions. They are
nothing but the operators of the previous section,

$$P_{j_\pi,m_\pi}^\dagger = P_{+,m_\pi}^{(j_\pi)},$$

$$\tilde{P}_{j_\nu,m_\nu} = P_{-,m_\nu}^{(j_\nu)}. \qquad (7.36)$$

The operators for β^+-decay are similar, except for the indices π
and ν which are interchanged.

The coefficients $\eta_{j_\pi,j_\nu}^{\mathrm{F}}$ and $\eta_{j_\pi,j_\nu}^{\mathrm{GT}}$ in (7.35) depend on the form of
the beta-decay operators. If the beta decay takes place between
proton and the neutron orbits with the same orbital quantum
numbers, the lowest order term of the decay operator can be used.
In that case, the η-coefficients are given by

$$\eta_{j_\pi,j_\nu}^{\mathrm{F}} = -\sqrt{2j_\pi+1}\, \delta_{j_\pi j_\nu},$$

$$\eta_{j_\pi,j_\nu}^{\mathrm{GT}} = (-)^{l_\pi+j_\pi-3/2} \sqrt{\frac{(2j_\pi+1)(2j_\nu+1)}{2}} \left\{ \begin{array}{ccc} 1/2 & 1/2 & 1 \\ j_\pi & j_\nu & l_\pi \end{array} \right\} \delta_{l_\pi l_\nu}. \qquad (7.37)$$

Transitions between orbits with different orbital quantum numbers $(l_\pi \neq l_\nu)$ are forbidden in lowest order. In this case, non-zero
beta-decay matrix elements arise from higher-order terms involving radial integrals and give more complicated expressions for the
η-coefficients. Beta-decay transitions of this type have also been
considered in the interacting boson–fermion model by Navrátil
and Dobeš (1988). Finally, we mention that in an approach similar to that leading to (7.37) the boson image of the double-beta
decay operator can be derived (Scholten and Yu, 1985).

8

Algebras

8.1 Introduction

The algebraic structure of the interacting boson–fermion model-2 is a combination of the algebraic structures discussed previously and those of Part II of Volume 1. There are now four parts corresponding to proton and neutron bosons and fermions. By combining these four pieces one can obtain a large number of possible couplings. Since these are simple extensions of the couplings described in Chapter 5 of Volume 1 and Chapter 2 of this volume, only a few selected examples will be discussed here.

8.2 Boson and fermion algebras

From the bilinear products of boson and fermion operators one can form now four algebras,

$$
\begin{aligned}
g_\pi^B &: B_{\pi,\alpha\beta} = b_{\pi,\alpha}^\dagger b_{\pi,\beta}, \\
g_\nu^B &: B_{\nu,\alpha\beta} = b_{\nu,\alpha}^\dagger b_{\nu,\beta}, \\
g_\pi^F &: A_{\pi,ik} = a_{\pi,i}^\dagger a_{\pi,k}, \\
g_\nu^F &: A_{\nu,ik} = a_{\nu,i}^\dagger a_{\nu,k}.
\end{aligned}
\tag{8.1}
$$

The algebras in (8.1) are the unitary algebras discussed previously,

$$
\begin{aligned}
g_\pi^B &= u_\pi^B(6), \quad g_\nu^B = u_\nu^B(6), \\
g_\pi^F &= u_\pi^B(\Omega_\pi), g_\nu^F = u_\nu^B(\Omega_\nu),
\end{aligned}
\tag{8.2}
$$

where Ω_π and Ω_ν are the dimensions of the fermionic spaces, i.e. $\Omega_\pi = \sum_{j_\pi}(2j_\pi+1)$ and $\Omega_\nu = \sum_{j_\nu}(2j_\nu+1)$. The algebraic structure of the model is thus that of the direct sum of all four algebras, or, using the notation appropriate for groups, the product

$$G = U_\pi^B(6) \otimes U_\nu^B(6) \otimes U_\pi^F(\Omega_\pi) \otimes U_\nu^B(\Omega_\nu). \tag{8.3}$$

The main question here is how to reduce this product to the rotation group O(3). There are two main routes, which will now be illustrated with an example. The first route is that in which bosons are first coupled and so are fermions and subsequently the combinations of bosons and fermions are coupled. The second route is that in which protons first are coupled and so are neutrons and subsequently the combinations of protons and neutrons are coupled. To clarify this, consider the case in which $\Omega_\pi = \Omega_\nu = 4$. This case has been extensively investigated (Hübsch and Paar, 1984; Hübsch *et al.*, 1985; Balantekin and Paar, 1986b; Hübsch and Paar, 1987). The first route corresponds to the lattice of algebras

$$
\begin{array}{ccccccccccc}
U_\pi^B(6) & & \otimes & & U_\nu^B(6) & \otimes & U_\pi^F(4) & & \otimes & & U_\nu^F(4)\\
\downarrow & \searrow & & \swarrow & \downarrow & & \downarrow & \searrow & & \swarrow & \downarrow\\
O_\pi^B(6) & & U_{\pi\nu}^B(6) & & O_\nu^B(6) & & SU_\pi^F(4) & & U_{\pi\nu}^F(4) & & SU_\nu^F(4)\\
\downarrow & \searrow & \downarrow & \swarrow & \downarrow & & \downarrow & \searrow & \downarrow & \swarrow & \downarrow\\
O_\pi^B(5) & & O_{\pi\nu}^B(6) & & O_\nu^B(5) & & Sp_\pi^F(4) & & SU_{\pi\nu}^F(4) & & Sp_\nu^F(4)\\
\downarrow & \searrow & \downarrow & \swarrow & \downarrow & & \downarrow & \searrow & \downarrow & \swarrow & \downarrow\\
O_\pi^B(3) & & O_{\pi\nu}^B(5) \searrow & & O_\nu^B(3) & & SU_\pi^F(2) \swarrow & & Sp_{\pi\nu}^F(4) & & SU_\nu^F(2)\\
& \searrow & \downarrow & \swarrow & & & & \searrow & \downarrow & \swarrow\\
& & O_{\pi\nu}^B(3) \searrow & \searrow & & & & \swarrow & SU_{\pi\nu}^F(2)\\
& & & & & O_{\pi\nu}^{BF}(6) & & & &\\
& & & \searrow & \searrow & \downarrow & \swarrow & \swarrow & &\\
& & & & & O_{\pi\nu}^{BF}(5) & & & &\\
& & & & \searrow & \downarrow & \swarrow & & &\\
& & & & & O_{\pi\nu}^{BF}(3) & & & &\\
& & & & & \downarrow & & & &\\
& & & & & O_{\pi\nu}^{BF}(2) & . & & &\\
\end{array}
$$

$$\tag{8.4}$$

The second route corresponds to the lattice of algebras

$$
\begin{array}{ccccccc}
U^B_\pi(6) & \otimes & U^F_\pi(4) & \otimes & U^B_\nu(6) & \otimes & U^F_\nu(4) \\
\downarrow & & \downarrow & & \downarrow & & \downarrow \\
O^B_\pi(6) & & SU^F_\pi(4) & & O^B_\nu(6) & & SU^F_\nu(4) \\
\downarrow \searrow & & \nearrow \downarrow & & \downarrow \searrow & & \nearrow \downarrow \\
O^B_\pi(5) & \mathrm{Spin}^{BF}_\pi(6) & Sp^F_\pi(4) & & O^B_\nu(5) & \mathrm{Spin}^{BF}_\nu(6) & Sp^F_\nu(4) \\
\downarrow \searrow & \downarrow & \nearrow \downarrow & & \downarrow \searrow & \downarrow & \nearrow \downarrow \\
O^B_\pi(3) & \mathrm{Spin}^{BF}_\pi(5) & \searrow SU^F_\pi(2) & & O^B_\nu(3) \nearrow & \mathrm{Spin}^{BF}_\nu(5) & SU^F_\nu(2) \\
\searrow & \downarrow & \nearrow & & & \downarrow & \nearrow \\
& \mathrm{Spin}^{BF}_\pi(3) & \searrow & & \nearrow & \mathrm{Spin}^{BF}_\nu(3) &
\end{array}
$$

$$
\begin{array}{c}
\searrow \qquad \searrow \qquad O^{BF}_{\pi\nu}(6) \qquad \nearrow \qquad \nearrow \\
\downarrow \\
O^{BF}_{\pi\nu}(5) \\
\searrow \qquad \downarrow \qquad \nearrow \\
O^{BF}_{\pi\nu}(3) \\
\downarrow \\
O^{BF}_{\pi\nu}(2) \quad .
\end{array}
$$

$$\text{(8.5)}$$

The complexity of the problem is clear from (8.4) and (8.5).

8.3 Dynamic symmetries

The only dynamic symmetry that will be considered here in detail is one that has found useful applications in the description of odd–odd nuclei in the region of the Au isotopes. This symmetry corresponds to bosons described by $O(6)$, protons occupying a single-particle level with $j_\pi = 3/2$, $\Omega_\pi = 4$ and neutrons occupying single-particle levels with $j_\nu = 1/2, 3/2, 5/2$, $\Omega_\nu = 12$.

8.3.1 Lattice of algebras

The lattice of algebras considered (Van Isacker *et al.*, 1985) is intermediate between the two schemes discussed in Sect. 8.2,

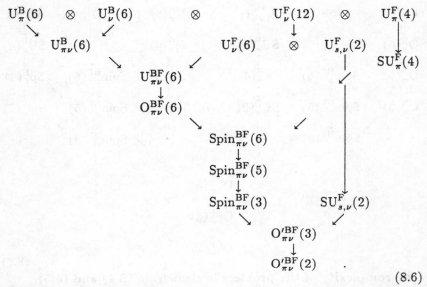

$$(8.6)$$

This lattice is a combination of those described in Sects. 3.2.1 and 3.2.3.

8.3.2 Energy eigenvalues

The usual procedure of writing the Hamiltonian in terms of Casimir operators gives

$$H = e_0 + e_1 \mathcal{C}_1(\mathrm{U}^{\mathrm{B}}_\pi 6) + e_2 \mathcal{C}_2(\mathrm{U}^{\mathrm{B}}_\pi 6) + e_3 \mathcal{C}_1(\mathrm{U}^{\mathrm{B}}_\nu 6) + e_4 \mathcal{C}_2(\mathrm{U}^{\mathrm{B}}_\nu 6)$$

$$+ e_5 \mathcal{C}_1(\mathrm{U}^{\mathrm{F}}_\pi 4) + e_6 \mathcal{C}_2(\mathrm{U}^{\mathrm{F}}_\pi 4) + e_7 \mathcal{C}_1(\mathrm{U}^{\mathrm{F}}_\nu 12) + e_8 \mathcal{C}_2(\mathrm{U}^{\mathrm{F}}_\nu 12)$$

$$+ a\mathcal{C}_1(\mathrm{U}^{\mathrm{B}}_{\pi\nu} 6) + a'\mathcal{C}_2(\mathrm{U}^{\mathrm{B}}_{\pi\nu} 6) + \eta \mathcal{C}_2(\mathrm{U}^{\mathrm{BF}}_{\pi\nu} 6)$$

$$+ \eta' \mathcal{C}_2(\mathrm{O}^{\mathrm{BF}}_{\pi\nu} 6) + \eta'' \mathcal{C}_2(\mathrm{Spin}^{\mathrm{BF}}_{\pi\nu} 6) + \beta \mathcal{C}_2(\mathrm{Spin}^{\mathrm{BF}}_{\pi\nu} 5)$$

$$+ \gamma \mathcal{C}_2(\mathrm{Spin}^{\mathrm{BF}}_{\pi\nu} 3) + \gamma' \mathcal{C}_2(\mathrm{SU}^{\mathrm{F}}_{s,\nu} 2) + \gamma'' \mathcal{C}_2(\mathrm{O}'^{\mathrm{BF}}_{\pi\nu} 3). \tag{8.7}$$

Taking the expectation value of H in the basis

$$
\begin{vmatrix}
U^B_\pi(6) & \otimes & U^B_\nu(6) & \otimes & U^F_\pi(4) & \otimes & U^F_\nu(12) \\
\downarrow & & \downarrow & & \downarrow & & \downarrow \\
[N_{B\pi} = N_\pi] & & [N_{B\nu} = N_\nu] & & \{N_{F\pi} = 1\} & & \{N_{F\nu} = 1\} \\[6pt]
\supset U^B_{\pi\nu}(6) \otimes U^F_\pi(4) \otimes U^F_\nu(6) \otimes U^F_{s,\nu}(2) & & & & & & \\
\quad\downarrow & & \quad\quad\downarrow & & & & \\
[N_1, N_2] & & [1] & & & & \\[6pt]
\supset U^{BF}_{\pi\nu}(6) \otimes U^F_\pi(4) \otimes U^F_{s,\nu}(2) \supset O^{BF}_{\pi\nu}(6) \otimes SU^F_\pi(4) \otimes U^F_{s,\nu}(2) & & & & & & \\
\quad\downarrow & & \quad\quad\downarrow & & & & \\
[N'_1, N'_2, N'_3] & & (\sigma_1, \sigma_2, \sigma_3) & & & & \\[6pt]
\supset \mathrm{Spin}^{BF}_{\pi\nu}(6) \otimes U^F_{s,\nu}(2) \supset \mathrm{Spin}^{BF}_{\pi\nu}(5) \otimes U^F_{s,\nu}(2) & & & & & & \\
\quad\downarrow & & \quad\downarrow & & & & \\
(\sigma'_1, \sigma'_2, \sigma'_3) & & (\tau_1, \tau_2) & & & & \\[6pt]
\supset \mathrm{Spin}^{BF}_{\pi\nu}(3) \otimes SU^F_{s,\nu}(2) \supset O'^{BF}_{\pi\nu}(3) \supset O'^{BF}_{\pi\nu}(2) & & & & & & \\
\quad\downarrow & & \quad\downarrow & & \downarrow & & \downarrow \\
\nu_\Delta, J & & S = 1/2 & & L & & M_L
\end{vmatrix}
$$

(8.8)

one obtains energy eigenvalues appropriate to describe odd–odd nuclei,

$$
\begin{aligned}
E(&N_{B\pi} = N_\pi, N_{B\nu} = N_\nu, N_{F\pi} = 1, N_{F\nu} = 1, [N'_1, N'_2, N'_3], \\
&(\sigma_1, \sigma_2, \sigma_3), (\sigma'_1, \sigma'_2, \sigma'_3), (\tau_1, \tau_2), \nu_\Delta, J, L, M_L) \\
=\ & e'_0 + a(N_1 + N_2) + a'[N_1(N_1 + 5) + N_2(N_2 + 3)] \\
&+ \eta[N'_1(N'_1 + 5) + N'_2(N'_2 + 3) + N'_3(N'_3 + 1)] \\
&+ 2\eta'[\sigma_1(\sigma_1 + 4) + \sigma_2(\sigma_2 + 2) + \sigma_3^2] \\
&+ 2\eta''[\sigma'_1(\sigma'_1 + 4) + \sigma'_2(\sigma'_2 + 2) + \sigma'^2_3] \\
&+ 2\beta[\tau_1(\tau_1 + 3) + \tau_2(\tau_2 + 1)] + 2\gamma J(J+1) + 2\gamma'' L(L+1),
\end{aligned}
$$

(8.9)

where the constant terms have been included in e'_0. Similar formulas can be obtained in the cases where the $U^F_\pi(4) \otimes U^F_\nu(12)$ representations are $\{0\} \otimes \{0\}$ (even–even nuclei), $\{1\} \otimes \{0\}$ (odd-proton nuclei) and $\{0\} \otimes \{1\}$ (odd-neutron nuclei).

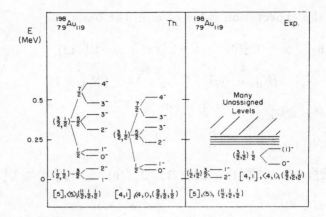

Fig. 8-1   An example of a nucleus with $U_\pi^B(6) \otimes U_\nu^B(6) \otimes U_\pi^F(4) \otimes U_\nu^F(12)$ symmetry: $^{198}_{79}\mathrm{Au}_{119}$ ($N_{B\pi} = 1, N_{B\nu} = 3, N_{F\pi} = 1, N_{F\nu} = 1$).

8.3.3 Examples of nuclei with $U_\pi^B(6) \otimes U_\nu^B(6) \otimes U_\pi^F(4) \otimes U_\nu^F(12)$ symmetry

Experimental examples of odd–odd nuclei which can be described by the expression (8.9) have been found in the Au region. One of these nuclei, $^{198}_{79}\mathrm{Au}_{119}$, is shown in Fig. 8.1. Recently, this nucleus has been remeasured by Warner *et al.* (1986) and many more states than shown in the figure have been established experimentally. However, due to the complexity of the odd–odd spectrum, it is difficult to establish a one-to-one correspondence between observed and calculated states. Thus, any assignment of quantum numbers to the observed levels in $^{198}_{79}\mathrm{Au}_{119}$ can only be viewed as tentative as long as they are not confirmed by nucleon transfer or electromagnetic decay properties. Simple analytic expressions are available for the former (Van Isacker, 1987) and will provide a test of proposed classifications of levels in nuclei in this mass region.

It is worthwhile commenting on the extreme difficulty, both experimental and theoretical, posed by odd–odd nuclei. From the experimental side, the high density of levels makes it very hard to assign spin and parity to individual states. This is reflected in the uncertainties shown in Fig. 8.1. A theoretical analysis of odd–odd nuclei, especially when many single-particle levels are included is hardly feasible. Dynamic symmetries offer here a unique opportunity. Despite the apparent complexity of the procedure described in this chapter, calculations are still feasible and straightforward. The only complication is in the bookkeeping aspect of the procedure, but this is greatly aided by the use of algebraic methods (group theory). It is in the treatment of these very complex cases that the full power of algebraic methods comes into play.

8.3.4 Examples of nuclei with $U_\pi^B(6) \otimes U_\nu^B(6) \otimes U_\pi^F(4) \otimes U_\nu^F(4)$ symmetry

Although not discussed here in detail, we note that examples of dynamic Bose–Fermi symmetries based on the chain (8.4) have been found in the spectra of the odd–odd Cu isotopes, in particular of $^{62}_{29}Cu_{33}$ (Hübsch and Paar, 1987). The odd–even Cu isotopes were discussed in Sect. 3.3.1.9 as examples of nuclei with $Spin^{BF}(5)$ symmetry. The odd–odd isotopes provide further examples of $Spin^{BF}(5)$ symmetry in the case in which the odd proton occupies an orbit with $j_\pi = 3/2$ and the odd neutron one with $j_\nu = 3/2$.

9

Superalgebras

9.1 Introduction

Superalgebras can also be used within the context of the interacting boson–fermion model-2. The corresponding algebraic structures are similar to those discussed in Chapter 4. The only difference is that proton and neutron indices now appear everywhere and that the number of routes possible in the reduction of the superalgebra to the rotation algebra increases considerably. Superalgebras based on the interacting boson–fermion model-2 are particularly useful in the description of odd–odd nuclei, since by fixing the parameters of the Hamiltonian and other operators from a study of even–even and odd–even nuclei, one is able to predict the structure of odd–odd nuclei. These predictions can be compared with experimental data (when they exist) or used as a guideline for future experiments. In this chapter only a few selected cases will be presented.

9.2 Supersymmetric chains

Supersymmetric chains can be obtained by embedding the algebras of Chapter 8 into superalgebras. There are again two main routes that will be illustrated with an example. Consider the case in which both protons and neutrons occupy a level with $j_\pi = j_\nu = 3/2$. The corresponding algebraic structure has been discussed in Sect. 8.2 and can be embedded into the superalgebra

$$G^* = U_\pi(6|4) \otimes U_\nu(6|4). \tag{9.1}$$

When considering the subalgebras of (9.1) one can either first combine the two subalgebras into their sum,

$$U_\pi(6|4) \otimes U_\nu(6|4) \supset U(6|4) \supset U^B(6) \otimes U^F(4), \qquad (9.2)$$

where the algebra $U(6|4)$ is obtained by adding the generators of $U_\pi(6|4)$ to the corresponding generators of $U_\nu(6|4)$, or one can go directly from the proton and neutron superalgebras to their maximal Lie subalgebras,

$$U_\pi(6|4) \otimes U_\nu(6|4) \supset U_\pi^B(6) \otimes U_\nu^B(6) \otimes U_\pi^F(4) \otimes U_\nu^F(4) \supset U^B(6) \otimes U^F(4).$$
$$(9.3)$$

The first alternative only exists if the proton and neutron spaces are identical, $\Omega_\pi = \Omega_\nu$. If the first alternative is possible, one can introduce a formalism similar to F-spin, but now applied to superalgebras. Proton bosons and fermions can be assigned to a supermultiplet with $F = 1/2$ and F-spin projection $F_z = +1/2$. Similarly, neutron bosons and fermions have $F = 1/2$ and $F_z = -1/2$, i.e.

$$|\pi\rangle = |\tfrac{1}{2}, +\tfrac{1}{2}\rangle,$$
$$|\nu\rangle = |\tfrac{1}{2}, -\tfrac{1}{2}\rangle. \qquad (9.4)$$

The supersymmetric multiplets now contain

$$|\pi\rangle = \begin{pmatrix} b_{\pi,\alpha}^\dagger \\ a_{\pi,i}^\dagger \end{pmatrix}; \quad |\nu\rangle = \begin{pmatrix} b_{\nu,\alpha}^\dagger \\ a_{\nu,i}^\dagger \end{pmatrix}; \quad (\alpha = 1, \dots, 6; i = 1, \dots, 4).$$
$$(9.5)$$

The F-spin basis can be obtained by constructing the Kronecker products of two $U(6|4)$ representations. The rules for this product, when expressed in terms of Young supertableaux, are identical to those of normal Lie algebras. For example,

$$\boxtimes \otimes \boxtimes = \boxtimes \boxtimes \oplus \begin{matrix} \boxtimes \\ \boxtimes \end{matrix}, \qquad (9.6)$$

or

$$[1\} \otimes [1\} = [2\} \oplus [1,1\}. \qquad (9.7)$$

One obtains in this case Young supertableaux which are not totally supersymmetric.

9.3   Dynamic supersymmetries

In heavy nuclei, the situation described in the previous section (the F-spin scheme) seldom occurs. One must therefore use the second possible reduction (9.3). Some examples of this kind have been found. One such example corresponds to the embedding of the chains discussed in Sect. 8.3 into

$$
\left|
\begin{array}{cccccc}
U_\pi(6|4) & \otimes & U_\nu(6|12) & \supset & U_\pi^B(6) \otimes U_\nu^B(6) \otimes U_\pi^F(4) \otimes U_\nu^F(12) \\
\downarrow & & \downarrow & & \downarrow \quad\; \downarrow \quad\;\; \downarrow \quad\;\;\; \downarrow \\
[\mathcal{N}_\pi\} & & [\mathcal{N}_\nu\} & & [N_{B\pi}] \;\; [N_{B\nu}] \;\; \{N_{F\pi}\} \;\; \{N_{F\nu}\}
\end{array}
\right\rangle .
$$

$$(9.8)$$

Use of supersymmetry now allows the construction of supermultiplets obtained by combining the proton supermultiplets with the neutron supermultiplets. Those can be constructed as discussed in Chapter 4. The important portion of the supermultiplet that can be accessed easily is that formed by an even–even nucleus, the adjoining odd–even and even–odd nuclei and the neighboring odd–odd nucleus, i.e. the nuclei with

$$
\begin{array}{llll}
N_{B\pi} = \mathcal{N}_\pi, & N_{F\pi} = 0, & N_{B\nu} = \mathcal{N}_\nu, & N_{F\nu} = 0, \\
N_{B\pi} = \mathcal{N}_\pi - 1, & N_{F\pi} = 1, & N_{B\nu} = \mathcal{N}_\nu, & N_{F\nu} = 0, \\
N_{B\pi} = \mathcal{N}_\pi, & N_{F\pi} = 0, & N_{B\nu} = \mathcal{N}_\nu - 1, & N_{F\nu} = 1, \\
N_{B\pi} = \mathcal{N}_\pi - 1, & N_{F\pi} = 1, & N_{B\nu} = \mathcal{N}_\nu - 1, & N_{F\nu} = 1.
\end{array}
$$

$$(9.9)$$

All these nuclei belong to the supermultiplet $[\mathcal{N}_\pi\} \otimes [\mathcal{N}_\nu\}$. The set of four nuclei (9.9) has been termed a magic square. An example of such a magic square is shown in Fig. 9.1. If a dynamic supersymmetry is present, all nuclei in the square should be described with the same Hamiltonian. For the nuclei shown in Fig. 9.1 the appropriate Hamiltonian is given by (8.7). One must insert in this formula the appropriate eigenvalues of the Casimir operators corresponding to the four cases in (9.9). The odd–odd formula is given by (8.9). The even–even and odd–even formulas are obtained in the manner discussed in Volume 1 and in Part I of this

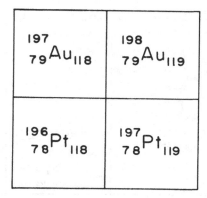

Fig. 9-1   An example of a magic square in the Pt–Au region.

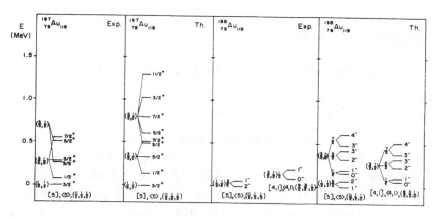

Fig. 9-2   Comparison between experimental and theoretical spectra of nuclei belonging to the magic square of Fig. 9.1: $^{197}_{79}\mathrm{Au}_{118}$ and $^{198}_{79}\mathrm{Au}_{119}$.

volume. A comparison of the spectra obtained in this way with those experimentally measured is shown in Figs. 9.2 and 9.3.

Another example of a dynamic supersymmetry including even-even, even–odd, odd–even and odd–odd nuclei has been presented by Hübsch and Paar (1987) in the region of the Cu isotopes.

A further generalization of these studies can be achieved by embedding the direct product of proton and neutron superalgebras into a single, larger superalgebra. This, in general, can be written

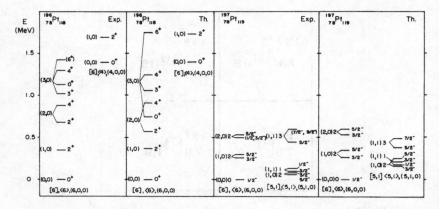

Fig. 9-3   Comparison between experimental and theoretical spectra of nuclei belonging to the magic square of Fig. 9.1: $^{196}_{78}\text{Pt}_{118}$ and $^{197}_{78}\text{Pt}_{119}$.

as (Jolie *et al.*, 1985a):

$$
\left| \begin{array}{ccc}
\mathrm{U}(12|\Omega_\pi + \Omega_\nu) & \supset & \mathrm{U}_\pi(6|\Omega_\pi) \otimes \mathrm{U}_\nu(6|\Omega_\nu) \\
\downarrow & & \downarrow \qquad\qquad \downarrow \\
[\mathcal{N}\} & & [\mathcal{N}_\pi\} \qquad\qquad [\mathcal{N}_\nu\}
\end{array} \right\rangle, \qquad (9.10)
$$

with $\mathcal{N}_\pi + \mathcal{N}_\nu = \mathcal{N}$. The single representation $[\mathcal{N}\}$ of $\mathrm{U}(12|\Omega_\pi + \Omega_\nu)$ now not only contains all even–even nuclei with $N_{\mathrm{B}\pi} + N_{\mathrm{B}\nu} = \mathcal{N}$ but also all associated even–odd, odd–even and odd–odd nuclei, as specified in (9.9). It is clear that, due to the large number of nuclei contained in one multiplet, such schemes have only a very limited applicability.

10

Numerical studies

10.1 Introduction

The degree of complexity when going from even–even to odd–even nuclei increases by at least one order of magnitude. It increases further by another order of magnitude when going from odd–even to odd–odd nuclei. Although the dynamic symmetries discussed in the previous chapter may give some insight into the structure of specific examples of nuclei, they cannot be used in all cases and one must resort to more realistic, numerical calculations. Many odd–even nuclei have been studied in this way with the interacting boson–fermion model-2, using a computer code written by Bijker (1983) and an example will be discussed in Sect. 10.2. Odd–odd nuclei, being more complex, have been studied less accordingly, but nevertheless a few calculations are available. Finally, we also discuss in this chapter an example of a broken-pair calculations for even–even nuclei.

10.2 Odd–even nuclei

In view of the large number of parameters appearing in the operators of Chapter 7, a semimicroscopic input is introduced (Alonso et al., 1984; Arias, 1985; Alonso, 1986) similar to the one discussed in Sect. 5.6. Here this input is more appropriate, since the interacting boson–fermion model-2 is directly related to the underlying shell model. Coefficients in the operators of Chapter 7 are written in terms of the occupation probabilities obtained through a BCS calculation (Bardeen, Cooper and Schrieffer, 1957). This calculation is done separately for protons and neutrons and provides the single-particle energies in the presence of several valence particles (quasi-particle energies, ϵ_{j_π} and ϵ_{j_ν}), in terms of the Fermi energies λ_π and λ_ν, the pairing gaps Δ_π and Δ_ν and the single-particle energies in the absence of other valence particles, E_{j_π} and E_{j_ν},

$$\epsilon_{j_\rho} = \sqrt{(E_{j_\rho} - \lambda_\rho)^2 + \Delta_\rho^2}, \qquad \rho = \pi, \nu. \tag{10.1}$$

The occupation probabilities are then given by

$$v_{j_\rho} = \left[\frac{1}{2} \left(1 - \frac{E_{j_\rho} - \lambda_\rho}{\epsilon_{j_\rho}} \right) \right]^{\frac{1}{2}},$$

$$u_{j_\rho} = (1 - v_{j_\rho}^2)^{\frac{1}{2}}, \qquad \rho = \pi, \nu. \tag{10.2}$$

The pairing gaps usually are taken as $\Delta = 12A^{-\frac{1}{2}}$ MeV (Bohr and Mottelson, 1975). The Fermi energies are obtained by solving (10.1) and (10.2) with the condition that the number of nucleons be

$$n_{j_\rho} = \sum_{j_\rho} v_{j_\rho}^2 (2j_\rho + 1), \qquad \rho = \pi, \nu. \tag{10.3}$$

For the calculation of energies and wave functions one needs the parameters of the boson–fermion interaction. On the basis of BCS theory one can write them in the form

$$\Gamma_{j_\rho j_\rho'} = (u_{j_\rho} u_{j_\rho'} - v_{j_\rho} v_{j_\rho'}) Q_{j_\rho j_\rho'} \Gamma_\rho,$$

$$Q_{j_\rho j_\rho'} = \langle l_\rho, \tfrac{1}{2}, j_\rho \| Y^{(2)} \| l_\rho', \tfrac{1}{2}, j_\rho' \rangle, \qquad \rho = \pi, \nu, \tag{10.4}$$

and

$$\Lambda_{j_\rho j_\rho'}^{j_\rho''} = -\beta_{j_\rho j_\rho'} \beta_{j_\rho'' j_\rho} \left(\frac{10}{N_\rho (2j_\rho + 1)} \right)^{\frac{1}{2}} \Lambda_\rho,$$

$$\beta_{j_\rho j_\rho'} = (u_{j_\rho} v_{j_\rho'} + v_{j_\rho} u_{j_\rho'}) Q_{j_\rho j_\rho'}, \qquad \rho = \pi, \nu. \tag{10.5}$$

All energies in odd–even nuclei are then calculated in terms of three coefficients, A_ρ, Γ_ρ and Λ_ρ. The BCS theory also provides a simple parametrization of the coefficients appearing in the one-nucleon transfer operators (7.33) and (7.34):

$$p_{j_\rho} = u_{j_\rho} \frac{1}{K'_{j_\rho}},$$

$$q_{j'_\rho}^{(j_\rho)} = -v_{j_\rho}\beta_{j'_\rho j_\rho} \left(\frac{10}{N_\rho(2j_\rho+1)}\right)^{\frac{1}{2}} \frac{1}{K_\rho K'_{j_\rho}},$$

$$p'_{j_\rho} = \frac{v_{j_\rho}}{\sqrt{N_\rho}} \frac{1}{K''_{j_\rho}},$$

$$q_{j'_\rho}^{'''(j_\rho)} = u_{j_\rho}\beta_{j'_\rho j_\rho} \left(\frac{10}{2j_\rho+1}\right)^{\frac{1}{2}} \frac{1}{K_\rho K''_{j_\rho}}, \qquad \rho = \pi,\nu, \quad (10.6)$$

where K_ρ, K'_{j_ρ} and K''_{j_ρ} are obtained from the three conditions:

$$K_\rho = \left(\sum_{j_\rho j'_\rho} \beta^2_{j_\rho j'_\rho}\right)^{\frac{1}{2}}, \qquad \rho = \pi,\nu, \qquad (10.7)$$

and

$$\sum_{\alpha J} \langle \text{odd}; \alpha J \parallel P_+^{(j_\rho)} \parallel \text{even}; 0_1^+\rangle^2 = (2j_\rho+1)u_{j_\rho}^2,$$

$$\sum_{\alpha J} \langle \text{even}; 0_1^+ \parallel P_+^{'(j_\rho)} \parallel \text{odd}; \alpha J\rangle^2 = (2j_\rho+1)v_{j_\rho}^2, \quad \rho = \pi,\nu.$$

$$(10.8)$$

The formulas (10.4)–(10.8) are valid when the odd nucleon is a particle. Corresponding formulas for a hole are obtained by interchanging u_{j_ρ} and v_{j_ρ}. From this parametrization one can obtain that of other operators, since these can be built from one-nucleon operators.

An example of a numerical calculation for a series of isotopes will now be discussed. Table 10.1 lists other, similar calculations that can found in the literature.

10.2.1 Energies

The calculation that will be described here concerns the odd–even isotopes of Xe ($Z = 54$) and Cs ($Z = 55$). The single-particle

Table 10-1 List of available numerical calculations of nuclear properties using the interacting boson–fermion model-2

Proton number	Odd particle	Neutron number	Single-particle orbits	Reference
43 Tc	π	54–60	$2d_{5/2}, 1g_{9/2}$ $2p_{1/2}, 2p_{3/2}, 1f_{5/2}, 1f_{7/2}$	Arias et al. (1987)
44 Ru	ν	55–61	$3s_{1/2}, 2d_{3/2}, 2d_{5/2}, 1g_{7/2},$ $1g_{9/2}$ $2f_{7/2}, 1h_{9/2}, 1h_{11/2}$	Arias et al. (1987)
45 Rh	π	54–60	$2d_{5/2}, 1g_{9/2}$ $2p_{1/2}, 2p_{3/2}, 1f_{5/2}, 1f_{7/2}$	Arias et al. (1987)
46 Pd	ν	55–61	$3s_{1/2}, 2d_{3/2}, 2d_{5/2}, 1g_{7/2},$ $1g_{9/2}$ $2f_{7/2}, 1h_{9/2}, 1h_{11/2}$	Arias et al. (1987)
52 Te	ν	53–79	$3s_{1/2}, 2d_{3/2}, 2d_{5/2}, 1g_{7/2}$ $1h_{11/2}$	Dellagiacoma (1988)
53 I	π	52–80	$3s_{1/2}, 2d_{3/2}, 2d_{5/2}, 1g_{7/2}$ $1h_{11/2}$	Dellagiacoma (1988)
54 Xe	ν	55–77	$3s_{1/2}, 2d_{3/2}, 2d_{5/2}, 1g_{7/2}$ $1h_{11/2}$	Arias et al. (1985)
55 Cs	π	54–78	$3s_{1/2}, 2d_{3/2}, 2d_{5/2}, 1g_{7/2}$ $1h_{11/2}$	Arias et al. (1985)
56 Ba	ν	53–79	$3s_{1/2}, 2d_{3/2}, 2d_{5/2}, 1g_{7/2}$ $1h_{11/2}$	Dellagiacoma (1988)
57 La	π	52–80	$3s_{1/2}, 2d_{3/2}, 2d_{5/2}, 1g_{7/2}$ $1h_{11/2}$	Dellagiacoma (1988)
63 Eu	π	84–92	$3s_{1/2}, 2d_{3/2}, 2d_{5/2}, 1g_{7/2}$ $1h_{11/2}$	Alonso et al. (1988)
77 Ir	π	108–118	$3s_{1/2}, 2d_{3/2}, 2d_{5/2}, 1g_{7/2}$ $1h_{9/2}, 1h_{11/2}$	Arias et al. (1986)
78 Pt	ν	107–117	$1i_{13/2}$ $3p_{1/2}, 3p_{3/2}, 2f_{5/2}, 2f_{7/2},$ $1h_{9/2}$	Arias et al. (1986)
79 Au	π	106–118	$1h_{9/2}$	Arias et al. (1986)

levels included in the calculation and their energies are given in Table 10.2. In addition, one needs the parameters appearing in the boson part of the Hamiltonian, H_B. These are determined by the energies of nuclei with no unpaired nucleons (even–even nuclei).

Table 10-2 Single-particle energies E_{j_ρ} (MeV) in the 50–82 shell

	$1g_{7/2}$	$2d_{5/2}$	$1h_{11/2}$	$3s_{1/2}$	$2d_{3/2}$
Proton levels	0.00	0.60	1.50	3.35	3.00
Neutron levels	0.00	0.80	2.00	2.10	2.50

Table 10-3 Boson parameters for the even–even Xe isotopes (Puddu et al., 1980)[a]

A	ϵ^b (MeV)	κ (MeV)	χ_ν	$c_0^{(\nu)}$ (MeV)	$c_2^{(\nu)}$ (MeV)
108	0.96	−0.255	−0.80	0.30	0.10
110	0.96	−0.215	−1.00	0.30	0.10
112	0.96	−0.185	−1.00	0.20	0.00
114	0.94	−0.155	−1.00	−0.20	−0.15
116	0.85	−0.135	−0.80	−0.25	−0.15
118	0.78	−0.130	−0.60	−0.25	−0.12
120	0.76	−0.130	−0.40	−0.20	−0.12
122	0.72	−0.137	−0.20	−0.05	−0.12
124	0.70	−0.145	0.00	0.05	−0.10
126	0.70	−0.155	0.20	0.10	−0.10
128	0.70	−0.170	0.33	0.30	0.00
130	0.76	−0.190	0.50	0.30	0.10
132	0.90	−0.210	0.90	0.30	0.10

[a] $\chi_\pi = -0.80$, $\xi_1 = \xi_2 = 0.12$ MeV and $\xi_3 = -0.09$ MeV for all nuclei. Other parameters not mentioned in the table are zero.

[b] $\epsilon_\pi = \epsilon_\nu = \epsilon$.

In the case discussed here these parameters are taken from the calculation of the even Xe isotopes discussed in Volume 1, Sect. 5.7. The appropriate parameters are shown in Table 10.3. The only new parameters needed for the calculation of odd–even nuclei, are the strengths of the monopole, quadrupole and exchange interactions, A_ρ, Γ_ρ and Λ_ρ. The calculation separates into two parts, one related to the negative-parity states and another part related

Table 10-4 Boson–fermion interaction
parameters (MeV) in the Xe–Cs isotopes
(Arias et al., 1985)

	Γ_ρ	Λ_ρ	A_ρ
(i) Negative-parity states			
Protons, $\rho = \pi$	0.60	1.30	−0.30
Neutrons, $\rho = \nu$	0.60	1.30	0.00
(ii) Positive-parity states			
Protons, $\rho = \pi$	0.60	0.10	−0.20
Neutrons, $\rho = \nu$	0.40	0.10	−0.50

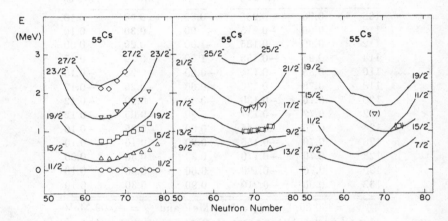

Fig. 10-1 Comparison between calculated (lines) and experimental (circles, triangles, squares) negative-parity spectra of odd–even $_{55}$Cs isotopes (Arias *et al.*, 1985). All energies are plotted relative to the lowest $11/2_1^-$ state.

to the positive-parity states. The corresponding values of the parameters are given in Table 10.4. The resulting energies are shown in Figs. 10.1, 10.2, 10.3 and 10.4.

It is of interest to contrast spectra of odd-proton nuclei with those of odd-neutron nuclei with the same mass number. One observes major differences. These differences arise from the fact that the boson–fermion coupling depends on the occupa-

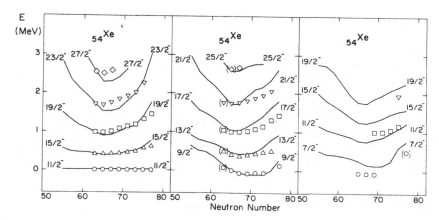

Fig. 10-2 Comparison between calculated (lines) and experimental (circles, triangles, squares) negative-parity spectra of odd–even $_{54}$Xe isotopes (Arias *et al.*, 1985). All energies are plotted relative to the lowest $11/2_1^-$ state.

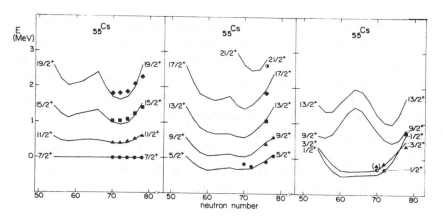

Fig. 10-3 Comparison between calculated (lines) and experimental (circles, triangles, squares) positive-parity spectra of odd–even $_{55}$Cs isotopes (Arias *et al.*, 1985). All energies are plotted relative to the lowest $7/2_1^+$ state.

tion probabilities which are different for protons and neutrons. The negative-parity states, for example, originate from the $1h_{11/2}$ single-particle level. The proton level $\pi 1h_{11/2}$ is almost completely empty in Cs, which has only five proton particles in the valence

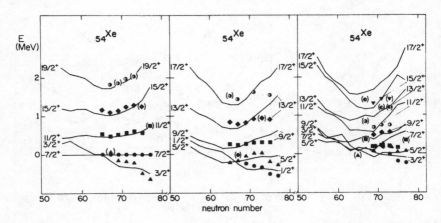

Fig. 10-4   Comparison between calculated (lines) and experimental (circles, triangles, squares) positive-parity spectra of odd–even $_{54}$Xe isotopes (Arias *et al.*, 1985). All energies are plotted relative to the lowest $7/2_1^+$ state.

shell. The emptiness of this level remains constant as neutrons are added. On the other hand, the neutron $\nu 1h_{11/2}$ in the odd–even Xe isotopes becomes increasingly full as more and more neutrons are added. As a result, the structure of the Cs isotopes is dominated by the quadrupole boson–fermion interaction, while the structure of the Xe isotopes arises from an interplay between the quadrupole and the exchange interactions. This is particularly evident in the lowering in Xe of a state with $J^P = 9/2^-$ in the middle of the shell. This feature arises from the exchange interaction as one can see by using arguments similar to those discussed in Sect. 5.2.

10.2.2 *Electromagnetic transitions and moments; E2*

Matrix elements of electromagnetic transition operators are calculated using the wave functions obtained from the numerical diagonalization of H and the operators discussed in Sect. 7.5.3. E2 transitions and moments are given in terms of the boson effective charges, e_π^B and e_ν^B. These are taken from the calculations reported in Volume 1, $e_\pi^B = e_\nu^B = 0.12\ eb$. The fermion part of the operator requires the fermion effective charges and radial integrals.

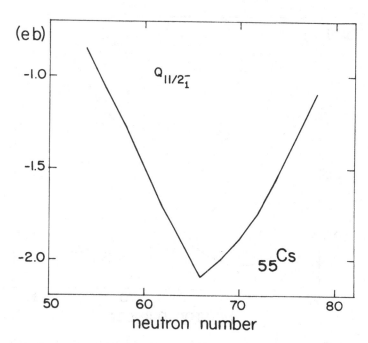

Fig. 10-5   Electric quadrupole moments of the lowest $11/2_1^-$ state in the $_{55}$Cs isotopes (Arias *et al.*, 1985).

The radial integrals are estimated to be $\langle n_\rho, l_\rho | r^2 | n_\rho, l_\rho \rangle = 0.0033\, b$ for the $1h_{11/2}$ level and $\langle n_\rho, l_\rho | r^2 | n_\rho, l_\rho \rangle = 0.0027\, b$ for the positive-parity levels. The fermion effective charges are taken as $e_\pi^{\rm F} = 1.5\,e$ and $e_\nu^{\rm F} = 0.5\,e$.

Without any further parameters, one can then compute all E2 transitions and moments. A portion of these results is shown in Figs. 10.5, 10.6, 10.7 and 10.8.   The experimental information on electromagnetic transitions and moments in odd–even nuclei in this mass region is rather meager. E2 transitions in odd–even nuclei are still dominated by the collective boson part. The fermion part contributes only 5–10% to the matrix elements. A study of the latter must thus await more accurate and systematic measurements of E2 transitions.

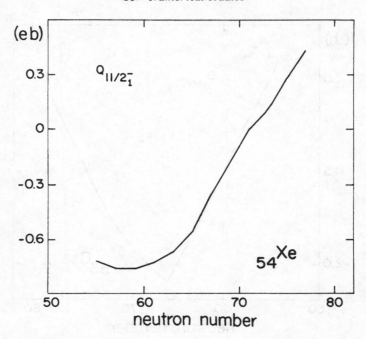

Fig. 10-6   Electric quadrupole moments of the lowest $11/2_1^-$ state in the $_{54}$Xe isotopes (Arias *et al.*, 1985).

10.2.3 Electromagnetic transitions and moments; M1

In contrast to E2 properties, M1 transitions and moments in odd–even nuclei are dominated by the fermion part of the M1 operator. Using the operator of Sect. 7.5.3, one can compute the corresponding transitions. The boson part of the operator requires a specification of g_π^B and g_ν^B. These can be taken from the calculations reported in Volume 1 for even–even nuclei (Sambataro *et al.*, 1984). The fermion part of the operator requires a specification of the fermion g-factors. The orbital g-factors are $g_{l,\pi}^F = 1\ \mu_N$ and $g_{l,\nu}^F = 0$. The spin g-factors are taken as the free values quenched by a factor of 0.7, i.e. $g_{s,\pi}^F = 0.7 \times 5.58\ \mu_N$ and $g_{s,\nu}^F = -0.7 \times 3.82\ \mu_N$. A portion of the results is shown in Figs. 10.9 and 10.10. Also here the experimental information is rather scarce. For those cases for which experimental data exist, the results of calculations

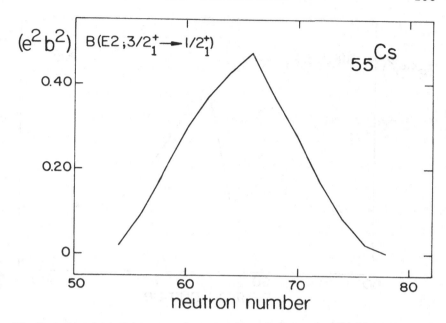

Fig. 10-7 $B(E2; 3/2_1^+ \rightarrow 1/2_1^+)$ values in the $_{55}$Cs isotopes (Arias *et al.*, 1985).

of M1 transitions agree in general less well with the data as compared to the corresponding calculations of E2 transitions. This indicates that while the collective degrees of freedom appear to be well described in odd–even nuclei, the single-particle degrees of freedom still require improvement.

10.2.4 Beta-decay probabilities

The image of beta-decay operators in the boson–fermion space has been described in Sect. 7.5.5. The transition probabilities are usually quoted through the ft value, defined by

$$ft = \frac{6163}{\langle M_F \rangle^2 + (G_A/G_V)^2 \langle M_{GT} \rangle^2} \text{ sec.} \qquad (10.9)$$

Here $\langle M_F \rangle$ and $\langle M_{GT} \rangle$ are the matrix elements of the operators $\hat{T}^{\beta F}$ and $\hat{T}^{\beta GT}$ and G_A and G_V are the axial vector and vector

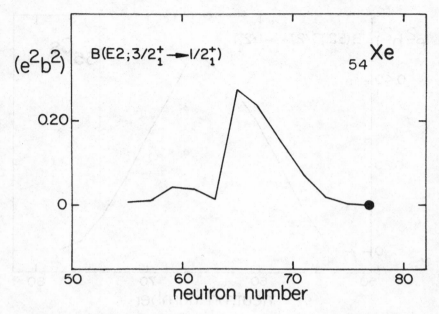

Fig. 10-8 $B(E2; 3/2_1^+ \to 1/2_1^+)$ values in the $_{54}$Xe isotopes (Arias *et al.*, 1985).

coupling constants, whose ratio in free space is $(G_A/G_V)^2 = 1.59$. Using (7.35)–(7.37), ft values can be calculated without introducing new parameters. The calculations typically reproduce the main features of the data, in particular, the trend with increasing proton or neutron number, but they overestimate the absolute value of the matrix elements by approximately a factor of 2. This result is also known from shell-model calculations There are two reasons for it. First, the value of the axial vector coupling constant is likely to be renormalized to a value $(G_A/G_V)^2 \approx 1.2$. Second, since the matrix elements of the operator $\hat{T}^{\beta\mathrm{GT}}$ are very sensitive to small components in the wave functions, the inclusion of only the valence space may not be sufficient to describe the data. An example of calculations of beta-decay probabilities for transitions from odd–even to even–odd nuclei is shown in Fig. 10.11. These calculations are done with an effective Gamow–Teller operator of

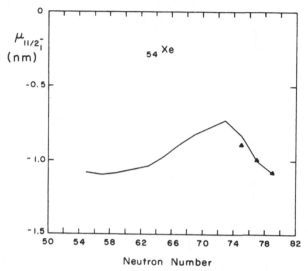

Fig. 10-9   Magnetic dipole moments of the lowest $11/2_1^-$ state in the $_{54}$Xe isotopes (Dellagiacoma, 1988).

the type

$$\eta_{j_\pi,j_\nu}^{'\mathrm{GT}} = \zeta^{\mathrm{GT}} \eta_{j_\pi,j_\nu}^{\mathrm{GT}}, \tag{10.10}$$

with $\eta_{j_\pi,j_\nu}^{\mathrm{GT}}$ given by (7.37) and $\zeta^{\mathrm{GT}} = 0.53$. The ft values can be obtained by inserting $\langle M_{\mathrm{GT}} \rangle^2$ in (10.9). In the case of transitions between low-lying states in medium-mass nuclei, the contribution of the Fermi matrix element, $\langle M_{\mathrm{F}} \rangle^2$, is absent since the low-lying states in the parent and daughter nucleus have different isospin.

In addition one can calculate matrix elements of the Gamow–Teller operator between states other than the ground states. These matrix elements can be extracted experimentally from a study of the (p,n) reaction. An example of such a calculation is shown in Fig. 10.12. The agreement between theory and experiment for transitions to excited states is usually only qualitative. This is due in part to the difficulty in extracting the matrix elements from the data and in part to the fact that beta decay is mostly a single-particle process, not a collective one.

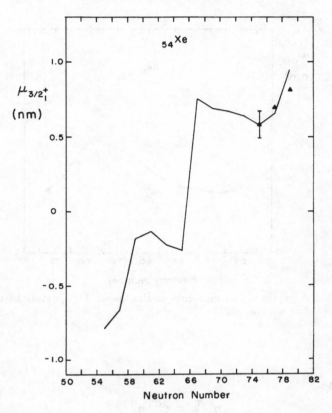

Fig. 10-10   Magnetic dipole moments of the lowest $3/2_1^+$ state in the $_{54}$Xe isotopes (Dellagiacoma, 1988).

10.3   Odd–odd nuclei

A brief discussion of odd–odd nuclei has been given in Chapters 8 and 9 within the context of Bose–Fermi symmetries and supersymmetries. In recent years considerable effort has gone into the study of these nuclei both from the experimental and theoretical points of view. Numerical calculations for odd–odd nuclei are rather difficult in view of the large size of the matrices that have to be diagonalized. Nonetheless, some calculations have been done (see Table 10.5). In Fig. 10.13 we show the results of one

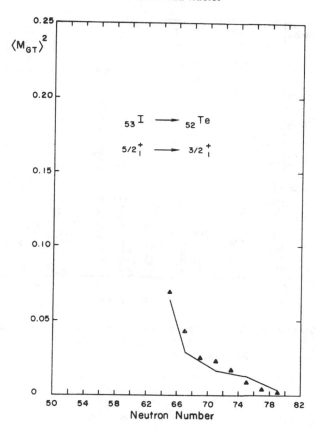

Fig. 10-11   Gamow–Teller matrix elements for the transitions $_{53}$I($5/2_1^+$) $\to$ $_{52}$Te($3/2_1^+$) (Dellagiacoma, 1988).

such a calculation (Lopac *et al.*, 1986) and compare it with the experimental data taken from Warner *et al.* (1986).

In addition to properties originating from the collective nature of the spectra, there is in odd–odd nuclei a new aspect originating from the residual interaction between the unpaired proton and neutron. This aspect is particular evident in the so-called parabolic rule of proton–neutron multiplets (Paar, 1979). In order to illustrate this rule it is convenient to use the coupling scheme (8.4) and to consider the case in which protons (neutrons) occupy only one orbit j_π (j_ν). The basis states for odd–odd nuclei can be

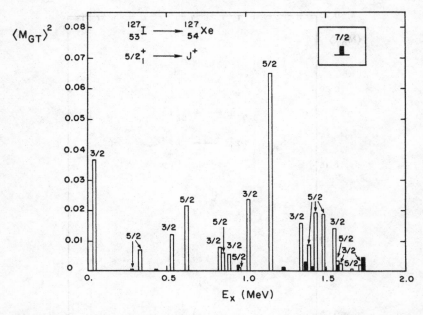

Fig. 10-12   Gamow–Teller matrix elements for transitions from the ground state of $^{127}_{53}\text{I}_{74}$ to all states of $^{127}_{54}\text{Xe}_{73}$ below 1.7 MeV (Dellagiacoma, 1988) as a function of excitation energy E_x.

Table 10-5 List of available numerical calculations for odd–odd nuclei using the interacting boson–fermion model-2

Proton number	Neutron number	Reference
57 La	83	Scholten *et al.* (1989)
59 Pr	83	Scholten *et al.* (1989)
61 Pm	83	Scholten *et al.* (1989)
63 Eu	83	Scholten *et al.* (1989)
79 Au	115	Blasi and Lo Bianco (1987)
79 Au	119	Lopac *et al.* (1986)

written as

$$\mathcal{BF}: \quad [[[N_{B\pi}, \varphi_\pi]^{(L_\pi)} \times [N_{B\nu}, \varphi_\nu]^{(L_\nu)}]^{(L_{\pi\nu})} \times [j_\pi \times j_\nu]^{(J_{\pi\nu})}]^{(J)}_{M_J} |o\rangle,$$

(10.11)

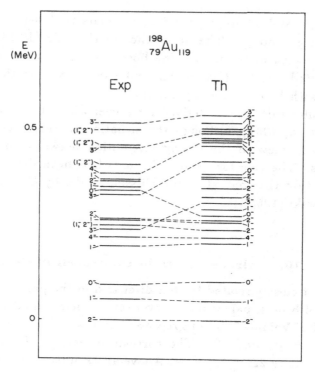

Fig. 10-13 Comparison between calculated (Th) and experimental (Exp) spectra of the odd–odd nucleus $^{198}_{79}\mathrm{Au}_{119}$ (Lopac *et al.*, 1986).

where φ_π and φ_ν generically denote boson quantum numbers. The parabolic rule arises from the dependence of the energies of the states on the angular momentum J. It is particularly simple in the case of nuclei with U(5) symmetry. Here, to a good approximation, the lowest states of odd–odd nuclei are characterized by wave functions of the type (10.11) with $L_{\pi\nu} = 0$ and thus $J = J_{\pi\nu}$. For these states the values of J are:

$$J = |j_\pi - j_\nu|, \ldots, j_\pi + j_\nu. \tag{10.12}$$

The dependence of the energies on J arises from the residual proton–neutron interaction in the orbits j_π and j_ν. To a good approximation, one obtains

$$E(J) = a + b\Big(J(J+1) - J_0(J_0+1)\Big)^2, \tag{10.13}$$

where a, b and J_0 are appropriate numbers that depend on the values of j_π and j_ν. The occurrence of the rule (10.13) within the framework of the interacting boson–fermion model has been investigated recently (Balantekin and Paar, 1986a) in the case of bosons with U(5) or O(6) symmetry.

A similar rule can be derived for nuclei with SU(3) symmetry (Paar et al., 1987), although the situation in this case is more involved in view of the strong interaction between bosons and fermions. The rule for deformed nuclei had been derived previously within the context of the Nilsson model by Gallagher and Moszkowski (1958).

10.4   Broken pairs in even–even nuclei

A subject closely related to that discussed in the previous section is that of broken pair states in even–even nuclei. As discussed in Sect. 4.2 of Volume 1, bosons represent correlated pairs of nucleons with $J^P = 0^+$ and 2^+. The correlation energy of s bosons is approximately $2\Delta = 1.6$ MeV. Above an excitation energy of this amount, one expects states corresponding to the breaking of a pair into two fermion states. These states are similar to those discussed in Sect. 10.3 except that now they are formed by two protons or two neutrons. Particularly important among these broken pair states are states with high angular momenta J since they appear at about the same energy as collective states with the same J value. In this case, they can be easily observed experimentally and indeed a fraction of the experimental activity in nuclear structure in the last few years has been devoted to the study of these states.

An analysis of broken pair states within the context of the interacting boson–fermion model was initiated several years ago (Gelberg and Zemel, 1980). As in odd–odd nuclei, one can attack this problem in two ways. The first is through the use of Bose–Fermi symmetries or supersymmetries. The states denoted by one star in Figs. 4.1, 4.7 and 4.10, e.g. $^{122}_{78}\mathrm{Pt}^*_{114}$ in Fig. 4.7, are indeed states of this type and some of their properties have been studied (Morrison and Jarvis, 1985; Baake et al., 1986). The second way

Table 10-6 List of available numerical calculations of broken pair states in even–even nuclei

Proton number	Neutron number	Broken pair	Reference
		Interacting boson–fermion model-1	
56 Ba	70	$(1h_{11/2})^2$	Gelberg and Zemel (1980)
	70–74	$(1h_{11/2})^2$	Faessler *et al.* (1985)
58 Ce	72–76	$(1h_{11/2})^2$	Faessler *et al.* (1985)
66 Dy	88–96	$(1h_{11/2})^2, (1i_{13/2})^2$	Chuu and Hsieh (1988)
78 Pt	104–114	$(1i_{13/2})^2$	Hsieh and Chuu (1987)
80 Hg	114–118	$(1i_{13/2})^2$	Morrison *et al.* (1981)
	110–118	$(1i_{13/2})^2$	Kuyucak *et al.* (1984)
		Interacting boson–fermion model-2	
32 Ge	34–40	$(\pi 1g_{9/2})^2$ $(\nu 1g_{9/2})^2$	Yoshida and Arima (1985)
54 Xe	70	$(\pi 1g_{7/2})^2$ $(\nu 1h_{11/2})^2$	Hanewinkel *et al.* (1983)
54 Xe	68–76	$(\nu 1h_{11/2})^2$	Kusakari and Sugawara (1984)
56 Ba	70, 72	$(\nu 1h_{11/2})^2$	Yoshida *et al.* (1982)
58 Ce	72–76	$(\nu 1h_{11/2})^2$	Yoshida *et al.* (1982)
66 Dy	88–92	$(\pi 1h_{11/2})^2$ $(\nu 1h_{9/2})^2, (\nu 1i_{13/2})^2$	Alonso *et al.* (1986)
84 Po	116–126	$(\pi 1h_{9/2})^2$	Zemel and Dobeš (1983)
86 Rn	116–126	$(\pi 1h_{9/2})^2$	Zemel and Dobeš (1983)
88 Ra	116–126	$(\pi 1h_{9/2})^2$	Zemel and Dobeš (1983)

is a numerical diagonalization of the interacting boson–fermion Hamiltonian with two fermions explicitly included. Several cases have been studied and Table 10.6 gives a summary of calculations available at the present time. These calculations reproduce details of properties of even–even nuclei at high angular momentum that cannot be obtained without the explicit breaking of pairs. One of these properties is the behavior of the $B(E2)$ values along the ground-state band, $B(E2; L \rightarrow L - 2)$. As discussed in Chapter 2 of Volume 1 (Fig. 2.14), one expects these $B(E2)$ values to be a smooth function of L, if only collective degrees of freedom play a

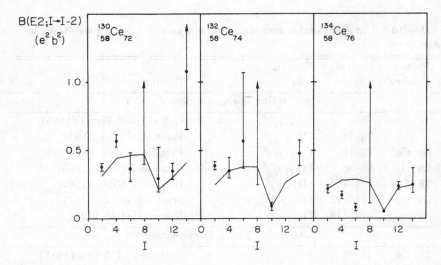

Fig. 10-14   Behavior of the $B(E2; L \rightarrow L-2)$ values in the ground-state band of $_{58}$Ce isotopes caused by the breaking of pairs (Yoshida *et al.*, 1982).

role. The explicit breaking of pairs causes sudden changes in this quantity. Figure 10.14 shows an example in the $_{58}$Ce nuclei.

We also mention here that an alternative approach exists for treating broken pair states, in which the fermion pair is treated as a boson with high angular momentum (Sorensen and Fowler, 1986).

When more than one pair is broken, calculations that include all states become impractical. Up to now the analysis of experimental data has been done by making use of the Nilsson model to determine the location of the intrinsic states and by subsequently cranking it in order to find the excitation energy of the corresponding rotational states. Similar calculations could be done, although they have not been done so far, within the context of the interacting boson–fermion model. Such calculations would become tractable in the large-N limit with the help of the techniques described in Chapter 6.

Part III

THE INTERACTING
BOSON–FERMION MODEL-k

Part III

THE INTERACTING
BOSON-FERMION MODELS

11

The interacting
boson–fermion models-3 and 4

11.1   Introduction

In light nuclei where protons and neutrons occupy the same single-particle levels, the use of the interacting boson–fermion model-1 or 2 is not appropriate, since one must take into account explicitly the isospin invariance of nuclear forces. This can be done by introducing more elaborate versions of the interacting boson–fermion model built on the isospin invariant forms of the interacting boson model (Elliott *et al.*, 1988). These versions are denoted by the interacting boson–fermion models-3 and 4.

11.2   The interacting boson–fermion model-3

In this model, one introduces a triplet of bosons describing correlated proton–proton (pp), neutron–neutron (nn) and proton–neutron (pn) pairs with isospin $T = 1$. The new bosons are called δ in Volume 1. Together with the π and ν bosons introduced earlier they form an isospin triplet,

$$
\begin{aligned}
|\pi\rangle &= |1, +1\rangle, \\
|\delta\rangle &= |1, 0\rangle, \\
|\nu\rangle &= |1, -1\rangle.
\end{aligned}
\tag{11.1}
$$

The corresponding creation and annihilation operators are denoted by

$$
\begin{aligned}
s_\pi^\dagger, d_{\pi,\mu}^\dagger, s_\delta^\dagger, d_{\delta,\mu}^\dagger, s_\nu^\dagger, d_{\nu,\mu}^\dagger, &\qquad (\mu = 0, \pm 1, \pm 2), \\
s_\pi, d_{\pi,\mu}, s_\delta, d_{\delta,\mu}, s_\nu, d_{\nu,\mu}, &\qquad (\mu = 0, \pm 1, \pm 2),
\end{aligned}
\tag{11.2}
$$

or, in a more compact notation,

$$b^\dagger_{\rho,l,m}; \qquad b_{\rho,l,m}; \qquad (\rho = \pi, \delta, \nu; l = 0, 2; -l \le m \le l). \qquad (11.3)$$

Since these operators form an isospin triplet, it is convenient to introduce explicitly isospin labels and denote the operators by

$$b^\dagger_{1,M_{T_B},l,m}; \qquad b_{1,M_{T_B},l,m}; \qquad (M_{T_B} = 0, \pm 1). \qquad (11.4)$$

Spherical tensors are built from creation and annihilation operators of the type

$$\tilde{b}_{1,M_{T_B},l,m} = (-)^{1-M_{T_B}+l-m} b_{1,-M_{T_B},l,-m}. \qquad (11.5)$$

The operators (11.4), together with the fermion operators (7.9), allow one to construct states in the isospin basis.

11.3   Isospin basis

In the isospin basis states can be classified as representations of

$$G = \mathrm{U}^B(6) \otimes \mathrm{U}^B_T(3) \otimes \mathrm{U}^F(\Omega) \otimes \mathrm{U}^F_T(2), \qquad (11.6)$$

with wave functions that are the product wave functions of a space-spin part, Φ, and an isospin part, X. Since these act on different spaces they cannot be combined together. The treatment of the space-spin part is identical to that discussed in Part I. The isospin part must be treated separately, but is relatively simple since it involves only the chain of algebras:

$$\mathrm{U}^B_T(3) \otimes \mathrm{U}^F_T(2) \supset \mathrm{SU}^B_T(3) \otimes \mathrm{SU}^F_T(2)$$
$$\supset \mathrm{SU}^B_T(2) \otimes \mathrm{SU}^F_T(2) \supset \mathrm{SU}^{BF}_T(2) \supset \mathrm{O}^{BF}_T(2). \qquad (11.7)$$

This chain simply describes the coupling of the boson and fermion isospins to the total isospin. The isospin wave functions, X, are thus labelled by the quantum numbers

$$
\left|
\begin{array}{ccccc}
\mathrm{U}^B_T(3) & \otimes & \mathrm{U}^F_T(2) & \supset & \mathrm{SU}^B_T(3) \otimes \mathrm{U}^F_T(2) \\
\downarrow & & \downarrow & & \downarrow \\
[N_{B1}, N_{B2}, N_{B3}] & & [N_{F1}, N_{F2}] & & [f_1, f_2] \\
\supset \mathrm{SU}^B_T(2) & \otimes & \mathrm{SU}^F_T(2) & \supset & \mathrm{SU}^{BF}_T(2) \supset \mathrm{O}^{BF}_T(2) \\
\downarrow & & \downarrow & & \downarrow \qquad\qquad \downarrow \\
T_B & & T_F & & T \qquad\qquad M_T
\end{array}
\right\rangle,
$$

$$(11.8)$$

Table 11-1 Partial classification scheme for isovector
bosons coupled to one fermion (Elliott et al., 1988)

N_B	$[f_1, f_2]$	n_s	n_d	T_B	N_F	T_F	T
0	[0,0]	0	0	0	1	1/2	1/2
1	[1,0]	1	0	1	1	1/2	1/2,3/2
		0	1	1	1	1/2	1/2,3/2
2	[2,0]	2	0	0	1	1/2	1/2
				2	1	1/2	3/2,5/2
		1	1	0	1	1/2	1/2
				2	1	1/2	3/2,5/2
		0	2	0	1	1/2	1/2
				2	1	1/2	3/2,5/2
	[1,1]	1	1	1	1	1/2	1/2,3/2
		0	2	1	1	1/2	1/2,3/2

with $f_1 = N_{B1} - N_{B3}$, $f_2 = N_{B2} - N_{B3}$ and $T_F = \frac{1}{2}(N_{F1} - N_{F2})$.
The representations of $SU_T^B(3)$ are labelled here by the partition
$[f_1, f_2]$ rather than by the Elliott quantum numbers (λ_T, μ_T) of
Volume 1. These two notations are related through:

$$\lambda_T = f_1 - f_2, \qquad \mu_T = f_2. \qquad (11.9)$$

The total wave functions are then given by

$$|\Psi\rangle = |\Phi\rangle \, |X\rangle$$
$$= |[N_{B1}, N_{B2}, N_{B3}]\alpha_B L_B; [N'_{F1}, \dots, N'_{F\Omega}]\alpha_F J_F; J M_J\rangle$$
$$\times |[f_1, f_2]T_B; T_F; T M_T\rangle, \qquad (11.10)$$

where $[N'_{F1}, \dots, N'_{F\Omega}]$ is the conjugate representation of $[N_{F1}, N_{F2}]$,
obtained by interchanging rows and columns. In addition to the
one described here, other coupling schemes are possible (Elliott *et
al.*, 1988). They all are obtained from (11.10) by isospin recoupling
transformations. A partial classification of states up to $N_B \leq 2$
and one fermion is presented in Table 11.1.

11.4   Physical operators

The structure of the Hamiltonian and other operators in the interacting boson–fermion model-3 is similar to that in the preceding

two models except that the coefficients now depend on $T_{\rm B}$, $T_{\rm F}$ and T. This is true not only for the terms describing bosons and fermions separately, but also for the boson–fermion interaction. In its multipole form, this interaction can now be written as

$$V_{\rm BF} = V'_{\rm BF} + (\vec{T}_{\rm B} \cdot \vec{T}_{\rm F})V''_{\rm BF}, \qquad (11.11)$$

where $\vec{T}_{\rm B}$ and $\vec{T}_{\rm F}$ denote the isospin vectors of bosons and fermions.

11.5   The interacting boson–fermion model-4

This model is a further generalization of the preceding one in which (pn) pairs with isospin $T = 0$ are added to the (pp), (nn) and (pn) pairs with isospin $T = 1$. The (pn) pairs with $T = 0$ are assigned an intrinsic spin $S = 1$, while those with $T = 1$ have $S = 0$ (Elliott and Evans, 1981). If one denotes the (pn) bosons with $T = 0$ and $S = 1$ by θ, one has now a sextet of bosons,

$$T = 1, S = 0: \quad \begin{pmatrix} \pi \\ \delta \\ \nu \end{pmatrix} ; \qquad T = 0, S = 1: \quad (\theta). \qquad (11.12)$$

The corresponding creation and annihilation operators are denoted by

$$s^\dagger_\pi, d^\dagger_{\pi,\mu}, s^\dagger_\delta, d^\dagger_{\delta,\mu}, s^\dagger_\nu, d^\dagger_{\nu,\mu}, s^\dagger_\theta, d^\dagger_{\theta,\mu}, \quad (\mu = 0, \pm 1, \pm 2),$$
$$s_\pi, d_{\pi,\mu}, s_\delta, d_{\delta,\mu}, s_\nu, d_{\nu,\mu}, s_\theta, d_{\theta,\mu}, \quad (\mu = 0, \pm 1, \pm 2), \quad (11.13)$$

or

$$b^\dagger_{\rho,l,m}; \qquad b_{\rho,l,m}; \qquad (\rho = \pi, \delta, \nu, \theta; l = 0, 2; -l \le m \le l). \tag{11.14}$$

If the spin and isospin labels are introduced explicitily, these operators are written as

$$b^\dagger_{T_{\rm B},M_{T_{\rm B}},S_{\rm B},l_{\rm B},j_{\rm B},m_{j_{\rm B}}}, \qquad b_{T_{\rm B},M_{T_{\rm B}},S_{\rm B},l_{\rm B},j_{\rm B},m_{j_{\rm B}}}. \qquad (11.15)$$

The bilinear products $b^\dagger_\alpha b_{\alpha'}$ $(\alpha, \alpha' = T_{\rm B}, M_{T_{\rm B}}, S_{\rm B}, l_{\rm B}, j_{\rm B}, m_{j_{\rm B}})$ span a 36-dimensional space, which is the representation space of

$$U^{\rm B}(36) \supset U^{\rm B}(6) \otimes U^{\rm B}_{ST}(6). \qquad (11.16)$$

Similarly, one can introduce the fermion operators in s–l coupling. These are labelled by

$$a^{\dagger}_{\frac{1}{2},m_t,\frac{1}{2},l_F,j_F,m_{j_F}}, \qquad a_{\frac{1}{2},m_t,\frac{1}{2},l_F,j_F,m_{j_F}}. \qquad (11.17)$$

The bilinear products $a^{\dagger}_i a_{i'}$ $(i,i' = \frac{1}{2},m_t,\frac{1}{2},l_F,j_F,m_{j_F})$ span a space of dimension $\Omega = 4\Omega_L$ with $\Omega_L = \sum_{l_F}(2l_F + 1)$. In the special case of the 8–20 shell, the single-particle levels have $l_F = 0,2$, and thus $\Omega = 24$, and the fermion representation space is

$$U^{F}(24) \supset U^{F}(6) \otimes U^{F}_{ST}(4). \qquad (11.18)$$

The orbital spaces of bosons and fermions are in this case equivalent.

11.6 Wigner basis

The particular case of the 8–20 shell offers the opportunity of constructing a boson–fermion basis which is related to the Wigner supermultiplet scheme (Wigner, 1937). This basis amounts to reducing the spin–isospin boson algebra to $O^{B}_{ST}(6) \approx SU^{B}_{ST}(4)$ and combining it with the fermion algebra $SU^{F}_{ST}(4)$, i.e.

$$U^{B}(36) \otimes U^{F}(24) \supset U^{B}(6) \otimes U^{B}_{ST}(6) \otimes U^{F}(6) \otimes U^{F}_{ST}(4)$$
$$\supset U^{B}(6) \otimes O^{B}_{ST}(6) \otimes U^{F}(6) \otimes U^{F}_{ST}(4) \supset U^{BF}(6) \otimes SU^{BF}_{ST}(4).$$

$$(11.19)$$

This scheme leads to the construction of several Bose–Fermi lattices of algebras of physical significance. Here, we discuss one of the possibilities (Szpikowski *et al.*, 1988) which corresponds to the lattice:

$$U^B(36) \quad \otimes \quad U^F(24)$$

$$
\begin{array}{ccccccc}
U^B(6) & \otimes & & U^F(6) & \otimes & U^B_{ST}(6) & \otimes & U^F_{ST}(4) \\
\downarrow & & \nearrow & \downarrow & & \downarrow & & \downarrow \\
O^B(6) & U^{BF}(6) & & O^F(6) & & O^B_{ST}(6) & \otimes & SU^F_{ST}(4) \\
\downarrow & \downarrow & \nearrow & \downarrow & & \searrow & \nearrow \\
O^B(5) & O^{BF}(6) & & O^F(5) & & SU^{BF}_{ST}(4) \\
\downarrow & \downarrow & \nearrow & \downarrow & & | \\
O^B(3) & O^{BF}(5) & & O^F(3) & & | \\
& \downarrow & \nearrow & & & | \\
& O^{BF}(3) & & & SU^{BF}_S(2) & \otimes & SU^{BF}_T(2) \\
& \searrow & \nearrow & & & \\
& \text{Spin}^{BF}(3) \\
& \downarrow \\
& \text{Spin}^{BF}(2) &.
\end{array}
$$

$$(11.20)$$

11.7   Dynamic symmetries

The classification scheme corresponding to the lattice (11.20) can be obtained via the usual methods discussed at length in previous chapters. We consider here only the spin–isospin part. The representations of $U^{BF}_{ST}(6)$ are six-rowed and characterized by six integers $[N_1, N_2, N_3, N_4, N_5, N_6]$. For a given boson number, N_B, the values of the N_is can be constructed as explained by Elliott and Evans (1981). Also the reduction from $U^B_{ST}(6)$ to $SU^B_{ST}(4) \approx O^B_{ST}(6)$ has been given and is reproduced in Volume 1, Table 7.2. The representations of $U^F_{ST}(4)$ are in general four-rowed. However, if only one odd fermion is considered, one has to deal only with the fundamental representation $[1,0,0,0]$ whose reduction is trivial, since it leads to the $(T = 1/2) \otimes (S = 1/2)$ representation of $SU^F_S(2) \otimes SU^F_T(2)$. In summary, the classification scheme is

$$
\left|
\begin{array}{ccc}
U^B_{ST}(6) & \otimes & U^F_{ST}(4) \supset O^B_{ST}(6) & \otimes & SU^F_{ST}(4) \\
\downarrow & & \downarrow & & \downarrow \\
[N_1,\ldots,N_6] & & [1,0,0,0] & & [\lambda_1, \lambda_2, \lambda_3] \\
& \supset & SU^{BF}_{TS}(4) \supset SU^{BF}_S(2) & \otimes & SU^{BF}_T(2) \\
& & \downarrow & & \downarrow \\
& & [\lambda'_1, \lambda'_2, \lambda'_3] & \quad S \quad & T
\end{array}
\right\rangle .
$$

$$(11.21)$$

Dynamic Bose–Fermi symmetries can be constructed by writing, as usual, the Hamiltonian in terms of Casimir operators. This procedure yields

$$
\begin{aligned}
H = {} & e_0 + e_1 C_2(\mathrm{O}^{\mathrm{B}}_{ST} 6) + e_2 C_2(\mathrm{SU}^{\mathrm{BF}}_{ST} 4) + e_3 C_2(\mathrm{U}^{\mathrm{BF}} 6) \\
& + \eta C_2(\mathrm{U}^{\mathrm{B}} 6) + 2\eta' C_2(\mathrm{O}^{\mathrm{BF}} 6) + 2\beta C_2(\mathrm{O}^{\mathrm{BF}} 5) + 2\gamma C_2(\mathrm{O}^{\mathrm{BF}} 3) \\
& + 2\gamma' C_2(\mathrm{SU}^{\mathrm{BF}}_S 2) + 2\gamma'' C_2(\mathrm{Spin}^{\mathrm{BF}} 3) + 2\omega C_2(\mathrm{SU}^{\mathrm{BF}}_T 2). \quad (11.22)
\end{aligned}
$$

Taking the expectation value of (11.22) in the basis provided by the lattice (11.20) produces energy formulas that can be used to analyze experimental data.

11.8 Dynamic supersymmetries

The Bose–Fermi algebras discussed in the previous section can be embedded into a superalgebra. The superalgebra appropriate to (11.20) is U(36|24) and can be constructed as in Chapters 4 and 9 by considering all bilinear products of creation and annihilation boson and fermion operators. Supermultiplets are now characterized by the total number of bosons plus fermions, $\mathcal{N}$. The situation here is more complex than in the previous cases, especially in odd–odd nuclei, since in these nuclei one has two types of states. The first type is formed by collective states constructed by means of (pn) bosons, δ and θ. These are the spin–isospin partners of the (pp) and (nn) bosons π and ν. The second type is formed by states with two fermions (one proton and one neutron) explicitly unpaired. An example of a supermultiplet is shown in Fig. 11.1.

11.9 Experimental examples

Several nuclei in the 8–20 shell have been analyzed recently in terms of the Bose–Fermi lattice (11.20). Figs. 11.2 and 11.3 show the set of four nuclei $^{30}_{14}\mathrm{Si}_{16}$, $^{30}_{15}\mathrm{P}_{15}$, $^{31}_{15}\mathrm{P}_{16}$ and $^{31}_{16}\mathrm{S}_{15}$ (Szpikowski *et al.*, 1988). All states shown in the figures belong either to the symmetric representation $[\mathcal{N}]$ or to the representation $[\mathcal{N} - 1, 1]$ of $\mathrm{U}^{\mathrm{BF}}(6)$. The excitation energy of other states is increased by

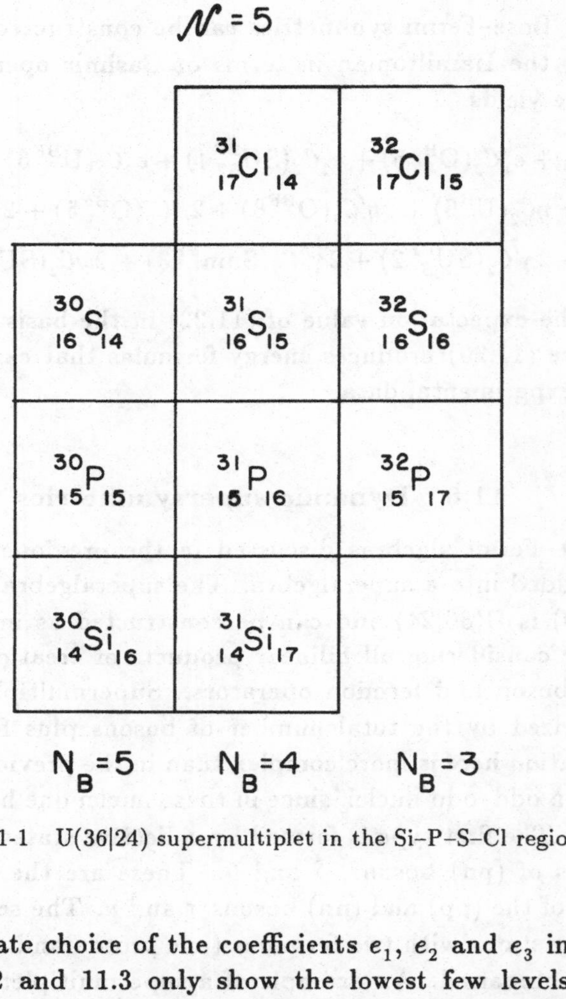

$$\mathcal{N} = 5$$

Fig. 11-1   U(36|24) supermultiplet in the Si–P–S–Cl region.

an appropriate choice of the coefficients e_1, e_2 and e_3 in (11.22). Figures 11.2 and 11.3 only show the lowest few levels in each nucleus. Many more levels are observed and also calculated at higher energies, but are omitted from the figures because of the ambiguities in the assignment of quantum numbers.

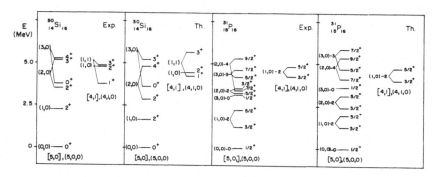

Fig. 11-2   An example of a (partial) U(36|24) supermultiplet in nuclei: $^{30}_{14}\text{Si}_{16}$ and $^{31}_{15}\text{P}_{16}$.

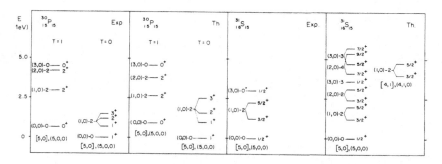

Fig. 11-3   An example of a (partial) U(36|24) supermultiplet in nuclei: $^{30}_{15}\text{P}_{15}$ and $^{31}_{16}\text{S}_{15}$.

Part IV

HIGH-LYING
COLLECTIVE MODES

12

Giant resonances

12.1 Introduction

In addition to low-lying collective modes extensively discussed in Volume 1 and in this book in terms of bosonic degrees of freedom, nuclei also display high-lying collective modes. The microscopic description of these modes is different from that of the low-lying modes, as shown schematically in Fig. 12.1. The latter are built from correlated pairs of nucleons in the valence shell, while the former are built from correlated particle–hole pairs, with one or more particles outside the valence shell. A description of high-lying modes in terms of bosons is also possible, although not particularly useful in itself since only one vibrational state of each mode is observed. It becomes useful only when coupling low-lying and high-lying modes. This coupling leads to the splitting and mixing of the high-lying modes which is often observed.

High-lying collective modes have been introduced in the interacting boson model by Morrison and Weise (1982) and, independently, by Scholtz and Hahne (1983). They proposed a description of the giant dipole resonance via a p boson coupled to a system of interacting s and d bosons and solved the resulting Hamiltonian numerically. Subsequently, Rowe and Iachello (1983) showed that, for deformed nuclei, a class of Hamiltonians exists that correspond to dynamic symmetries and that for such Hamiltonians analytic results can be obtained for energies and transition matrix elements. Since then the model has been applied to several (series of) isotopes (Maino *et al.*, 1984; 1985; Scholtz, 1985; Maino *et al.*, 1986a; Scholtz and Hahne, 1987; Nathan, 1988) and has been extended to include monopole and quadrupole giant reso-

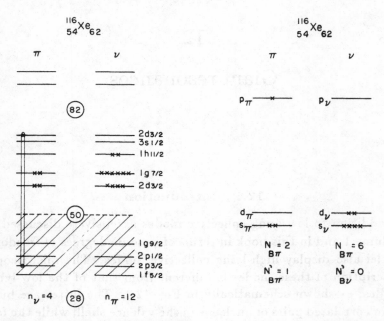

Fig. 12-1   Schematic illustration of the microscopic structure of high-lying collective modes in nuclei: the shell-model problem for the giant dipole resonance in $^{116}_{54}\mathrm{Xe}_{62}$ (left) and the boson-model problem that replaces it (right).

nances (Maino *et al.*, 1986b) and dipole resonances in light (Maino *et al.*, 1988) and odd–even nuclei (Maino, 1989).

The study of the coupling between high-lying and low-lying modes can be done in a way similar to that discussed in the preceding chapters where the coupling of a particle to the low-lying collective modes was treated. The coupling will be studied first by making use of symmetry considerations and later by numerical methods.

12.2   Giant resonances

The high-lying collective modes we consider here are those listed in Table 12.1. These modes will be treated in terms of boson operators. The dipole mode will be described in terms of the

Table 12-1 High-lying modes considered in this book

Mode	Quantum numbers	Boson label
Dipole	$J^P = 1^-, T = 1$	p
Isoscalar quadrupole	$J^P = 2^+, T = 0$	d'
Isoscalar monopole	$J^P = 0^+, T = 0$	s'
Isovector quadrupole	$J^P = 2^+, T = 1$	d''
Isovector monopole	$J^P = 0^+, T = 1$	s''

creation and annihilation operators for p bosons,

$$p_\mu^\dagger; \qquad p_\mu; \qquad (\mu = 0, \pm 1). \tag{12.1}$$

The quadrupole and monopole modes are described by the boson operators

$$d_\mu'^\dagger, s'^\dagger; \qquad d_\mu', s'; \qquad (\mu = 0, \pm 1, \pm 2),$$
$$d_\mu''^\dagger, s''^\dagger; \qquad d_\mu'', s''; \qquad (\mu = 0, \pm 1, \pm 2), \tag{12.2}$$

The prime superscript distinguishes these modes from the low-lying s and d bosons. Since high-lying modes are built from particle–hole pairs, their number need not to be conserved, in contrast to the low-lying modes which are built from particle–particle pairs. In the absence of any coupling between low-lying and high-lying modes, the latter modes are simply a single state at an energy ϵ_p or $\epsilon_{d'}, \epsilon_{s'}, \epsilon_{d''}, \epsilon_{s''}$. The presence of a coupling produces a pattern of splitting and mixing, as described in the sections below.

12.3   Mode–mode coupling.   Dipole

We consider first the coupling of the giant dipole resonance to the low-lying modes. The Hamiltonian for this coupling can be written as:

$$H = H_{sd} + H_p + V_{sd,p}, \tag{12.3}$$

where H_{sd} describes the low-lying modes, H_p the dipole mode and $V_{sd,p}$ their interaction. The Hamiltonian H_{sd} can be taken in one of its forms, interacting boson model-1, -2, -3 or -4. The Hamiltonian H_p is given by

$$H_p = \epsilon_p \hat{n}_p,$$
$$\hat{n}_p = \sum_\mu p^\dagger_\mu p_\mu. \tag{12.4}$$

The coupling can be written in terms of the products

$$V_{sd,p} = c_0 [d^\dagger \times \tilde{d}]^{(0)} \cdot [p^\dagger \times \tilde{p}]^{(0)} + c_1 [d^\dagger \times \tilde{d}]^{(1)} \cdot [p^\dagger \times \tilde{p}]^{(1)}$$
$$+ c_2 \left([s^\dagger \times \tilde{d} + d^\dagger \times \tilde{s}]^{(2)} + \chi [d^\dagger \times \tilde{d}]^{(2)} \right) \cdot [p^\dagger \times \tilde{p}]^{(2)}, \tag{12.5}$$

or, alternatively,

$$V_{sd,p} = c'_0 \hat{n}_d \hat{n}_p + c'_1 \hat{L}_{sd} \cdot \hat{L}_p + c'_2 \hat{Q}^\chi_{sd} \cdot \hat{Q}_p, \tag{12.6}$$

where the operators $\hat{n}_d$, $\hat{L}_{sd}$ and $\hat{Q}^\chi_{sd}$ are the multipole operators for s and d bosons and $\hat{n}_p$, $\hat{L}_p$ and $\hat{Q}_p$ are the corresponding ones for p bosons. The forms (12.5) and (12.6) assume an interacting boson model-1 description of low-lying states.

In studying transition properties of the high-lying dipole mode, an important operator is the electric dipole operator, given by

$$T^{(E1)}_\mu = \alpha'_1 [p^\dagger + \tilde{p}]^{(1)}_\mu. \tag{12.7}$$

The presence of a single operator p in (12.7) instead of a bilinear product, stresses the difference between the high-lying modes (particle–hole) and the low-lying modes (particle–particle).

12.4 Dynamic symmetries. Dipole

As in all algebraic problems, there are cases in which the eigenvalues of H can be found in closed form. In the problem discussed here there is a case of particular importance that corresponds to giant dipole states in deformed nuclei. Consider in fact an s,d-boson Hamiltonian with SU(3) symmetry. Since a p boson transforms as the fundamental representation of SU(3) one can combine the two parts of the wave function into a single one corresponding to the lattice of algebras (Rowe and Iachello, 1983)

$$
\begin{array}{ccc}
U^{sd}(6) & \otimes & U^{P}(3) \\
\downarrow & & \downarrow \\
SU^{sd}(3) & & SU^{P}(3) \\
& \searrow \quad \swarrow & \\
& SU(3) & \\
& \downarrow & \\
& O(3) & \\
& \downarrow & \\
& O(2) & .
\end{array}
\tag{12.8}
$$

By writing the Hamiltonian as

$$
H^{(II)} = e_0 + \delta C_2(SU^{sd}3) + \delta' C_2(SU^{P}3) + \delta'' C_2(SU3) + \gamma C_2(O3),
\tag{12.9}
$$

and evaluating its eigenvalues in the basis

$$
\left|
\begin{array}{ccccccc}
U^{sd}(6) & \otimes & U^{P}(3) \supset & SU^{sd}(3) & \otimes SU^{P}(3) \supset & SU(3) \supset & O(3) \supset & O(2) \\
\downarrow & & \downarrow & \downarrow & \downarrow & \downarrow & \downarrow & \downarrow \\
[N] & & [1] & (\lambda_B,\mu_B) & (1,0) & (\lambda,\mu) & \tilde{\chi},L & M_L
\end{array}
\right\rangle,
\tag{12.10}
$$

one obtains

$$
\begin{aligned}
E^{(II)}&(N,(\lambda_B,\mu_B),(\lambda,\mu),\tilde{\chi},L,M_L) \\
&= e_0 + \tfrac{2}{3}\delta\,(\lambda_B^2 + \mu_B^2 + \lambda_B\mu_B + 3\lambda_B + 3\mu_B) + \tfrac{8}{3}\delta' \\
&\quad + \tfrac{2}{3}\delta''\,(\lambda^2 + \mu^2 + \lambda\mu + 3\lambda + 3\mu) + 2\gamma L(L+1).
\end{aligned}
\tag{12.11}
$$

The allowed values of (λ,μ) can be obtained from those of (λ_B,μ_B) by multiplication. For axially symmetric nuclei, $\mu_B = 0$,

$$
(\lambda_B,0) \otimes (1,0) = (\lambda_B + 1,0) \oplus (\lambda_B - 1,1).
\tag{12.12}
$$

Thus, the single dipole state splits into two parts with energies

$$
\begin{aligned}
E_{\parallel} &= E - \kappa(2\lambda_B + 4), \\
E_{\perp} &= E + \kappa(\lambda_B - 1),
\end{aligned}
\tag{12.13}
$$

where $\kappa = -\frac{2}{3}\delta''$ and

$$E = e_0 + \tfrac{2}{3}(\delta + \delta'')(\lambda_B^2 + 3\lambda_B) + \tfrac{8}{3}\delta' + 4\gamma. \qquad (12.14)$$

The notation $\parallel$ is used in (12.13) to indicate a collective oscillation of the protons against the neutrons parallel with the axis of symmetry of the nucleus. Likewise, $\perp$ denotes an oscillation perpendicular to the symmetry axis. The former corresponds to the SU(3) quantum numbers $(\lambda_B + 1, 0)$, while $\perp$ corresponds to $(\lambda_B - 1, 1)$. When the dynamic symmetry is present, one can also evaluate the matrix elements of the dipole operator (12.7) between the ground state

$$|[N][0](\lambda_B, \mu_B = 0)(0,0)(\lambda = \lambda_B, \mu = 0)\tilde{\chi} = 0, L = 0, M_L = 0\rangle$$

and the two states with $L = 1$ in (12.10). Denoting

$$S_{1,i} = |\langle 1_i^- \parallel T^{(E1)} \parallel 0_1^+ \rangle|^2, \qquad (12.15)$$

one finds

$$\frac{S_{1,\perp}}{S_{1,\parallel}} = \frac{\left\langle \begin{matrix} (\lambda_B, 0) & (1,0) & | & (\lambda_B - 1, 1) \\ 0 & 1 & | & 1 \end{matrix} \right\rangle^2}{\left\langle \begin{matrix} (\lambda_B, 0) & (1,0) & | & (\lambda_B + 1, 0) \\ 0 & 1 & | & 0 \end{matrix} \right\rangle^2} = \frac{2\lambda_B}{\lambda_B + 3}, \qquad (12.16)$$

where we have used the fact that the operator $p^\dagger$ transforms as the $(1,0)$ representation of SU(3) and thus that its matrix elements are given by the SU(3) isoscalar factors in (12.16). For large λ_B, the dynamic symmetry result reduces to

$$E_\parallel \overset{\lambda_B \to \infty}{\Longrightarrow} E - 2\kappa\lambda_B,$$

$$E_\perp \overset{\lambda_B \to \infty}{\Longrightarrow} E + \kappa\lambda_B, \qquad (12.17)$$

and

$$\frac{S_{1,\perp}}{S_{1,\parallel}} \overset{\lambda_B \to \infty}{\Longrightarrow} 2. \qquad (12.18)$$

12.5 Numerical studies. Dipole

The study of the dipole mode is particularly interesting in view of the large number of available experimental data. Before comparing the results of calculations of splitting and mixing of high-lying states with experiments, one must keep in mind that, since these states lie at high excitation energies, they do not appear as isolated states, but rather acquire a width due to the coupling to more complex states, such as two-particle–two-hole states or unbound states, which lie at the same energy. In absence of a microscopic calculation, the coupling to the underlying states is taken into account by assigning a width. This width is usually taken to be energy-dependent, reflecting the variation of the density of states with energy. A commonly used form is (Eisenberg and Greiner, 1978):

$$\Gamma(E) = \Gamma_0(E/E_0)^\gamma. \tag{12.19}$$

12.5.1 Dipole resonances in the U(5) limit

This limit is characterized by the s,d-boson Hamiltonian of Eq. (2.75) of Volume 1. The actual structure of the spectrum in the presence of a p boson must be found numerically. However, in view of the fact that the spacings of the states with different values of n_d are large, the effect of the sd–p coupling is minor. The dipole state is thus, to a good approximation, not split much in this limit, as shown in Fig. 12.2(a).

12.5.2 Dipole resonances in the SU(3) limit

This case can be studied by means of the dynamic symmetry discussed in Sect. 12.4. In the exact limit, the dipole mode is split into two states with strengths given by Eq. (12.16). This situation is depicted in Fig. 12.2(b).

12.5.3 Dipole resonances in the O(6) limit

This case requires a full numerical study. The strength here is fragmented in several pieces, as shown in Fig. 12.2(c).

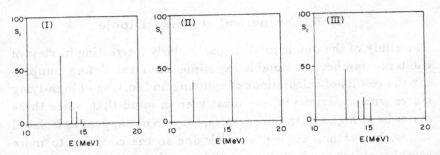

Fig. 12-2 Schematic illustration of the splitting of the dipole strength S_1 in the three limiting cases (a) $U^B(5)$, (b) $SU^B(3)$ and (c) $O^B(6)$ of the interacting boson model.

12.5.4 Dipole resonances in transitional regions

Several transitional regions have been studied. An example of results in the transitional region A, the Sm isotopes, is shown in Fig. 12.3 (Maino *et al.*, 1985). The dipole strengths have been spread by using (12.19) with $\gamma = 1.6$, $\Gamma_0/E_0^\gamma = 0.06$ MeV$^{(1-\gamma)}$.

12.6 Mode–mode coupling.
Monopole and quadrupole

The coupling of monopole and quadrupole giant resonances to the low-lying modes can be studied in a way similar to that described above for the dipole. We consider here the case of the isoscalar monopole and quadrupole modes. The Hamiltonian for the coupling can be written as (Maino *et al.*, 1986b):

$$H = H_{sd} + H_{s'd'} + V_{sd,s'd'}. \qquad (12.20)$$

The study in this case is made easy by the fact that the angular momenta of the s' and d' bosons are precisely the same as those of the s and d bosons. The problem here is thus analogous to that encountered in the study of the interacting boson model-2. The role of proton and neutron bosons is taken here by the s,d and s',d' bosons. It is, however, easier than in that case since one

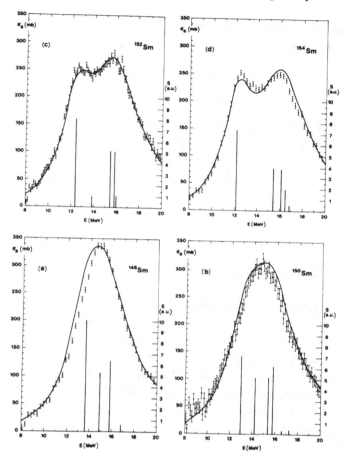

Fig. 12-3   Splitting of the dipole strength S_1 in the transitional class A of the interacting boson model. In addition to the strength (right scale and bars), the total photo-absorption cross section σ_a (left scale) is plotted against energy E. The spreading of the strength is done using (12.19) (Maino *et al.*, 1985).

considers only one s' or d' boson. The Hamiltonian $H_{s'd'}$ is then simply given by:

$$H_{s'd'} = \epsilon_{s'}\hat{n}_{s'} + \epsilon_{d'}\hat{n}_{d'}, \qquad (12.21)$$

where $\hat{n}_{s'}$ and $\hat{n}_{d'}$ are the number operators for s' and d' bosons. It is interesting to point out that in this case the interaction $V_{sd,s'd'}$ can have two types of terms. The first type conserves the number of s,d and s′,d′ bosons separately. This type causes splitting and mixing of the s′,d′ states. The structure of this term is:

$$V_{sd,s'd'}^{(I)} = \sum_{\alpha\beta,\alpha'\beta'} w_{\alpha\beta,\alpha'\beta'} b_\alpha^\dagger b_\beta b_{\alpha'}^\dagger b_{\beta'}, \qquad (12.22)$$

where α, β and α', β' generically denote s,d and s′,d′ bosons. Particularly important in (12.22) are the terms

$$V_{sd,s'd'}^{\prime(I)} = w_0 \left(\hat{n}_s + \hat{n}_d\right)\left(\hat{n}_{s'} + \hat{n}_{d'}\right) + w_1 \hat{L}_{sd} \cdot \hat{L}_{s'd'} + w_2 \hat{Q}_{sd}^\chi \cdot \hat{Q}_{s'd'}^{\chi'}, \qquad (12.23)$$

with

$$\hat{Q}_{sd,\mu}^\chi = [s^\dagger \times \tilde{d} + d^\dagger \times \tilde{s}]_\mu^{(2)} + \chi [d^\dagger \times \tilde{d}]_\mu^{(2)},$$
$$\hat{Q}_{s'd',\mu}^{\chi'} = [s'^\dagger \times \tilde{d}' + d'^\dagger \times \tilde{s}']_\mu^{(2)} + \chi' [d'^\dagger \times \tilde{d}']_\mu^{(2)}. \qquad (12.24)$$

The second type of terms does not conserve the number of s,d and s′,d′ bosons separately. This term causes a mixing of the giant monopole and quadrupole resonances into the low-lying states. This mixing causes a renormalization of the properties of the low-lying states as has been repeatedly emphasized within the context of the symplectic model (Rosensteel and Rowe, 1977; 1980). Its effects have also been studied within the context of the interacting boson model (Park and Elliott, 1986). The structure of these terms is

$$V_{sd,s'd'}^{(II)} = \sum_{\alpha\alpha',\beta\beta'} u_{\alpha\alpha',\beta\beta'} b_\alpha^\dagger b_{\alpha'} b_\beta^\dagger b_{\beta'} + \sum_{\alpha,\alpha'\beta'\gamma'} u'_{\alpha\alpha',\beta'\gamma'} b_\alpha^\dagger b_{\alpha'} b_{\beta'}^\dagger b_{\gamma'}$$
$$+ \sum_{\alpha\beta\gamma,\alpha'} u''_{\alpha\beta,\gamma\alpha'} b_\alpha^\dagger b_\beta b_\gamma^\dagger b_{\alpha'}. \qquad (12.25)$$

We consider here only the first type of terms.

In addition to the Hamiltonian, one is also interested in transitions. Especially important here are the electric monopole and quadrupole transitions. The corresponding operators contain a part related to s and d, and a part related to s′ and d′,

$$T_\mu^{(L)} = T_{sd,\mu}^{(L)} + T_{s'd',\mu}^{(L)} \qquad (12.26)$$

The s,d part is the usual one, extensively discussed in Volume 1. The s',d' part is:

$$T_{s'd',0}^{(E0)} = \alpha_0'[s'^\dagger + \tilde{s}']_0^{(0)},$$
$$T_{s'd',\mu}^{(E2)} = \alpha_2'[d'^\dagger + \tilde{d}']_\mu^{(2)}. \tag{12.27}$$

This form is a consequence of the assumed particle–hole character of the s' and d' bosons.

12.7 Dynamic symmetries. Monopole and quadrupole

Since the monopole and quadrupole giant resonances have the same quantum numbers of the low-lying collective modes, it is possible to construct dynamic symmetries for all three limits U(5), SU(3) and O(6). This is done in a way similar to that presented in Volume 1, Chapter 5, for the coupling of proton and neutron bosons. We discuss here only the case of SU(3) which has extensive applications in medium-mass and heavy nuclei. The appropriate chain is:

$$\left|\begin{matrix} U^{sd}(6) \otimes U^{s'd'}(6) \supset SU^{sd}(3) \otimes SU^{s'd'}(3) \supset SU(3) \supset O(3) \supset O(2) \\ \downarrow \qquad\quad \downarrow \qquad\qquad \downarrow \qquad\qquad \downarrow \qquad\qquad \downarrow \qquad \downarrow \qquad \downarrow \\ [N] \qquad\quad [1] \qquad\quad (\lambda_B,\mu_B) \qquad (2,0) \qquad (\lambda,\mu) \quad \tilde{\chi}, L \quad M_L \end{matrix}\right\rangle. \tag{12.28}$$

By writing the Hamiltonian as

$$H^{(II)} = e_0 + \delta C_2(SU^{sd}3) + \delta' C_2(SU^{s'd'}3) + \delta'' C_2(SU3) + \gamma C_2(O3), \tag{12.29}$$

and evaluating its eigenvalues in the basis (12.28), one obtains

$$\begin{aligned} E^{(II)}&(N, (\lambda_B, \mu_B), (\lambda, \mu), \tilde{\chi}, L, M_L) \\ &= e_0 + \tfrac{2}{3}\delta\,(\lambda_B^2 + \mu_B^2 + \lambda_B\mu_B + 3\lambda_B + 3\mu_B) + \tfrac{20}{3}\delta' \\ &+ \tfrac{2}{3}\delta''\,(\lambda^2 + \mu^2 + \lambda\mu + 3\lambda + 3\mu) + 2\gamma L(L+1). \end{aligned} \tag{12.30}$$

The allowed values of (λ, μ) can be obtained from those of (λ_B, μ_B) by multiplication. For axially symmetric nuclei, $\mu_B = 0$,

$$(\lambda_B, 0) \otimes (2, 0) = (\lambda_B + 2, 0) \oplus (\lambda_B, 1) \oplus (\lambda_B - 2, 2). \tag{12.31}$$

The product contains four states with $L = 2$ and two states with $L = 0$. The energies of the four states with $L = 2$ are given by

$$E_1^{(2)} = E - \kappa(4\lambda_B + 10) + 12\gamma,$$
$$E_2^{(2)} = E - \kappa(\lambda_B + 4) + 12\gamma,$$
$$E_3^{(2)} = E_4^{(2)} = E + \kappa(2\lambda_B - 4) + 12\gamma, \qquad (12.32)$$

where $\kappa = -\frac{2}{3}\delta''$ and

$$E = e_0 + \tfrac{2}{3}(\delta + \delta'')\,(\lambda_B^2 + 3\lambda_B) + \tfrac{20}{3}\delta', \qquad (12.33)$$

while the energies of the two $L = 0$ states are given by

$$E_1^{(0)} = E - \kappa(4\lambda_B + 10),$$
$$E_2^{(0)} = E + \kappa(2\lambda_B - 4), \qquad (12.34)$$

where $\kappa = -\frac{2}{3}\delta''$. The ratios of the strengths of E2 and E0 can be obtained by considering matrix elements of the operators (12.27). These are in turn given by the isoscalar factors:

$$\left\langle \begin{matrix} (\lambda_B,0) & (2,0) & | & (\lambda,\mu) \\ 0 & 2 & | & 2 \end{matrix} \right\rangle \quad \text{and} \quad \left\langle \begin{matrix} (\lambda_B,0) & (2,0) & | & (\lambda,\mu) \\ 0 & 0 & | & 0 \end{matrix} \right\rangle.$$

12.8   Numerical studies. Monopole and quadrupole

The dynamic symmetry discussed in the previous section gives the general features of the splitting of the giant monopole and quadrupole resonances in SU(3) nuclei. However, it assumes that the s′ and d′ bosons are degenerate in energy, $\epsilon_{s'} \approx \epsilon_{d'}$. This is not the case in practice, since typical values of $\epsilon_{s'}$ and $\epsilon_{d'}$ in medium-mass nuclei are $\epsilon_{s'} \approx 14$ MeV and $\epsilon_{d'} \approx 11$ MeV. One thus needs to perform numerical calculations. Furthermore, in order to compare these with experiment, one must introduce a spreading width similar to (12.19) but with different values of E_0 and γ. In doing numerical studies, one must choose a mode–mode coupling interaction. A commonly used interaction, which contains the main features of the coupling, is that given by (12.23). Since

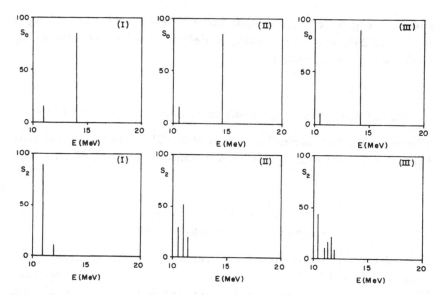

Fig. 12-4   Schematic illustration of the splitting of the monopole, S_0, and quadrupole, S_2, strength in the three limiting cases (a) $U^B(5)$, (b) $SU^B(3)$ and (c) $O^B(6)$ of the interacting boson model.

the terms w_0 and w_1 have only a minor effect, they are usually neglected.

12.8.1 Monopole–quadrupole resonances in the U(5) limit

The giant monopole and quadrupole resonances are only slightly affected by the coupling to the low-lying states in the limit in which these states are described by a U(5) Hamiltonian. They appear as single states with $J^P = 0^+$ and 2^+, Fig. 12.4(a).

12.8.2 Monopole–quadrupole resonances in the SU(3) limit

In this limit, the coupling of s,d and s′,d′ modes causes a large splitting and mixing. As discussed in Sect. 12.7, the splitting of $J = 2$ states is concentrated in four fragments while that of $J = 0$ states is concentrated in two fragments. A typical situation is shown in Fig. 12.4(b). The splitting is determined by the strength w_2 in (12.23), while the mixing is determined by the ratio of w_2 to the energy difference $\epsilon_{s'} - \epsilon_{d'}$.

12.8.3 Monopole–quadrupole resonances in the $O(6)$ limit

The situation here is similar to that of the SU(3) limit, except that the splitting and mixing has a more complex pattern, Fig. 12.4(c).

12.8.4 Monopole–quadrupole resonances in transitional regions

Several transitional regions have been studied. An example of the results obtained in the transitional region A is shown in Fig. 12.5. The monopole and quadrupole strengths have been spread in a way similar to (12.19).

12.9   Giant resonances in light nuclei

As discussed in Chapter 11, in light nuclei where protons and neutrons occupy the same single-particle levels, isospin conservation plays an important role and one must introduce all bosons of the $T = 1$ triplet in order to describe low-lying states. Isospin has also a major effect in the coupling of the low-lying and high-lying levels. We consider here specifically the coupling of the giant dipole resonance, $J^P = 1^-, T = 1$, to bosons described by the interacting boson model-3. The problem can again be attacked either by using symmetry considerations or in a numerical way.

12.9.1 Isospin coupling

We consider in particular the case of bosons with SU(3) symmetry. The coupling of the space–spin part can be done as discussed in previous sections. For the isospin part, one must couple the isospin

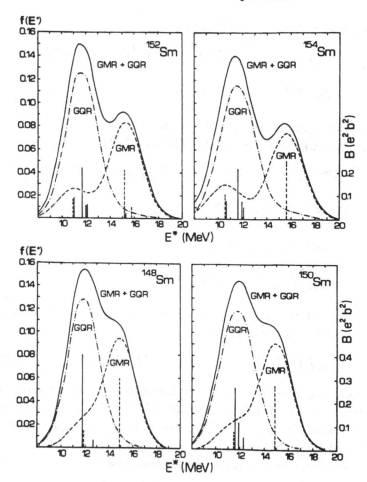

Fig. 12-5 Splitting of the monopole, S_0, and quadrupole, S_2, strength in the transitional class A of the interacting boson model. The $B(E0; 0_1^+ \rightarrow 0_i^+)$ (dashed bars) and $B(E2; 0_1^+ \rightarrow 2_i^+)$ (full bars) values are plotted on the right-hand scale against energy E. The spreading of the monopole (dashed), quadrupole (dash–dotted) and summed (full) strength is done using (12.19) (Maino *et al.*, 1986b).

of the s and d bosons, described by:

$$\left| \begin{array}{cccc} U_T^{sd}(3) & \supset & SU_T^{sd}(3) & \supset & SU_T^{sd}(2) & \supset & O_T^{sd}(2) \\ \downarrow & & \downarrow & & \downarrow & & \downarrow \\ [N_1, N_2, N_3] & & [f_1, f_2] & & T_B & & M_{T_B} \end{array} \right\rangle,$$

$$\text{(12.35)}$$

with the isospin of the giant dipole resonance, $T = 1$, which corresponds to the representation [1] of $U_T^p(3)$. One thus has:

$$\left| \begin{array}{cccccc} U_T^{sd}(3) & \otimes & U_T^p(3) & \supset & SU_T^{sd}(3) & \otimes & SU_T^p(3) \\ \downarrow & & \downarrow & & \downarrow & & \downarrow \\ [N_1, N_2, N_3] & & [1] & & [f_1, f_2] & & [1,0] \\[2ex] \supset & SU_T(3) & \supset & SU_T(2) & \supset & O_T(2) \\ & \downarrow & & \downarrow & & \downarrow \\ & [f_{1T}, f_{2T}] & & \tilde{\chi}, T & & M_T \end{array} \right\rangle.$$

$$\text{(12.36)}$$

The values of the quantum numbers $[f_{1T}, f_{2T}]$ can be obtained from the multiplication rule:

$$[f_1, f_2] \otimes [1, 0] = [f_1 + 1, f_2] \oplus [f_1, f_2 + 1]. \qquad \text{(12.37)}$$

The coupling scheme (12.36) is, however, not particularly useful since states with different values of T_B belonging to the same representation $[f_1, f_2]$ of $SU_T^{sd}(3)$ are considerably split, as shown Fig. 7.2 of Volume 1.

A much more appropriate isospin coupling (Maino *et al.*, 1988) is that at the level of $SU_T(2)$, i.e. the coupling

$$\left| \begin{array}{cccccc} U_T^{sd}(3) & \otimes & U_T^p(3) & \supset & SU_T^{sd}(3) & \otimes & SU_T^p(3) \\ \downarrow & & \downarrow & & \downarrow & & \downarrow \\ [N_1, N_2, N_3] & & [1] & & [f_1, f_2] & & [1,0] \\[2ex] \supset & SU_T^{sd}(2) & \otimes & SU_T^p(2) & \supset & SU_T(2) & \supset & O_T(2) \\ & \downarrow & & \downarrow & & \downarrow & & \downarrow \\ & T_B & & T_p = 1 & & T & & M_T \end{array} \right\rangle.$$

$$\text{(12.38)}$$

For each set of states with isospin T_B one obtains states with $T = T_B + 1$, T_B, $T_B - 1$. However, since in a given nucleus the value

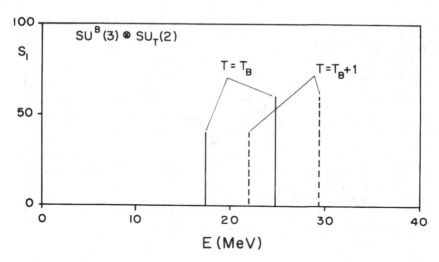

Fig. 12-6 Schematic illustration of the splitting of the dipole strength S_1 in the $SU^B(3)$ limit of the interacting boson model with isospin. The boson isospin is denoted by T_B.

of M_T is $\frac{1}{2}(A - 2Z)$, only states with $T = T_B + 1$ and T_B survive. As a consequence the giant dipole resonance built on nuclei with $T_B \neq 0$ is further split into two isospin parts. Introducing an isospin-dependent Hamiltonian of the type:

$$H_T = \omega C_2(SU_T^{sd}2) + \omega' C_2(SU_T^p 2) + \omega'' C_2(SU_T 2), \qquad (12.39)$$

one obtains the energy formula

$$E_T(T_B, T) = 2\omega T_B(T_B + 1) + 4\omega' + 2\omega'' T(T + 1). \qquad (12.40)$$

The splitting between the states with $T = T_B$ and $T = T_B + 1$ is then given by

$$\Delta E_T(T_B) = 4\omega''(T_B + 1), \qquad (12.41)$$

and depends on the single parameter ω''. The situation in light nuclei with an SU(3) symmetry is then that illustrated in Fig. 12.6.

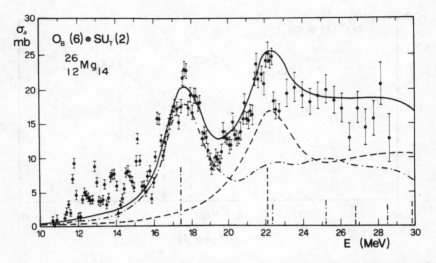

Fig. 12-7   Splitting of the dipole strength S_1 in the interacting boson model with isospin. The two isospin components are denoted by dash-dotted and dashed lines. The total photo-absorption cross section σ_a is also shown. The spreading of the strength is done using (12.19) (Maino *et al.*, 1988).

12.9.2 Numerical studies

Several cases of light nuclei both in the s–d and the p–f shell have been studied using the interacting boson model-3. It is convenient to start the investigation from nuclei with the same number of protons and neutrons and with the ground state having $T_B = 0$. The study of these nuclei fixes the coupling of the p boson to the s and d bosons. Analysis of nuclei with $T_B \neq 0$ determines then the isospin part of the interaction. An example is shown in Fig. 12.7.

12.10   Giant resonances in odd–even nuclei

The fragmentation of giant resonances in odd–even nuclei is a very difficult subject that has not been investigated much. Numerical analyses are very involved, especially when several single-particle orbits play an important role. In these complex cases, symmetry considerations offer the possibility to study at least the general features of the coupling. We consider here the case of the giant dipole resonance. Since the p boson transforms as the representation $(1,0)$ of $SU^P(3)$, as discussed in Sect. 12.4, one can study giant resonances in odd nuclei by coupling this $SU^P(3)$ to the chain of algebras discussed in Sect. 3.4. We consider here in particular the case in which the single particle occupies states with $j = 1/2, 3/2$. The lattice of algebras here is (Maino, 1989):

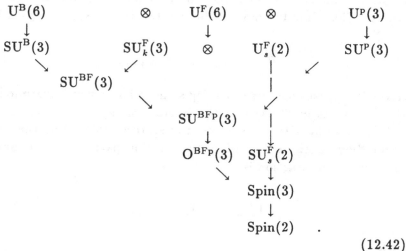

$$(12.42)$$

The basis states for the $SU^{BF}(3)$ part are discussed in Sect. 3.4. By taking the product of $(\lambda_B, \mu_B) = (2N, 0)$ with $(\lambda_F, \mu_F) = (1, 0)$ one has

$$(2N, 0) \otimes (1, 0) = (2N + 1, 0) \oplus (2N - 1, 1). \qquad (12.43)$$

These representations must then be combined with the representation $(\lambda_p, \mu_p) = (1, 0)$ to yield

$$(2N + 1, 0) \otimes (1, 0) = (2N + 2, 0) \oplus (2N, 1),$$
$$(2N - 1, 1) \otimes (1, 0) = (2N, 1) \oplus (2N - 2, 2). \qquad (12.44)$$

Thus, assuming the ground state of the odd–even nucleus has $SU^{BF}(3)$ quantum numbers $(2N + 1, 0)$, the giant dipole resonance built on it will have $SU^{BFP}(3)$ quantum numbers $(2N + 2, 0)$ and

$(2N, 1)$. Furthermore, by reducing $\text{SU}^{\text{BF}}(3) \otimes \text{SU}^{\text{F}}_s(2)$ to $\text{Spin}^{\text{BF}}(3)$ one finds that the ground state has $J = 1/2$. The dipole operator (12.7) can only excite states with $J = 1/2$ and $3/2$ when acting on the ground state and five such states are contained in the $\text{SU}^{\text{BFP}}(3) \otimes \text{SU}^{\text{F}}_s(2)$ representations with $(2N + 2, 0)$ and $(2N, 1)$ quantum numbers. Of these five states, two have $J = 1/2$ and three $J = 3/2$. In order to find the energies of these states one must introduce the Hamiltonian coupling low-lying and high-lying states. This Hamiltonian can be written in the same way as in (12.3):

$$H = H_{\text{sd,a}} + H_{\text{p}} + V_{\text{sd,a,p}}. \tag{12.45}$$

Here, $H_{\text{sd,a}}$ describes the low-lying s and d bosons and a fermions, $H_{\text{p}} = \epsilon_{\text{p}} \hat{n}_{\text{p}}$ as in (12.4) and the interaction $V_{\text{sd,a,p}}$ can be taken as in (12.6) but with the s,d-boson operators replaced by the corresponding operators including the fermion part. The dominant term in this interaction is

$$V'_{\text{sd,a,p}} = c'_2 \hat{Q}_{\text{sd,a}} \cdot \hat{Q}_{\text{p}}, \tag{12.46}$$

with

$$\hat{Q}_{\text{sd,a},\mu} = [s^\dagger \times \tilde{d} + d^\dagger \times \tilde{s}]^{(2)}_\mu - \frac{\sqrt{7}}{2} [d^\dagger \times \tilde{d}]^{(2)}_\mu$$
$$- \frac{1}{\sqrt{2}} \left([a^\dagger_{1/2} \times \tilde{a}_{3/2}]^{(2)}_\mu - [a^\dagger_{3/2} \times \tilde{a}_{1/2}]^{(2)}_\mu - [a^\dagger_{3/2} \times \tilde{a}_{3/2}]^{(2)}_\mu \right). \tag{12.47}$$

Diagonalization of (12.45) produces the desired energies.

A simpler result can be obtained by considering the case in which a dynamic symmetry exists. This occurs when the Hamiltonian can be written in the form

$$H^{(\text{II})} = e_0 + \delta C_2(\text{SU}^{\text{BF}}3) + \delta' C_2(\text{SU}^{\text{P}}3) + \delta'' C_2(\text{SU}^{\text{BFP}}3)$$
$$+ \gamma C_2(\text{O}^{\text{BFP}}3) + \gamma' C_2(\text{Spin3}). \tag{12.48}$$

Evaluating the expectation value of (12.48) in the state

$$
\left|
\begin{array}{c}
U^{BF}(3) \quad \otimes \quad U^{P}(3) \otimes U^{F}_{s}(2) \supset SU^{BFP}(3) \otimes U^{F}_{s}(2) \\
\downarrow \qquad\qquad \downarrow \qquad\qquad\qquad\qquad \downarrow \\
(\lambda_{BF},\mu_{BF}) \qquad (1,0) \qquad\qquad\qquad (\lambda,\mu) \\[4pt]
\supset O^{BFP}(3) \otimes SU^{F}_{s}(2) \supset Spin(3) \supset Spin(2) \\
\downarrow \qquad\qquad \downarrow \qquad\quad \downarrow \qquad\quad \downarrow \\
\tilde{\chi}, L \qquad S = 1/2 \qquad\; J \qquad\;\; M_{J}
\end{array}
\right\rangle ,
$$

$$(12.49)$$

one obtains

$$
\begin{aligned}
E^{(II)} & (N, (\lambda_{BF}, \mu_{BF}), (\lambda, \mu), \tilde{\chi}, L, J, M_J) \\
&= e_0 + \tfrac{2}{3}\delta\,(\lambda_{BF}^2 + \mu_{BF}^2 + \lambda_{BF}\mu_{BF} + 3\lambda_{BF} + 3\mu_{BF}) + \tfrac{8}{3}\delta' \\
&\quad + \tfrac{2}{3}\delta''\,(\lambda^2 + \mu^2 + \lambda\mu + 3\lambda + 3\mu) + 2\gamma L(L+1) + 2\gamma' J(J+1).
\end{aligned}
$$

$$(12.50)$$

In the special case discussed above, $(\lambda_{BF}, \mu_{BF}) = (2N+1, 0)$, the two $J = 1/2$ states have energies:

$$
\begin{aligned}
E_1^{(1/2)} &= E - \kappa(2\lambda_{BF} + 4), \\
E_2^{(1/2)} &= E + \kappa(\lambda_{BF} - 1),
\end{aligned}
$$

$$(12.51)$$

where $\kappa = -\tfrac{2}{3}\delta''$ and

$$
E = e_0 + \tfrac{2}{3}(\delta + \delta'')\,(\lambda_{BF}^2 + 3\lambda_{BF}) + \tfrac{8}{3}\delta' + \tfrac{3}{2}\gamma'. \qquad (12.52)
$$

The three $J = 3/2$ states have energies:

$$
\begin{aligned}
E_1^{(3/2)} &= E - \kappa(2\lambda_{BF} + 4) + 6\gamma', \\
E_2^{(3/2)} &= E + \kappa(\lambda_{BF} - 1) + 4\gamma + 6\gamma', \\
E_3^{(3/2)} &= E + \kappa(\lambda_{BF} - 1) + 12\gamma + 6\gamma'.
\end{aligned}
$$

$$(12.53)$$

This situation is schematically illustrated in Fig. 12.8. A study similar to that reported in the previous sections can also be done for the dipole strengths and for the coupling of other giant resonances to the low-lying collective modes in odd–even nuclei.

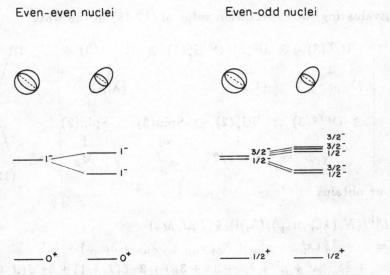

Fig. 12-8 Schematic illustration of the splitting of the dipole strength S_1 in even–even and odd–even nuclei with $SU^B(3)$ symmetry.

The guiding principle in all these studies is the one systematically used in this book and consists of writing all operators in terms of the relevant degrees of freedom, studying the general features of the solutions by means of the concept of dynamic symmetry and finally performing more detailed numerical calculations in order to obtain a good description of the experimental data.

This guiding principle has been used to study the interplay of the three basic degrees of freedom of nuclear structure:

(i) Cooper pairs, described in Volume 1 (s, d,... bosons);

(ii) single particles, described in Chapters 1 to 11 of this volume (j fermions); and

(iii) phonons, described in Chapter 12 of this volume (p, s', d',... bosons).

References

Alonso, C. E. (1986), *Estudio de la Estructura de Nucleos Medios y Pesados en el Modelo de los Bosones en Interaccion*, PhD. Thesis University of Sevilla.
Alonso, C. E., Arias, J. M., Bijker, R. and Iachello, F. (1984), *Phys. Letters* **144B**, 141.
Alonso, C. E., Arias, J. M. and Lozano, M. (1986), *Phys. Letters* **177B**, 130.
Alonso, C. E., Arias, J. M. and Lozano, M. (1988), *J. Phys.* **G14**, 877.
Arias, J. M. (1985), *Desarollo y Applicaciones del Modelo de Bosones (Protonicas y Neutronicas) en Interaccion par Nucleos de A Impar*, PhD. Thesis, University of Sevilla.
Arias, J. M., Alonso, C. E. and Bijker, R. (1985), *Nucl. Phys.* **A445**, 333.
Arias, J. M., Alonso, C. E. and Lozano, M. (1986), *Phys. Rev.* **C33**, 1482.
Arias, J. M., Alonso, C. E. and Lozano, M. (1987), *Nucl. Phys.* **A466**, 295.
Arima, A., Gelberg, A. and Scholten, O. (1987), *Phys. Letters* **185B**, 259.
Arima, A., Harvey, M. and Shimizu, K. (1969), *Phys. Letters* **30B**, 517.
Arima, A. and Iachello, F. (1975), *Phys. Rev. Letters* **35**, 1069.
Arima, A. and Iachello, F. (1976), *Ann. Phys.* (NY) **99**, 253.
Arima, A. and Iachello, F. (1978), *Ann. Phys.* (NY) **111**, 201.
Arima, A. and Iachello, F. (1979), *Ann. Phys.* (NY) **123**, 468.
Arima, A., Otsuka, T., Iachello, F. and Talmi, I. (1977), *Phys. Letters* **66B**, 205.
Auble, R. L. (1979), *Nucl. Data Sheets* **28**, 559.
Baake, M., Reinicke, P. and Gelberg, A. (1986), *Phys. Letters* **166B**, 10.
Balantekin, A. B. (1982), *A Study of Dynamical Supersymmetries in Nuclear Physics*, PhD. Thesis, Yale University.
Balantekin, A. B. and Bars, I. (1981), *J. Math. Phys.* **22**, 1149.
Balantekin, A. B., Bars, I., Bijker, R. and Iachello, F. (1983), *Phys. Rev.* **C27**, 1761.
Balantekin, A. B., Bars, I. and Iachello, F. (1981), *Nucl. Phys.* **A370**, 284.
Balantekin, A. B. and Paar, V. (1986a), *Phys. Letters* **169B**, 9.
Balantekin, A. B. and Paar, V. (1986b), *Phys. Rev.* **C34**, 1917.
Bardeen, J., Cooper, L. N. and Schrieffer, R. (1957), *Phys. Rev.* **108**, 1175.
Biedenharn, L. C. (1963), *Phys. Letters* **3**, 254.
Bijker, R. (1983), *Computer Program ODDPAR*, University of Groningen.
Bijker, R. (1984), *Dynamical Boson–Fermion Symmetries in Nuclei*, PhD. Thesis, University of Groningen.
Bijker, R. and Dieperink, A. E. L. (1982), *Nucl. Phys.* **A379**, 221.
Bijker, R. and Iachello, F. (1985), *Ann. Phys.* (NY) **161**, 360.
Bijker, R. and Kota, V. K. B. (1984), *Ann. Phys.* (NY) **156**, 110.
Bijker, R. and Kota, V. K. B. (1988), *Ann. Phys.* (NY) **187**, 148.

Bijker, R. and Scholten, O. (1985), *Phys. Rev.* **C32**, 591.

Blasi, N. (1984), *Single Proton Transfer Reactions on Odd-Even Nuclei*, PhD. Thesis, University of Groningen.

Blasi, N., Bijker, R., Harakeh, M. N., Iwasaki, Y., Sterrenburg, W. A., van der Werf, S. Y. and Vergnes, M. (1982), *Nucl. Phys.* **A388**, 77.

Blasi, N. and Lo Bianco, G. (1987), *Phys. Letters* **185B**, 254.

Bohr, A. (1952), *K. Dan. Vidensk. Selsk. Mat. Fys. Medd.* **26**, No. 14.

Bohr, A., and Mottelson, B. R. (1975), *Nuclear Structure*, Vol. 2, Benjamin.

Bolotin, H. H., Kennedy, D. L., Linard, B. J., Stuchbery, A. E., Sie, S. H., Katayama, I. and Sakai, H. (1979), *Nucl. Phys.* **A321**, 231.

Bruce, A. M., Gelletly, W., Lukasiak, J., Phillips, W. R. and Warner, D. D. (1985), *Phys. Letters* **165B**, 43.

Brussaard, P. J. and Glaudemans, P. W. M. (1977), *Shell-Model Applications in Nuclear Spectroscopy*, North-Holland.

Bucurescu, D., Căta, G., Cutoiu, D., Constantinescu, G., Ivaşcu, M. and Zamfir, N. V. (1983), *Nucl. Phys.* **A401**, 22.

Bucurescu, D., Căta, G., Cutoiu, D., Constantinescu, G., Ivaşcu, M. and Zamfir, N. V. (1985), *Nucl. Phys.* **A443**, 217.

Bucurescu, D., Căta, G., Ivaşcu, M., Zamfir, N. V., Liang, C. F. and Paris, P. (1988), *J. Phys.* **G14**, L175.

Cartan, E. (1894), *Sur la Structure des Groupes de Transformation Finis et Continues*, PhD. Thesis, Nony, Paris.

Casta nos, O., Chacón, E., Frank, A. and Moshinsky, M. (1979), *J. Math. Phys.* **20**, 35.

Casten, R. F., Kleinheinz, P., Daly, P. T. and Elbek, B. (1972), *K. Dan. Vidensk. Selsk. Mat. Fys. Medd.* **38**, No. 13.

Casten, R. F. and Smith, G. J. (1979), *Phys. Rev. Letters* **43**, 337.

Casten, R. F. and von Brentano, P. (1985), *Phys. Letters* **152B**, 22.

Chuu, D. S. and Hsieh, S. T. (1988), *Phys. Rev.* **C38**, 960.

Chuu, D. S. and Hsieh, S. T. (1989), *Nucl. Phys.* **A496**, 45.

Cizewski, J. A., Burke, D. G., Flynn, E. R., Brown, R. E. and Sunier, J. W. (1981), *Phys. Rev. Letters* **46**, 1264.

Cizewski, J. A., Burke, D. G., Flynn, E. R., Brown, R. E. and Sunier, J. W. (1983), *Phys. Rev.* **C27**, 1040.

Cizewski, J. A., Colvin, G. G., Börner, H. G., Hoyler, F., Kerr, S. A. and Schreckenbach, K. (1987), *Phys. Rev. Letters* **58**, 10.

Cunningham, M. A. (1982), *Nucl. Phys.* **A385**, 204, 221.

De Gelder, P., De Frenne, D., Heyde, K., Kaffrell, N., van den Berg, A. M., Blasi, N., Harakeh, M. N. and Sterrenburg, W. A. (1983), *Nucl. Phys.* **A401**, 397.

De Gelder, P., De Frenne, D. and Jacobs, E. (1982), *Nucl. Data Sheets* **35**, 443.

Dellagiacoma, F. (1988), *Beta Decay of Odd-Mass Nuclei in the Interacting Boson-Fermion Model*, PhD. Thesis, Yale University.

de-Shalit, A. and Talmi, I. (1963), *Nuclear Shell Theory*, Academic Press.

Dukelsky, J. and Lima, C. (1986), *Phys. Letters* **182B**, 116.

Eisenberg, J. M. and Greiner, W. (1978), *Nuclear Structure*, Vol. 1, North-Holland.

Elliott, J. P. (1958), *Proc. Roy. Soc.* **A245**, 128, 562.

Elliott, J. P. and Evans, J. A. (1981), *Phys. Letters* **101B**, 216.

Elliott, J. P., Evans, J. A. and Van Isacker, P. (1988), *Nucl. Phys.* **A481**, 245.

Faessler, A., Kuyucak, S., Petrovici, A. and Petersen, L. (1985), *Nucl. Phys.* **A438**, 78.

Flowers, B. H. (1952), *Proc. Roy. Soc.* **A212**, 248.

Frank, A., Pittel, S., Warner, D. D. and Engel, J. (1986), *Phys. Letters* **182B**, 233.

Frank, A., Van Isacker, P. and Warner, D. D. (1987), *Phys. Letters* **197B**, 474.

Gallagher, C. J. and Moszkowski, S. A. (1958), *Phys. Rev.* **111**, 1282.

Gelberg, A. (1983), *Z. Phys.* **A310**, 117.

Gelberg, A. (1984), *Z. Phys.* **A315**, 119.

Gelberg, A. and Zemel, A. (1980), *Phys. Rev.* **C22**, 937.

Gilmore, R. (1974), *Lie Groups, Lie Algebras and Some of Their Applications*, Wiley.

Hamermesh, M. (1962), *Group Theory and Its Application to Physical Problems*, Addison-Wesley.

Hanewinkel, H., Gast, W., Kaup, U., Harter, H., Dewald, A., Gelberg, A., Reinhardt, R., von Brentano, P., Zemel, A., Alonso, C. E. and Arias, J. M. (1983), *Phys. Letters* **133B**, 9.

Harmatz, B. (1977), *Nucl. Data Sheets* **22**, 433.

Harmatz, B. (1981), *Nucl. Data Sheets* **34**, 101.

Hecht, K. T. and Adler, A. (1969), *Nucl. Phys.* **A137**, 129.

Hecht, K. T. (1987), *Nucl. Phys.* **A475**, 276.

Hsieh, S. T. and Chuu, D. S. (1987), *J. Phys.* **G13**, L241.

Hübsch, T. and Paar, V. (1984), *Z. Phys.* **A319**, 111.

Hübsch, T. and Paar, V. (1987), *Z. Phys.* **A327**, 287.

Hübsch, T., Paar, V. and Vretenar, D. (1985), *Phys. Letters* **151B**, 320.

Iachello, F. (1980), *Phys. Rev. Letters* **44**, 772.

Iachello, F. and Arima, A. (1987), *The Interacting Boson Model*, Cambridge University Press.

Iachello, F. and Kuyucak, S. (1981), *Ann. Phys.* (NY) **136**, 19.

Iachello, F. and Levine, R. D. (1982), *J. Chem. Phys.* **77**, 3046.

Iachello, F. and Scholten, O. (1979), *Phys. Rev. Letters* **43**, 679.

Iachello, F. and Talmi, I. (1987), *Rev. Mod. Phys.* **59**, 339.

Iwasaki, Y., Aarts, E. H. L., Harakeh, M. N., Siemssen, R. H. and van der Werf, S. Y. (1981), *Phys. Rev.* **C23**, 1477.

Janssen, D., Jolos, R. V. and Dönau, F. (1974), *Nucl. Phys.* **A224**, 93.

Jolie, J. (1986), *The Interacting Boson-Fermion Model: Bose-Fermi Symmetries and Supersymmetries*, PhD. Thesis, University of Gent.

Jolie, J., Heyde, K., Van Isacker, P. and Frank, A. (1987), *Nucl. Phys.* **A466**, 1.

Jolie, J., Van Isacker, P., Heyde, K. and Frank, A. (1985a), *Phys. Rev. Letters* **55**, 1457.

Jolie, J., Van Isacker, P., Heyde, K., Moreau, J., Van Landeghem, G., Waroquier, M. and Scholten, O. (1985b), *Nucl. Phys.* **A438**, 15.

Kac, V. G. (1975), *Functional Anal. Appl.* **9**, 91.

Kac, V. G. (1977), *Comm. Math. Phys.* **53**, 31.

Kaup, U. and Gelberg, A. (1979), *Z. Phys.* **A293**, 311.

Kaup, U., Gelberg, A., von Brentano, P. and Scholten, O. (1980), *Phys. Rev.* **C22**, 1738.

Kölbl, W. R., Billowes, J., Burde, J., De Raedt, J. A. G., Grace, M. A. and Pakou, A. (1986), *Nucl. Phys.* **A456**, 349.

Kota, V. K. B. (1986), *Phys. Rev.* **C33**, 2218.

Kusakari, H. and Sugarawa, M. (1984), *Z. Phys.* **A317**, 287.

Kuyucak, S. (1982), *Study of Spinor Symmetries in Nuclear Structure*, PhD. Thesis, Yale University.

Kuyucak, S., Faessler, A. and Wakai, M. (1984), *Nucl. Phys.* **A420**, 83.

Leander, G. (1976), *Nucl. Phys.* **A273**, 286.

Lederer, C. M. (1982), *Nucl. Data Sheets* **35**, 525.

Leviatan, A. (1987), *Ann. Phys.* (NY) **179**, 201.

Leviatan, A. (1988), *Phys. Letters* **209B**, 415.

Leviatan, A. and Shao, B. (1990), *Phys. Letters* **B243**, 313.

Ling, Y. S., Zhang, M., Xu, J. M., Vallières, M., Gilmore, R., Feng, D. H. and Sun, H. Z. (1984), *Phys. Letters* **148B**, 13.

Lopac, V., Brant, S., Paar, V., Schult, O. W. B., Seyfarth, H. and Balantekin, A. B. (1986), *Z. Phys.* **A323**, 491.

Maino, G. (1989), *Phys. Rev.* **C40**, 988.

Maino, G. and Mengoni, A. (1988), *Phys. Rev.* **C38**, 2520.

Maino, G., Ventura, A., Van Isacker, P. and Zuffi, L. (1986a), *Phys. Rev.* **C33**, 1089.

Maino, G., Ventura, A., Van Isacker, P. and Zuffi, L. (1986b), *Europhys. Letters* **2**, 345.

Maino, G., Ventura, A. and Zuffi, L. (1988), *Phys. Rev.* **C37**, 1379.

Maino, G., Ventura, A., Zuffi, L. and Iachello, F. (1984), *Phys. Rev.* **C30**, 2101.

Maino, G., Ventura, A., Zuffi, L. and Iachello, F. (1985), *Phys. Letters* **152B**, 17.

Mauthofer, A., Stelzer, K., Gerl, J., Elze, Th. W., Happ, Th., Eckert, G., Faestermann, T., Frank, A. and Van Isacker, P. (1986), *Phys. Rev.* **C34**, 1958.

Meyer-ter-Vehn, J. (1975), *Nucl. Phys.* **A249**, 111.

Miyazawa, H. (1966), *Progr. Theor. Phys.* (Kyoto) **36**, 1266.

Morrison, I., Faessler, A. and Lima, C. (1981), *Nucl. Phys.* **A372**, 13.

Morrison, I. and Jarvis, P. D. (1985), *Nucl. Phys.* **A435**, 461.

Morrison, I. and Weise, J. (1982), *J. Phys.* **G8**, 687.

Mundy, S. J., Lukasiak, J. and Phillips, W. R. (1984), *Nucl. Phys.* **A426**, 144.

Murnaghan, F. D. (1938), *The Theory of Group Representations*, Johns Hopkins University Press.

Nathan, A. M. (1988), *Phys. Rev.* **C38**, 92.

Navrátil, P. and Dobeš, J. (1988), *Phys. Rev.* **C37**, 2126.

Nilsson, S. G. (1955), *K. Dan. Vidensk. Selsk. Mat. Fys. Medd.* **29**, No. 16.

Otsuka, T. and Arima, A. (1978), *Phys. Letters* **77B**, 1.

Otsuka, T., Arima, A., Iachello, F. and Talmi, I. (1978), *Phys. Letters* **76B**, 139.

Otsuka, T., Yoshida, N., Van Isacker, P., Arima, A. and Scholten, O. (1987), *Phys. Rev.* **C35**, 328.

Paar, V. (1979), *Nucl. Phys.* **A331**, 16.

Paar, V. (1980), *Inst. Phys. Conf. Ser.* **49**, 53.

Paar, V. and Brant, S. (1981), *Phys. Letters* **105B**, 81.

Paar, V., Brant, S. and Kraljević, H. (1982), *Phys. Letters* **110B**, 181.

Paar, V., Sunko, D. K. and Vretenar, D. (1987), *Z. Phys.* **A327**, 291.

Park, P. and Elliott, J. P. (1986), *Nucl. Phys.* **A448**, 381.

Pittel, S. and Frank, A. (1986), *Nucl. Phys.* **A454**, 226.

Price, R. H., Burke, D. G. and Johns, M. W. (1971), *Nucl. Phys* **A176**, 338.

Puddu, G., Scholten, O. and Otsuka, T. (1980), *Nucl. Phys.* **A348**, 109.

Racah, G. (1949), *Phys. Rev.* **76**, 1352.

Ramond, P. (1971), *Phys. Rev.* **D3**, 2415.

Ring, P. and Schuck, P. (1980), *The Nuclear Many-Body Problem*, Springer-Verlag.

Rosensteel, G. and Rowe, D. J. (1977), *Phys. Rev. Letters* **38**, 10.

Rosensteel, G. and Rowe, D. J. (1980), *Ann. Phys.* (NY) **126**, 343.

Rowe, D. J. and Iachello, F. (1983), *Phys. Letters* **130B**, 231.

Sambataro, M., Scholten, O., Dieperink, A. E. L. and Piccitto, G. (1984), *Nucl. Phys.* **A423**, 333.

Scholten, O. (1979), *Computer Program ODDA*, University of Groningen.

Scholten, O. (1980), *The Interacting Boson Approximation Model and Applications*, PhD. Thesis, University of Groningen.

Scholten, O. (1985), *Prog. Part. Nucl. Phys.* **14**, 189.

Scholten, O. and Blasi, N. (1982), *Nucl. Phys.* **A380**, 509.

Scholten, O. and Dieperink, A. E. L. (1981) *in* F. Iachello (ed.), *Interacting Bose–Fermi Systems in Nuclei*, Plenum.

Scholten, O. and Ozzello, T. (1984), *Nucl. Phys.* **A424**, 221.

Scholten, O. and Warner, D. D. (1984), *Phys. Letters* **142B** (1984), 315.

Scholten, O., Wu, H. C. and Dieperink, A. E. L. (1989), *Z. Phys.* **A332**, 1.

Scholten, O. and Yu, Z. R. (1985), *Phys. Letters* **161B**, 13.

Scholtz, F. G. (1985), *Phys. Letters* **151B**, 87.

Scholtz, F. G. and Hahne, F. J. W. (1983), *Phys. Letters* **123B**, 147.

Scholtz, F. G. and Hahne, F. J. W. (1987), *Nucl. Phys.* **A471**, 545.

Semmes, P. B., Leander, G. A., Lewellen, D. and Dönau, F. (1986) *Phys. Rev.* **C33**, 1476.

Sofia, H. M. and Vitturi, A. (1989), *Phys. Letters* **222B**, 317.

Sorensen, R. A. and Fowler, K. (1986), *in* R. A. Meyer and V. Paar (eds.), *Nuclear Structure, Reactions and Symmetries*, World Scientific.

Sun, H. Z., Frank, A. and Van Isacker, P. (1983), *Phys. Rev.* **C27**, 2430.

Szpikowski, S., Kłosowski, P. and Próchniak, L. (1988), *Nucl. Phys.* **A487**, 301.

Talmi, I. (1981) *in* F. Iachello (ed.), *Interacting Bose–Fermi Systems in Nuclei*, Plenum.

Tanaka, Y., Steffen, R. M., Shera, E. B., Reuter, W., Hoehn, M. V. and Zumbro, J. D. (1983), *Phys. Rev. Letters* **51**, 1633.

Van der Jeugt, J. (1985), *J. Phys.* **A18**, L745.

Van Isacker, P. (1987), *J. Math. Phys.* **28**, 957.

Van Isacker, P., Frank, A. and Sun, H. Z. (1984), *Ann. Phys.* (NY) **157**, 183.

Van Isacker, P., Jolie, J., Heyde, K. and Frank, A. (1985), *Phys. Rev. Letters* **54**, 653.

Vergados, J. D. (1968), *Nucl. Phys.* **A111**, 681.

Vergnes, M., Berrier-Ronsin, G. and Bijker, R. (1983), *Phys. Rev.* **C28**, 360.

Vergnes, M., Berrier-Ronsin, G. and Rotbard, G. (1987), *Phys. Rev.* **C36**, 1218.

Vergnes, M., Berrier-Ronsin, G., Rotbard, G., Vernotte, J., Maison, J. M. and Bijker, R. (1984), *Phys. Rev.* **C30**, 517.

Vergnes, M., Rotbard, G., Kalifa, J., Berrier-Ronsin, G., Vernotte, J., Seltz, R. and Burke, D. G. (1981), *Phys. Rev. Letters* **46**, 584.

Vervier, J. (1981), *Phys. Letters* **B100**, 383.

Vervier, J. (1987), *Rivista Nuovo Cimento* **10**, No. 9.

Vervier, J. and Janssens, R. V. F. (1982), *Phys. Letters* **108B**, 1.
Vervier, J., Van Isacker, P., Jolie, J., Kota, V. K. B. and Bijker, R. (1985), *Phys. Rev.* **C32**, 1406.
Warner, D. D. (1984), *Phys. Rev. Letters* **52**, 259.
Warner, D. D. and Bruce, A. M. (1984), *Phys. Rev.* **C30**, 1066.
Warner, D. D., Casten, R. F. and Frank, A. (1986), *Phys. Letters* **180B**, 207.
Warner, D. D., Casten, R. F., Stelts, M. L., Börner, H. G. and Barreau, G. (1982), *Phys. Rev.* **C26**, 1921.
Wigner, E. (1937), *Phys. Rev.* **51**, 106.
Wood, J. L. (1981a), *Phys. Rev.* **C24**, 1788.
Wood, J. L. (1981b), *in* F. Iachello (ed.), *Interacting Bose–Fermi Systems in Nuclei*, Plenum.
Wybourne, B. G. (1974), *Classical Groups for Physicists*, Wiley.
Yamazaki, Y., Sheline, R. K. and Shera, E. B. (1978), *Phys. Rev.* **C17**, 2061.
Yoshida, N. and Arima, A. (1985), *Phys. Letters* **164B**, 231.
Yoshida, N., Arima, A. and Otsuka, T. (1982), *Phys. Letters* **114B**, 86.
Yoshida, N., Sagawa, H., Otsuka, T. and Arima, A. (1988), *Phys. Letters* **215B**, 15.
Zemel, A. and Dobeš, J. (1983), *Phys. Rev.* **C27**, 2311.

Index